Winds of Change

Winds of Change

A Comparative Study of the Politics of Wind Energy Innovation in California and Denmark

ACADEMISCH PROEFSCHRIFT

ter verkrijging van de graad van doctor aan de
Universiteit van Amsterdam
op gezag van de Rector Magnificus prof. dr. J. J. M. Franse
ten overstaan van een door het College voor Promoties ingestelde
commissie in het openbaar te verdedigen in de Aula der Universiteit

op woensdag 8 september 1999 te 15.00 uur

door Quirinus Cornelis van Est

geboren te Fijnaart en Heijningen

promotor: prof. dr. R. Hoppe
co-promotores: dr. H. van de Graaf, dr. J. Grin

Faculteit der Maatschappij- en Gedragswetenschappen

Cover design: Marjo Starink, Amsterdam
Cover photo: Danish Vestas wind turbines at Zond's Victory Garden site in the Tehachapi Pass near Mojave, California. California state flower (poppy) blooming in Spring. Photo by Paul Gipe.
DTP: Henk Pel, Zeist
Printing: Haasbeek, Alphen aan den Rijn

Keywords: Wind energy – political aspects, technological innovation, policy sciences, political culture
ISBN: 90 5727 027 7

International Books, A. Numankade 17, 3572 KP Utrecht
tel.: ++31 30 273 18 40, fax: ++ 31 30 273 36 14
e-mail: i-books@antenna.nl, http://www.antenna.nl/i-books

Contents

Acknowledgements

Because the past is just a goodbye, teach your children well. *Graham Nash, 1970*

It is the end of the world as we know it (and I feel fine). *R.E.M., 1988*

Technological change constantly challenges the world in both minor and fundamental ways. It is essential, therefore, to have a proper view on the role of technology within our society. Unfortunately, it is still popular to consider technology as an autonomous force, something out of human control. This fatalist perspective – whether optimistic, pessimistic or neutral – is dangerous because it takes away responsibility from individuals, organizations, and society as a whole, for the development, use and impact of technology. A more worldly-wise perspective on the role of technology is badly needed in order to take up the challenges of the coming century. A sophisticated understanding should not be reserved to a select group of science and technology scholars. In order to make some dreams come true it should become common knowledge in the next two decades to politicians, decision makers within business and government and to a large segment of the population.

Hopefully, this book presents a small step towards that goal. It tries to make sense of technology by seeing it as a result of human endeavor, as the outcome of the work of people, who are part of a family and a circle of friends. Instead of looking at technology as something coming from outer space, artifacts are considered to be our 'children.' Not children of flesh and blood of course, but made out of nuts and bolts and human imagination and ingenuity. The book describes how the 'child' wind energy technology grew up in California and Denmark, and which codes it was given by its parents and family (policy makers, technologists, business managers and consumers) to live by. It explores the question of how to teach our technology-children well and feed them on our dreams.

As this book is finally reaching the light of day, I would like to thank: The Social and Behavioral Sciences Faculty of the University of Amsterdam, for giving me the opportunity to do my Ph.D.-research. The Danish Ministry of Education and Research, for funding a two-month stay at the International People's College (IPC) at Helsingør in Denmark. The Netherlands Organization for Scientific Research (NOW), for providing me with scholarships for my research visits to California. The Rathenau Institute, for partly funding this book.

My promotor Prof. Rob Hoppe, for his intelligent critic. My co-promotors Dr. John Grin, for being the best coach one could wish for, and Dr. Henk van de Graaf, for being an inspiring and true master of policy sciences. My colleagues at the University of Amsterdam and the Rathenau Institute, for supporting me with their kindness and concern. In particular, Jan Eberg, Martin Ket and Michiel van Well, for keeping up my spirit with their sense of humor.

Prof. Paul Sabatier, for inspiring me with his Advocacy Coalition Framework, and inviting me to stay at the University of California, Davis at the Division of Environmental Studies. His family, for its hospitality during my research visits in California. Matt Zafonte, for making me feel at ease in the Golden State. Paul Gipe, for commenting on my case descriptions and allowing me to use his picture for the cover of this book.

Ole Skou, for kindly putting his apartment at my free disposal during my work in Copenhagen. Richard Schalburg, for helping me during my stay at Roskilde University, Department of Environment, Technology, and Social Studies.

Pé, my dear wife, for believing in me and letting me give a huge amount of myself to the making of this book. I love you. Our four beautiful children, Berne, Chris, Ries, and Krijn, for stealing my time, and giving me precious love in return. My sisters, brothers and friends, for putting this work in perspective and giving me joy and recreation. Last but not least, my parents, for throwing me in a kettle of love potion when I was young and their infinite support. This book is dedicated with endless love and thanks to you.

List of Abbreviations

AB	Assembly Bill
AC	Alternating current
ACF	Advocacy Coalition Framework
AE-plan	Alternative Energy Plan
ATV	Danish Academy of Technical Sciences (*Akademiet for de Tekniske Videnskaber*)
AWEA	American Wind Energy Association
AWT	Advanced Wind Turbine
BRPU	Biennial Resource Plan Update
CAL SEIA	California Solar Energy Industries Association
CEC	California Energy Commission
CEERT	Center for Energy Efficiency and Renewable Technologies
CEO	Chief Executive Officer
CEQA	California Environmental Quality Act of 1970
CO_2	Carbon Dioxide
CPUC	California Public Utilities Commission
D	Democratic Party
DC	Direct current
DEA	Danish Energy Agency (*Energistyrelsen*)
DEF	Association of Danish Electric Utilities (*Danske Elværkers Forening*)
DEFU	Electric Utilities' Joint Research Institute (*Danske Elværkers Forenings Udredningsafdeling*)
Difko	Danish Investment Fund (*Dansk Investerings Fond*)
DKP	Denmark's Communist Party
DOE	Department of Energy
DSM	Demand Side Management
DtH	Danish Technical University (*Danmarks tekniske Højskole*)
DV	Association of Danish Wind Turbine Owners (*Danmarks Vindmølleforening – Danske Vindkraftværker*)
DVES	Danish Wind Power Association (*Dansk Vind Elektricitets Selskab*)
DWR	Department of Water Resources
DWT	Danish Wind Technology (*Dansk Vindteknik A/S*)
EC	European Community
EDF	Environmental Defend Fund
EM	Ministry of Energy (*Energiministeriet*)
EPA	Energy Policy Act of 1992
EPC	Energy Policy Committee
EPIC	End Poverty in California

EPRI	Electric Power Research Institute
ERDA	Energy Research and Development Administration
ETA	Energy Tax Act of 1978
FDV	Association of Danish Wind Turbine Manufacturers (*Foreningen af Danske Vindmøllefabrikanter*)
FERC	Federal Energy Regulatory Commission
FSO4	Final Standard Offer Number 4
FWEP	Federal Wind Energy Program
GOP	Grand Old Party (Republican Party)
GRP	Glass fibre reinforced polyester
HAWT	Horizontal axis wind turbine
HM	Ministry of Trade (*Handelsministeriet*)
IEP	Independent Energy Producers Association
IOU	Investor-Owned Utilities
IPP	Independent Power Producer
IRP	Integrated Resource Planning
ISO4	Interim Standard Offer Number 4
KF	Conservative People's Party (*Konservative Folkeparti*)
kW	Kilowatt
kWh	Kilowatt-hour
LADWP	Los Angeles Department of Water and Power
m	Metre
MIT	Massachusetts Institute of Technology
MOU	Memorandum of Understanding
MM	Ministry of Environmental Affairs (*Miljøministeriet*)
MW	Megawatt
NASA	National Aeronautics and Space Administration
NEA	National Energy Act of 1978
NEPA	National Environmental Policy Act of 1970
NO_x	Oxides of Nitrogen
NRC	Nuclear Regulatory Commission
NRDC	Natural Resource Defense Council
NREL	National Renewable Energy Laboratory, before 1992 known as SERI
NSF	National Science Foundation
NSI	National System of Innovation
OAT	Office of Appropriate Technology
O&M	Operation and Maintenance
OOA	Organization for Information about Nuclear Power (*Organisationen til Oplysining om Atomkraft*)
OPEC	Organization of Petroleum Exporting Countries
OVE	Organization for Renewable Energy Energy (*Organisationen for Vedvarende Energi*)
PEPA	Private Energy Producers Act of 1976
PG&E	Pacific Gas and Electric
PURPA	Public Utilities Regulatory Policies Act of 1978

QF	Qualifying Facility
RANN	Research Applied to National Needs
R&D	Research and Development
RD&D	Research, Development and Demonstration
RPS	Renewable Portfolio Standard
RV	Social Liberal Party (*Radikale Venstre*)
SCE	Southern California Edison
SD	Social Democratic Party
SDG&E	San Diego Gas and Electric
SERI	Solar Energy Research Institute, since 1992 known as NREL
SF	Socialist People's Party (*Socialistik Folkeparti*)
SMUD	Sacramento Municipal Utility District
SNOX	Danish law adapted in 1989 to reduce SO_2 and NO_x emissions
SO_2	Sulphur Dioxide
SUNRAE	Solar Use Now for Resources And Employment
TEN	Techno-Economic Network
TURN	Toward Utility Rate Normalization
UCS	Union of Concerned Scientists
USW	U.S. Windpower, after 1988 Kenetech Windpower
V	Liberal Party (*Venstre*)
VAWT	Vertical axis wind turbine
VS	Left-Wing Socialist Party (*Venstre Socialister*)
WEC	Wind Energy Committee (*Vindenergiudvalget*)
WECS	Wind Energy Conversion System
WEIN	Wind energy innovation networks
WFL	Windfarms Limited
WPP	Wind Power Program
WPRS	Wind Performance Reporting System

List of Figures

1 Introduction

> Our view of politics, technology and social choice is one in which their clear separation is impossible. They are, rather, an entanglement: an inchoate mass. And, if inchoateness is the very essence of politics, technology and social choice, then to try to get to grips with them by making each of them choate (which ... is what current approaches set out to do) is inevitably to distance yourself from them. So, we begin, not by getting rid of the inchoate, but by analyzing it. Politics, technology and social choice, we observe, are all entangled because they all hinge crucially upon cognition: upon seeing *and* knowing. *Michiel Schwarz and Michael Thompson, 1990: 1*

Everyone seems to agree that technology is a subject of increasing importance. It is commonplace to argue that technology shapes the ways of modern life and changes the contours of our society. Some even contend that technology has its internal, unstoppable dynamic and moulds society to fit its patterns. Within this so-called technological determinist perspective, social change is seen as something produced by technologies that have developed, so to speak, outside society. When the relationship between technology and society is interpreted in determinist terms, posing the question of the purpose or end of all the technological enterprises of the last half of the twentieth century (nuclear weapons, the pill, pesticides, television, space exploration, artificial intelligence, the Internet, and genetic engineering) becomes senseless and ridiculous. For better or worse, man becomes the slave of technology.

Like most social scientists who study the technology-society relationship, we do not adhere to the image of technology as the driving force behind much of societal history. It is recognized that technologies not only shape society, but are also shaped by society. In this view, technologies result from a complex interplay between different types of actors, including technologists, consumers, firm managers and policy makers. Seeing technology as something that both produces and is produced by social change makes the question of how society can shape technological development a relevant one. Instead of being the slave to technology, is it possible to become the master of technology? Can we mobilize technology in such a way that it comes to agree with our social needs and political aims? In this book we will take up a policy science perspective on the social shaping of technology. This means that we are particularly interested in the question of how policy makers might influence techno-economic innovation. Consequently, the book's central question is: *How can policy makers guide techno-economic innovation into a democratically legitimate direction?*

1.1 Why Wind Energy Innovation?

This study presents a modest, enquiring attempt to find proper ways to address this question and to find appropriate means of meeting the challenge of the democratic guidance of techno-economic innovation. We will stick to one specific example which we will describe and analyze in great detail. Since this inquiry is about the political control of techno-economic innovation, we will focus on the role policy makers have played within the innovation process.[1] Of all the possible techno-economic developments one can choose to study, we selected the wind energy innovation process in both California and Denmark.

It is clear that a comprehensive study on wind energy innovation in just two countries cannot provide a complete answer to the above fundamental question. Nevertheless, the study delivers some relevant insights and provides an opportunity to investigate some ideas about how policy makers could keep innovation processes in line with social and ecological needs. There are several good reasons why this particular research focus presents a good starting point for discussing the theme of democratic guidance of technology. The choice of wind energy innovation is related to the fact that governmental support is crucial for the development of the technology and the industry, and the fact that, both in California and Denmark, governments have actively encouraged the development of wind energy for most of the period under study. The selection of wind energy innovation is also inspired by the relatively simple nature of wind turbine technology[2] and the survivable size of the wind power industry. The development of the modern wind industry makes the selection of California and Denmark almost self-evident. California grew to dominate worldwide wind energy development during the early 1980s, and the boom in sales to California produced a thriving Danish wind turbine industry. About fifty percent of the fifteen thousand wind turbines installed in California were built by Danish firms, such as Micon, Bonus, Nordex, Vestas, and Danwin.[3] However, what really makes the story of wind energy in California and Denmark interesting from a political perspective is the fact that a large variety of political, technical and economical strategies played a role.

The two oil crises of the 1970s triggered a severe political controversy over the future of the energy supply system. The two oil shocks posed serious challenges to the energy and economic policies of all the industrialized nations, and prompted a debate on who and what was to blame and what kind of change was required. Some blamed utility and energy companies for the energy shortages and higher prices, others accused the government or environmentalists. The electric utility firms viewed nuclear power as their main answer to the 'oil supply crisis.' Their critics saw energy conservation as most important, and the use of renewable energy technologies such as wind turbines. Despite the resentment of the electric utility industry, over 15,000 wind turbines in California and 2,800 in Denmark have now been integrated into existing utility grids.[4] By contrast, the development of nuclear power has been curtailed in both states. It is clear that from the beginning wind turbine technology presented a politically charged artefact. In particular, it is interesting to compare the political nature of wind energy innovation in California and Denmark. There was a marked difference between the two. In Denmark, groups of neighbours stimulated the decentralized, small-scale use and smooth development of wind power, while futuristic-looking large-scale wind power plants – or wind farms – sprouted out like mushrooms on the California hills. These elements provide a number of politically interesting questions. Did public policies influence the way wind energy developed? If yes, how and to what extent did it control this development? How can we understand or explain the differences between the Californian and Danish developments in the field of wind energy?

The way in which the wind energy industry developed in both countries also provides some fascinating puzzles. The first is that small wind turbine producers came to control the wind energy market. Initially, both large multinational corporations, as well as small and medium-sized firms, were active within the field of wind energy. While the former aimed at building multi-megawatt size wind turbines, smaller companies focused on making small and medium-sized machines. At the time that most of the larger companies retreated from the wind turbine business the smaller firms produced thousands of 50 to 200 kilowatt turbines for the California market. In addition, while in California the wind industry was building wind power plants (so-called wind farms) and generating electricity with wind energy, a wind turbine manufacturing industry developed in Denmark. Supported by government subsidies in both states, a market for wind turbines was developed. Although the Californian market was much larger than the Danish one, only one American wind turbine manufacturer resulted from the effort of the Californian government in the early 1980s. To make it even worse, the only surviving company[5] went bankrupt in May 1996. In contrast, the Danish incentive program seems to have led to a thriving wind turbine industry that in the 1990s produced more than half of the world market.[6]

This leaves the reader with an intriguing question. How could the California market thrive without resulting in a successful American wind turbine industry and/or technology? It becomes even more puzzling when it is realized that "California is the science center of this planet."[7] Modern physics is built upon the pioneering work at the University of California. Apollo, and the author's beloved microcomputer, all bear the 'Made in California' stamp. Why is one of the most vibrant economies in the world, which holds the intellectual capacity to unravel the secrets of the atom, put a man on the moon, and invent a machine that beats the world's best chess player, not able to build a proper wind turbine? Moreover, how does it come about that Denmark – a country that "definitely qualifies for a place in the league of small nations of the world"[8] – did succeed?

With respect to the Danish and American approaches to technological development there also existed several remarkable differences. The Danish social scientists, Peter Karnøe (1991, 1995, 1996) and the German historian Matthias Heymann (1995), have made it clear that the Danes and Americans initiated different ways of developing wind turbine technology. Danish firms started with a small, simple, and robust, wind turbine design, and then attempted to improve it gradually, based on practical experience and craftsmanship. The American approach was characterized by the pursuit of "more advanced technological designs, and it relied on and sought to develop more formalized, theoretically based knowledge." (Karnøe 1995: 244)

1.2 The Seemingly Seamless Innovation Web

> [A] research program that aims at contributing to a greater understanding of the social processes involved in technological development [should be] respecting the seamless web character of technology and society. *Wiebe E. Bijker, Thomas P. Hughes and Trevor J. Pinch, 1987: 10*

The above two authors argue that the difference in technical practices explains the 'failure' of the American wind turbine companies and the 'success' of the Danish industry. Such an explanation smacks of technological determinism. In this book, a distinct and broader, or more holistic, perspective is chosen to study wind energy innovation. We promote the integrated inquiry into political, technical and economic aspects of innovation. This implies that techno-economic innovation has to be understood as a result of the interaction between policy making institutions, technological communities and the management of firms (cf. Grin and Van de Graaf 1996). Such a view takes the shine off Karnøe's and Heymann's conclusion that the rather simple technological design and the hands-on, practical, problem-solving approach of the Danes led to the 'success' of their wind turbine industry. Within the interactive systems view of innovation, there is no linear relationship between the knowledge strategy and the success of the industry since the success or failure of a technical approach is itself dependent on the success or failure of the industry. Innovative 'success' or 'failure' do not depend solely on technological development, but should be seen as the outcome of the complex interplay between technologists, managers, policy makers, and other relevant actors. Interdependence should be seen, therefore, as a natural starting point of every inquiry into the 'heterogeneous' process of technological innovation.[9]

The multitude of interactions among the various dimensions of innovation have led several science, technology and society scholars to use the metaphor of a *'seamless web'* of society and technology (Hughes 1986; Bijker *et al.* 1987, 1992). 'Seamless' denotes the idea that the interdependence of the various aspects of innovation is so strong that it no longer makes sense to make analytic distinctions between technical, social, economic, and political aspects (Bijker *et al.* 1987: 3). As Bijker (1993: 120) explains, "The web of modern society is not made up of distinct pieces of scientific, technical, social, cultural, and economic cloth; rather, whatever creases can be seen are made by the actors or by the analyst." He adds that one needs "to recognize that a successful engineer is not solely a technical wizard but an economic, political, and social one as well." (*ibid.*)

The image of the seamless web is useful for conceiving the interdependence of the different aspects of innovation. But when one is interested in the political guidance of techno-economic innovation the metaphor becomes less useful. In that case, one would like to study the influence of the political process on techno-economic development and vice versa. Bijker's (1992: 95) suggestion to try to avoid altogether the categories of 'society' and 'technology' would make such attempt impossible. Sweeping all aspects of innovation into a big heap carries the danger of disregarding the political character of innovation. As Horn (1994: 387) aptly comments, "[T]he weaving together of the social and the technical often works to displace the former by privileging the latter." In this study, we will try to articulate both interdependence as well as difference between the various facets of innovation. For this, we will use the metaphor of games. We will depict the innovation process as an event in which three games – the technical, the political, and the market game – are played simultaneously. Each game is defined by a unique set of rules, styles of play, and intellectual and emotional demands on the players. But each game influences and is influenced by the other two games. We think if one is interested in the issue of the democratic guidance of techno-economic innovation it is fruitful to unravel the 'seamless' innovation web in such a way that it becomes clear that the various convictions and interests of social and techno-economic actors are different, but still form a '*seemingly seamless web*.'

The aim of this study is to analyze how the political game in the field of (wind) energy is played, how it is influenced by the technical and market games, and how it influences the innovation game. Focusing on the role of the policy makers does not imply that the policy maker is seen as the dominant actor within the development of wind energy technology and business. Firm managers, technologists, consumers, and many other actors, have done their bit. Some even lost their lives during the construction, operation or maintenance of wind turbines (Gipe 1995: 364-365).[10] The policy maker thus represents one of the main actors, equal among others. Each actor has a different perspective. The policy maker may look at a wind turbine as a means to produce clean electricity in order to stop the depletion of the ozone layer; the firm manager probably sees it as a product to make profit. Which perspective is best is irrelevant. But since the development of wind energy is examined from a policy science perspective we will primarily focus on the policy maker's roles.

One might argue that policy makers play two interrelated roles. First, they democratically legitimize wind energy innovation. Apart from acting as a political watchdog, interpreter, or advocate, they may also play a constructive or stimulating role in the development of wind energy technology and industry. To do this successfully, they have to take up a meta-perspective. This implies that they need an insight into the variety of different actors and perspectives that exist within the field of wind energy, and the way in which these actors and views interact, and promote or hinder joint action that would foster the achievement of the chosen political goal. The policy actor becomes a public manager of interdependence; the interdependence between the political and techno-economic system and the interdependence between the various actors within the field of wind energy. For example, with respect to wind energy innovation, the policy maker's role is to co-ordinate the action of scientists, engineers, wind turbine manufacturers and users in such a way that some correspondence comes to exist between the technology that is developed, produced, and used. In this study, we will examine the two roles, the way they interact, and the way they influence and are influenced by techno-economic developments.

1.3 Outline of the Book

The book consists of four Parts. In Part One "*The Story of Wind Energy*," a comprehensive description is given of the development of wind energy in both California (Chapter 2) and Denmark (Chapter 3). The two case study descriptions demonstrate the dynamic and often unpredictable character of innovation, and illustrate the heterogeneous nature of the involved actors. The two stories show how changes within one institutional dimension induce and are induced by changes within other dimensions. Since we are particularly interested in the political shaping of wind energy innovation, we will first examine the political process behind the development of wind energy before we study the political features of wind energy innovation in interaction with its technical and economic aspects.

The politics of wind energy is described in Part Two "*The Political Game.*" To analyze the Californian and Danish political debate related to wind energy, a theoretical framework is developed in Chapter 4. Stimulated by the work of Sabatier and Jenkins-Smith (1993), the policy making process is conceptualized in terms of interacting *ideological coalitions*, each consisting of policy actors who share a set of basic political beliefs that guide their behaviour.

In Chapter 5, the policy analysis starts. First, the relevant ideological coalitions and their policy beliefs are determined. Next, it is studied how the various coalitions have interacted with each other, and how this affected their policy beliefs and influenced the public policy in the field of (wind) energy. In short, Chapter 5 deals with the following compound question:

> What were the politically significant ideological coalitions and their associated policy beliefs within the political debate related to wind energy innovation in California and Denmark, and how did the interaction between the various ideological coalitions change their policy beliefs over the last two decades?

The debates around energy policy in California and Denmark show remarkable similarities as well as differences. In Chapter 6, we examine whether the differences can be related to a difference in national political culture. A brief historical description of the national political cultures in the United States and Denmark is given. Next, the impact of the national political culture – which is not regarded as a single political culture, but as a mix of political traditions – on the interaction of the various ideological coalitions within the energy policy subsystem is studied. Chapter 6, therefore, deals with the next two questions:

> What political traditions constitute the national political culture in the United States and in Denmark? How does the mix of political traditions enable and constrain political learning processes within the energy subsystem?

The thorough political analysis in Part Two allows us to investigate the interdependence between the political game and technical and market games in an accurate and sophisticated way. This is done in Part Three '*The Innovation Game.*' To analyze the interactions between policy makers and other actors, such as consumers, technologists and firm managers, a theoretical framework is developed in Chapter 7. Inspired by the work of Callon *et al.* (1992), the notion of innovation network is introduced, which consists of a scientific, a technical, a production, consumption, and a political pole. Consequently, the innovation process will be analyzed in terms of interacting innovation networks, each populated by different types of innovation actors (policy makers, technologists, firm managers, and consumers) who are all guided by different (professional) frames of meaning.

In Chapter 8, the relationship between national political and techno-economic traditions is studied historically. It is found that these different types of traditions rely on each other and, as such, have jointly developed into connected wholes, which are named national innovation traditions. This seems to imply that different ideological coalitions have different ideal types of innovation networks in mind. In this way, we hope to answer the next question:

> What kind of wind energy innovation networks do the various ideological coalitions in California and Denmark ideally promote?

After having distinguished the various politically desired wind energy innovation networks in both countries, the dynamics and interaction of these networks is scrutinized in Chapter 9. As we noted before, the success of a network depends on the capability of innovation actors to integrate or converge their perspectives and interests. For establishing congruent meanings among the various innovation actors, a so-called joint innovation learning process is required. The central question in Chapter 9 is:

> How did the various politically desired wind energy innovation networks in California and Denmark develop over the last two decades, and what role has joint innovation learning (or the lack thereof) played within those developments?

In Part Four '*Conclusion,*' we will look back on our analysis of the political guidance of wind energy innovation and see what this means for the general challenge of the democratic guidance of techno-economic innovation. In addition, we will look ahead and discuss the future of wind energy innovation in California and Denmark. In Chapter 10, we will summarize the comparative findings of our examination of the case studies. The central question will be:

> How can policy makers guide techno-economic innovation into a democratically legitimate direction?

Finally in Chapter 11, we will speculate on the likely future of wind energy in California and Denmark on the basis of the theories and insights developed in this book.

1.4 Methodological Approach

> I believe that, encapsulated within many everyday events – and, yes, possibly any event whatsoever – lies the essence of an entire century. *Peter Høeg, 1996: 2*

Another way of framing the above quote is 'there are no details, only essentials.' This basic principle has guided this research venture. Every occurrence, every situation, every idea and perspective has been studied in its totality and (historical and cultural) context. Wind energy innovation is not only analyzed through its technical, managerial, and political constituents, but studied as a whole; as a 'seemingly seamless web'.

In the course of the past five years, the research project has focused on reconstructing and revealing the dynamic of relations within the wind energy innovation process over the last two decades. In addition, the historical (techno-economic and political) context of wind energy innovation in Denmark and California has been examined. Studying broadly and deeply, and over a long period of time, the interaction and change of several aspects of wind energy that cover different professional disciplines, is a risky, uncommon venture that has to do without the guidance of an existing comprehensive theoretical framework which covers all the various facets. The qualitative case study research method that guided the inquiry resembles the grounded theory approach, a methodology that was originally developed by Glaser and Strauss (1967).[11] Theoretical concepts have been further developed on the basis of our case study research. In their turn, the improved 'sensitizing concepts' (Blumer 1969) enhanced the quality of the collection and analysis of data. Three principal cycles of data collection and analysis, reflection and verification can be distinguished in the research project.

In the first round, the Advocacy Coalition Framework developed by Sabatier and Jenkins-Smith (1993),[12] together with Fischer's method (1980, 1985) for the evaluation of policies, provide the theoretical concepts for organizing the research and collecting and analyzing the research material. In the first year, both California and Denmark have been visited, each for a couple of months. In this period some exploratory interviewing took place, but the dominant activity was to photocopy and acquire data: administrative documents, white papers, annual reports, proceedings of conferences on wind power, professional journals, media information, *et cetera*. These collected data were ordered within two preliminary case study descriptions. This first reconstruction of the wind energy story formed the basis for both evaluating and reformulating the initial research set up.

In the second round, our theoretical perspective was sharpened by the work of two faculty members, Grin and Van de Graaf (1996), on the interplay between policy makers, corporate managers and technologists and the work done by Callon *et al.* (1992). The preliminary case descriptions and the elaborated theoretical view represented the crucial preparation for the second two-month research period in both countries. In this second round, registered gaps in information were filled in through acquiring new data. Interviewing key actors from policy making institutions, technological communities and the management of firms was the main task within this phase of the research project.[13] Most key informants were discovered and selected on the basis of the proverbial 'snowball method.' Since the wind energy community in California and Denmark is fairly small this method works efficiently and effectively. The in-depth interviewing is not done in a prestructured manner. The interviews are not considered separately, but as inextricably bound up with the complete hermeneutic circle of interviews. This means that interviewees were regularly confronted with remarks of former informants, and vice versa, through personal correspondence afterwards. In this way, an effort was made to reveal the dynamic of relations in a consistent and controllable fashion.

While the proof of the pudding is in the eating, the proof of the research is in the writing. In this elaborate third round, the content and structure of the case descriptions and the theoretical frames for analyzing policy change and the interaction between policy and techno-economic change took up their definite shape. In this last cycle, a lot of extra desk research was done on the political and techno-economic traditions of the United States of America and Denmark. We would like to invite the reader to join the research project and enter the fourth round.

Part One

The Story of Wind Energy

2 California: Common Sense Wind Energy

We have it in our power to begin the world all over again. *Thomas Paine (1737-1809) in Common Sense, 1774*

After the Second World War, the pent-up demand for consumer goods, combined with America's efforts to rebuild Europe and Japan, resulted in a tremendous expansion of the U.S. economy. California benefited. The state became a major tourist center, agricultural producer and food processor, as well as a base for light and heavy manufacturing. California's three private and two municipal electric utilities entered into a golden age of steady growth. Each company possessed a monopoly franchise to construct facilities and provide power to a specific geographic region. Pacific Gas and Electric (PG&E) – one of the world's largest private utilities – served most of the northern two-thirds of California. The investor-owned utilities (IOUs) Southern California Edison (SCE), and San Diego Gas and Electric (SDG&E) operated in the South. The two public systems were the Sacramento Municipal Utility District (SMUD), and the Los Angeles Department of Water and Power (LADWP) – the largest municipal electric utility in the nation.

In the first half of the century, competing companies had fought for service areas, and these five electric companies had remained from the territorial battle. Now that it become clear that a company could no longer grow by expanding its service area, the utilities campaigned even more vigorously to increase electricity demand, and investor-owned electric utility sales escalated at an annual rate of more than eight percent (Hyman 1988: 193). In return for their monopoly position, the private utilities were subject to regulatory authority. The administrative body responsible for regulating utilities – the California Public Utilities Commission (CPUC)[1] – required that the utilities provide safe and reliable service to all customers in its service area at "just and reasonable" prices. At the same time, the rates were supposed to provide the owners and stockholders of these utilities a fair return on their investment (CPUC 1993: 7-10). In essence, the Commission's role was to balance the public interest and utility shareholders' expectation of a fair profit on investment (Bell and Price 1980: 248).

The primary objective of the electric industry was the achievement of lower costs and greater reliability of supply. The private utilities added the profit motive. Within the industry, an implicit consensus maintained that the proper business strategy was to expand electricity sales and construct new, bigger and more efficient central power plants. Both scale economies[2] and improvements in thermal efficiency rendered gradual cost reduction, and provided utilities with a way to both raise their profits as well as lowering electricity costs. Besides, by the economics of new power plant construction, this so-called *grow-and-build strategy* was supported by the industry's regulatory system (cf. Fenn 1984: 26; CPUC 1993: 10, 26). First, the trend toward lower cost electricity caused regulators to sit quietly on the sidelines and interfere as little as possible. From time to time, usually at the request of the utility, they allocated the benefits of declining costs. By preventing lower costs from being passed on immediately to consumers through lower rates, regulatory lags served to fatten company profit margins. At the same time infrequent rate proceedings prompted utilities to be efficient and market their product. Utility growth was further motivated by the rate design philosophy. Utilities were allowed a rate of return on their investment. Consequently, if a utility was already earning its allowed profit margin, it could increase profits only by building new power plants.[3] Finally, the utility industry was encouraged to grow and build by its "duty" to serve. This feature of the traditional regulatory compact required utilities to maintain sufficient generating capacity to meet all anticipated customer demand. Utilities traditionally chose to be conservative in their forecasts and err on the side of overbuilding rather than risk being caught short of generating capacity by an unexpected upswing in electricity demand growth. Industry regulators routinely approved the utilities' plans, as "it was a matter of faith that the utility industry was the best judge of the state's need for electricity." (Mazmanian 1992: 197)

2.1 The Arrival of the *E*-word

Mercy Mercy (The Ecology)

Mercy mercy me
Things ain't what they used to be, no no
Where did all the blue skies go?
Poisson is the wind that blows from the north and south and east
Mercy, mercy me, mercy father
Things ain't what they used to be, no no
Oil wasted on the ocean and upon our seas, fish full of mercury
Mercy, mercy me
Things ain't what they used to be, no no
Radiation under ground and in the sky
Animals and birds who live near by are dying
Mercy, mercy me
Things ain't what they used to be
What about this over crowded land
How much more abuse from man can she stand

Marvin Gaye, 1971

Around the time California became the nation's most populous state, the benefits of unbridled growth began to be challenged. The issues of expansion – more schools, freeways, houses, industrial parks, power plants,[4] and shopping centers – became a matter of public debate. In 1962, the impact of pesticide and other chemical pollution on the biota were first documented for public consumption by Rachel Carson. The publication year of *Silent Spring* is probably the best starting point for modern environmentalism in America (Dowie 1995: 23). In the same year California Tomorrow printed a little book called *California Going, Going ...*[5] The pamphlet was the first real outcry about what was happening to the landscape, water, air and civilization of the Golden State. California Tomorrow served as a sort of think-tank and agenda-setter for the planning and conservation movement in California (Hart 1984).

The old conservation movement, represented by the Sierra Club and the Audubon Society, had worked to preserve forests and establish national parks. California Tomorrow broadened the scope of those concerns in the state and claimed as its subject the 'whole environment.' By the late 1960s, the environmental crusade peaked in California as it did across the country. The first Earth Day in 1970 – the largest one-day outpouring of public support for any social cause in American history (Dowie 1995: 24) – demonstrated that environmental quality had become a broad-based concern. Several environmental disasters, such as the Santa Barbara oil spill off the coast of California in 1969, had advanced the environmental initiative. Lawmakers responded to polls that showed strong public support for both state and federal protection of environmental health. In 1970, the National Environmental Policy Act (NEPA) became the ultimate landmark of the new movement. Slightly stronger legislation was enacted in California. Governor Ronald Reagan, no friend of the harp seal or the redwoods, nonetheless signed the California Environmental Quality Act (CEQA) into law.

Not only did California Tomorrow raise an outcry about the assault on the California environment, it also pushed the idea of better planning and better government machinery as the means of turning things around. With the establishment of the Coastal Commission and California Energy Commission, in the early 1970s, the planning movement seemed on the verge of achieving much of its program. In the energy field, environmental and planning aspects had concentrated on the power plant siting issue. During the 1960s, the electric utilities met difficulties in implementing what they considered vital building plans. They complained about the size and poorly organized process of licensing. As many as thirty federal, state and local permits could be required before a utility could begin construction of a new plant (CEC 1977: 1). In 1965, the State Power Plant Committee was established in the California State Resources Agency to promulgate siting guidelines. But the permits and approvals needed, continued to increase as the word *environment* entered the American vocabulary.

The apparent need for new power plants led to popular protest. The electric utilities planned to build new nuclear power plants near the coast, in order to use ocean cooling water.[6] The magnificent coastline, however, is one of the most earthquake-prone parts of the American continent. Conservationists – concerned about coastal land use, seismic safety of nuclear plants, and water pollution – argued that the rapid rise in electricity should be dampened and required stringent regulation of utility companies. Although expressing different grounds for dissatisfaction, both conservationists and the electric utilities called for coordinated planning. The Assembly Committee on Planning and Land Use responded to these demands and contracted the Rand Corporation to carry out a comprehensive study. In late 1972, three reports were released.[7] Rand strongly criticized the grow-and-build strategy of the electric utilities and questioned the magnitude of the utilities' growth forecast. It captured the specter of electricity growth, by estimating that if the utilities effected their building plans, California's thousand-mile coastline would harbor a power plant every eight miles by the year 2000. This became a screamer headline across the front pages of California newspapers. The consultancy firm – having no ties with the power business – advised the creation of an agency to coordinate power plant siting, estimate future electricity demand, implement measures to slow the growth in electricity consumption, and manage a research and development (R&D) program for new power technologies (Ball 1972: xiii).

The three Rand reports had an appreciable impact on the Assembly, particularly on Charles Warren (D, San Jose). As a member of the Planning and Land Use Committee, Warren received responsibility for chairing an Energy Subcommittee and soon began hearings on long-term energy policy. The Public Utilities Commission's bias in favor of the growth-happy utilities became strongly criticized and its ability to fulfill its responsibilities was questioned. In early 1973, a bill emerged from these hearings proposing the creation of a demand-side planning authority, that would substantially reduce the power of the CPUC. At the same time, pro-nuclear Senator Alfred Alquist (D, Santa Clara) introduced a siting bill for creating an efficient "one-stop" siting procedure for new power plants. Eventually, Alquist's supply-side bill and Warren's conservation bill were combined. This marriage of convenience at best, was passed by the Legislature, but vetoed by Governor Ronald Reagan in October 1973. The veto reflected the concerns of local government officials and industry about the creation of a powerful new state agency that could override existing power plant siting authority (CEC 1977: 2).[8]

2.1.1 The 1973 WECS Workshop[9]

> As you also know, a wind machine built in 1940 at Grandpa's Knob, Vermont, generated about 1.25 megawatts of electricity but had to shut down because of the fatigue failure of one of the blades. We've come a long way since that time in our materials and structural design capability. I am sure we have methods available today that would have solved those problems. So there is a clear opportunity to pick up where the technology stopped and move forward with deliberate haste. Let's concentrate on keeping the capital cost down. *Alfred J. Eggers, Jr. (National Science Foundation), 1973*[10]

On the national level, the National Science Foundation (NSF) had become worried about America's energy future. In the early 1970s, projections of the U.S. energy demand showed a strong growth, and it was generally expected that a significant portion of this increase would be derived from nuclear energy and fossil fuels. Concerns about increasing dependence on imported petroleum, and about negative environmental, health, and safety impacts of this forecast, had caused the NSF to stress the importance of assessing the potential of solar energy – including wind energy – as a national, inexhaustible, and non-polluting resource. For this purpose, the Solar Energy Panel was organized jointly by the NSF and the National Aeronautics and Space Administration (NASA), in January 1972 (NSF/NASA 1972: 1). One of the Panel's conclusions was that solar energy could economically provide up to 20 percent of the electricity requirements by the year 2020. In case the Federal government took the lead in implementing a research and development program, wind energy systems could play a major role in achieving this (*ibid.* 5).[11] It had been decided to hold a 'wind energy conversion systems' (WECS) workshop in 1973 "to try to determine what was the state of the art of wind energy systems technology and what direction the future efforts should take." (NSF/NASA 1973: iii)

The June workshop had a mixed audience: utility and industry representatives, university members, and government policy makers with active interest in the field of wind power, but also pioneers. Among the wind power pioneers were figureheads of both small-scale farm windmills, and large-scale wind generators. Marcellus L. Jacobs, founder of one of the most successful manufacturers of homestead-size wind generators in America, attended the workshop.[12] Beauchamp E. Smith spoke about the 1250 kW experimental Smith-Putnam wind turbine at Grandpa's Knob in the mountains of central Vermont. He told the audience that on October 19, 1941 – 17 months after his company had accepted the risk of manufacturing the test unit – for the first time in history, synchronous wind power was fed directly to a utility electricity grid and that the project thus proved "by actual demonstration the feasibility of generating electricity in useful quantity from the wind." (*ibid.* 6)[13]

In his welcoming address, Seymour C. Himmel set the tone for the workshop. The NASA representative stated "In the development of practical wind energy systems ... there are no technological barriers to overcome. There are no breakthroughs that have to be made other than a very vital one: the economics of wind energy systems." He added, "The problem must be approached in a systems engineering fashion, which I always like to describe as solving your problems subject to a number of boundary conditions."[14] (*ibid.* 4) Looking back in wisdom, these two points of view characterized the outlook of most of the wind energy proponents who attended the workshop. There was a general feeling that developing the technology would be straightforward, and that the essential question was how to harness the wind economically.

It was 'enthusiastically received' that the "best way to get the most power per dollar out of the wind" was by "selecting and trading off configuration candidates." (*ibid.* 215) The two basic balancing factors were the height of the tower and the diameter of the rotor. William Heronemus, for example, was indifferent about the size of the rotors as long as they were mounted on very high towers. He reasoned that to get the wind "we must go where it is, namely high up, more than 500 feet above the ground." (*ibid.* 198) In contrast, Ulrich Hütter claimed that "the costs are minimum for the shortest tower and that, since the energy extracted varies as the square of the rotor diameter, the rotor should be as large as possible." (*ibid.* 207) Professor Hütter's mathematical argument had the beauty of simplicity and was met with general agreement.

It was also the consensus of the workshop that a 100-kW machine was state-of-the-art technology. An R&D program should start off with that size, and then "scaling laws" could be used "to generate larger prototypes from the performance of smaller machines." (cf. *ibid.* 211, 216) It was expected that in such a "normal engineering development program" it "should be feasible in the next few years" to develop a wind turbine with a rotor diameter larger than a hundred meters (*ibid.* 215-216). Based on the workshop recommendations, Joseph M. Savino from NASA Lewis Research Center, Louis V. Divone and a few others drew up a five-year wind energy proposal for the NSF. The objective of this program was "to develop reliable and cost competitive wind energy conversion systems that are capable of rapid commercial expansion to produce significant quantities of energy on a national scale." (*ibid.* 242) At the time, few politicians showed interest in the energy topic, let alone in sponsoring the development of a wind energy technology, and the draughtsmen anticipated that their plan would soon be forgotten. The oil crisis, however, saved the R&D proposal from oblivion.

2.1.2 The Oil Embargo's Breath of Life

> I've lived through both of these oil-crises in this country. It fundamentally changed the way Americans thought about themselves, not only in terms of energy use. We were vulnerable for the first time. I mean, standing in a gas line for an hour waiting to fill your car was about the most un-American thing imaginable. It had never happened. It was the loss of, apparently a false, sense of security. *Jan Smutny-Jones (IEP), Interview Oct 31, 1995*

A few weeks after Governor Reagan had vetoed the Warren-Alquist bill, the Arab oil export embargo hit the American public by surprise, and there was a sense of panic. In California, lines snaked down city streets for blocks and tempers rose around the gas pumps. Energy suddenly preoccupied every level of government. In Washington, President Nixon unveiled 'Project Independence', a policy aimed at making the United States absolutely independent of any source of energy outside it borders by 1985. The policy appealed to patriotism and the ideal of self-sufficiency. It was designed to stimulate the development of domestic energy sources and to relax environmental regulations that might hinder the extraction or the utilization of these resources. Wind energy also received a piece of the action. Divone[15] remembers "we laid out a plan, not expecting too many people to pay too much attention to it, and then the Arab oil embargo hit. Anything that had the word energy in it – it didn't matter what the adjective in front was – got funded."[16]

In 1975, Congress launched the Energy Research and Development Administration (ERDA). Its Division of Solar Energy carried on NSF's Solar Program. In the meanwhile, NASA[17] was testing a 5-kW commercial farm windmill, and Westinghouse had finished the construction of the advised 100-kW prototype (McCormack 1975: 75). The latter turbine with a rotor 38 m in diameter was designated Mod-0 and was ready to be used as an engineering test bed for evaluating advanced-design concepts and validating the analytical methods and computer codes used to design later-generation utility machines. Westinghouse, was subsequently hired to build an uprated version, the Mod-0A (200 kW, 38 m diameter). Four of these engineering test machines were planned to be installed to provide information on long-term performance, component reliability, and maintenance requirements of utility-size machines. A year later, NASA had scaled up this configuration and hired General Electric to build the Mod-1 (2 MW, 61 m diameter). The main goal of the Mod-1 was to obtain early operating experience with a megawatt-scale machine. Furthermore, a test facility started at Rocky Flats to test farm and rural use wind systems.

California's immediate response to the energy problem was less dramatic than the Federal one. The heightened general awareness of an 'energy crisis,' revitalized the comprehensive energy planning theme. The Warren-Alquist bill was revived in the Legislature and Governor Reagan – not cool to the idea of state planning – reluctantly signed it into law in May 1974. It created the Energy Resources Conservation and Development Commission (later called the California Energy Commission). The Commission – a by American standards "unprecedented experiment in centralized energy planning" (Hart 1984: 282) – was given the responsibility to research and forecast future energy needs, to decide on power-plant siting, to regulate the electric power consumption of appliances, to set standards for energy-saving features of new buildings, and to speed along the development of 'alternative' energy technologies (CEC 1977: 3-4).

2.1.3 The Nuclear Power Issue

In the aftermath of the first oil price shock, orders for light water reactors reached a new peak in the United States. The utility industry considered nuclear power as necessary and cheap, and welcomed the Rasmussen report as the final confirmation of the safety of nuclear power.[18] By 1976, however, a dramatic decline in the fortunes of the nuclear industry took place and evidence accumulated that neither the economic nor the safety case was as clear-cut as the 'nuclear camp' indicated. The turnaround stemmed from the downturn in growth of electricity consumption, significant time and cost overruns, and increases in regulatory requirements, which frequently necessitated costly retrofitting. In addition, there was significant public opposition to nuclear power in the United States (Evans and Hope 1984: ch. 2).

In 1976, California nuclear opponents – consumer and environmental advocacy groups, with Ralph Nader at the fore – organized an anti-nuclear initiative.[19] The focus of the campaign was on plutonium, high level wastes, and the risks of sabotage and nuclear war. The problems of reactor safety receded, but reactor closure and a moratorium on new building remained a chief object of agitation (Burn 1978: 91). Among the nuclear supporters were the leaders of organized labor, who saw the anti-nuclear vote as "a grave threat to our national security, to our precious energy supplies, and to this state's expanding employment." In addition to the utility companies, the majority of the Energy Commission publicly opposed the proposition. Commission Chairman Maullin and Governor Edmund G. "Jerry" Brown, Jr. kept a cautious silence on the matter (Hart 1984: 284).

The result in California – a 2:1 majority in favor of nuclear energy – was in line with national polls, but it had not been taken for granted. Beforehand, many legislators feared the Proposition would pass and nuclear development be curtailed. As a result, pronuclear representatives joined ranks with environmentalists – including Warren again – to author three nuclear safety bills. Although generally less stringent than Proposition 15, the bills did require real safeguards. They required that, before authorizing construction of a nuclear plant, the California Energy Commission must determine that a demonstrated technology for the disposal of high-level nuclear wastes exists (Rubin 1979: 204). After the statutes were adopted, the initiative failed. Pronuclear legislators expressed regret that they had supported the three "nuclear fuel cycle" bills, but the laws were on the books.

2.2 Solar California

The first five commissioners of the California Energy Commission (CEC), appointed by newly elected Governor Brown in the spring of 1975, had been primarily development oriented, and by and large pronuclear. Shortly, however, after signing the fuel-cycle bills in 1976, Governor Brown began to respond to complaints from environmentalists about the pro-development stance of the Energy Commission, and at the end of 1977, both the CPUC's and CEC's list of friends and enemies had turned topsy-turvy. Three political key events help to explain this shift of policy. The first major event was the coincidental outcome of the anti-nuclear struggle. The second and probably main factor was the growing public complaint about increasing electricity costs. The decade before, declining electricity prices had put antitrust sentiments on ice, but as soon as electricity rates rose and inflation struck again during the 1970s this argument against the monopoly position of the electric utilities took on new force and received massive media coverage. As a result, the utilities were not only blamed for being unresponsive to public demand for environmental protection, and for choosing risky technologies, but were also accused of being callous in passing along fuel costs' increases without the accountability of efficiency, and sometimes even of endangering California's supply security.

Although Jerry Brown had few hard and fast notions about energy policy, the political crisis atmosphere surrounding the energy issue convinced him that it was necessary to experiment and improvise. In search of intellectual stimulation and creativity, he became inspired by the book *Energy for Survival* (Clark 1974), and hired its author Wilson Clark to become his personal energy advisor. Wilson Clark and Amory Lovins had very similar philosophical ideas.[20] The latter had become the focal point of the national energy debate since the fall of 1976, when *Foreign Affairs* published his watershed essay, "Energy Strategy: The Road Not Taken?" In this article Lovins lumped together the above wave of criticism in terms of opposing the 'hard' energy path – which consisted of conventional energy sources, and was favored by the electric utilities –, and a proposal for a 'soft' energy path. Like Clark, Lovins held that the United States should begin a transition to an exclusive reliance on renewable energy sources. In 1977, the 'energy whiz kid' spent most of his time in the Golden State, where his work became widely known. As an invited lecturer at UC Berkeley, he attended debates across the state, and personally briefed Governor Brown about his energy strategy. Jerry Brown became inspired by the potential of renewable technologies to make new nuclear power plants redundant, to lower California's dependence on foreign fuel, and to generate electricity at competitive prices.

The problems connected with conventional energy sources, combined with the potential bright future of the soft energy path, led Brown decide to pay lip service to conservation and renewable energy supply technologies, to appoint 'soft energy' advocates to the CEC and CPUC, and to establish the Office of Appropriate Technology (OAT). Moreover, he encouraged his administrators to try new things. Jan Hamrin recalls that the Governor said to them "I want you to try anything new. Whatever we have been doing hasn't worked that well. If you can think of some new way to do something, try it. That's what I want in my Administration. I want people to try new things."[21] This innovative style attracted young PhD progressive types into government, and, combined with Brown's *common sense* approach, created an atmosphere of positive support for renewables, and led to a chaos of experimentation and a flood of legislation and state programs that, however, lacked much ideological consistency.

2.2.1 Disagreement about the Utilities' Role in Promoting Renewables

As Brown appointees trickled into the CPUC, its position towards utility policy and on conservation slowly altered. The sense that changes were on the way, was first signaled in a September 1975 decision. Although PG&E was naturally given its demanded rate increase, the still Reagan-dominated, pro-utility Commission nevertheless found itself saying in its official voice, "We regard conservation as the most important task facing utilities today." It went on to say, that "unchecked proliferation of power plants," could not be allowed to continue (CPUC 1993: 46). By 1978, California IOUs had still made little efforts to promote conservation. Furthermore, they had made little progress toward building cogeneration power plants, let alone toward building renewable power plants. There were many forces against the utility-use of these alternative sources.

First, the utilities did not take much interest in the new options. The dramatic changes in the electric utility business environment during the 1970s had little impact on their basic strategies. The California utilities continued to follow a strategy based on expected growth in electricity usage and additional investment in new central-station generating facilities. Pursuing conservation, renewables and cogeneration was not in line with their grow-and-build strategy, since it was expected that conservation and cogeneration would result in a loss of revenue, and small-scale electricity generating systems were perceived as costly.

Secondly and more surprisingly, an intense argument over the role of the utilities in the development of new energy forms, spread discord within the Brown Administration and thwarted concrete political action. Soft-energy proponents held that utilities should be barred from the business of small-scale benign energy sources. They championed the view that solar and conservation should rest on community effort, and should be reserved to small entrepreneurs or self-service.[22] Stressing the importance of a new social outlook, these advocates assumed that in due time, an informed public would get involved in solar power, to free itself from utility dominance. The stance of mainstream environmental groups was quite different. Major organizations, like the Sierra Club and the Natural Resource Defense Council (NRDC), were so intent on achieving conservation that they applauded utility intervention for its scale and efficiency. The solar industry took the view in the middle. The California Solar Energy Industries Association (CAL SEIA) was interested neither in undermining the utilities' monopoly, nor in their involvement in conservation. Its sole concern was that the utilities would 'take over' the solar industry.

All three opinions found some support within the Brown Administration. The orthodox theory was thought to have the Governor's ear (Roe 1984: 121). The basic challenge of his OAT was to empower people with the knowledge, resources, and freedom required to solve their own energy problems. Initially, also the CPUC was fearful of letting utilities get involved with financing solar installations, and "wanted that young, evolving industry to make it on its own." (Leary 1980: 223) In contrast, CEC's wind program attempted to stimulate utility participation and include utility funding in the wind project.

2.2.2 The CEC's Commitment to Wind Power

> The most compelling reason for an aggressive State of California wind program is the critical element of time. *James Lerner and Mat Ginosar (CEC), 1979: 132*

When the CEC wind energy program was established in 1976, California had no historical connection with wind energy and lacked industrial activity in this field. As a result, wind energy was viewed as a relatively minor element of the solar energy program (CEC 1980: 1). This judgement drastically changed over the next two years. The CEC became curious about the 'overwhelmingly optimistic' federal projections on the potential market for wind turbines. And when an initial wind resource assessment indicated that California had some excellent wind power sites (cf. Miller and Simon 1978) its enthusiasm quickly rose. The Commission realized that the federal wind energy program in several aspects might have great relevance to the CEC program.

First, California's rationale for becoming involved in wind energy was similar to that of the national government. Wind technology was seen as environmentally benign, and capable in the short term of reducing the severity of oil price escalation and ultimate scarcity. Also, the ideas on how best to achieve this corresponded. Both DOE and CEC envisioned "large scale wind systems located at remote high wind sites, integrated with the utility network, and owned and operated by utilities" (Lerner and Ginosar 1979: 139). Within this view, wind energy comprised just another generating option open to the utilities, and implied no changes in the utility system itself. Accordingly, it was self-evident that the CEC showed interest in, for example, the Mod-2 project, which had just started and aimed at developing the first wind machine for general utility use with a competitive cost-of-energy.[23]

Secondly, the last few years the federal program had met with growing political enthusiasm and financial support. In August 1977, the Carter Administration's new Department of Energy (DOE) absorbed the ERDA and gave wind power its own Federal Wind Energy Program (FWEP). Congressmen showed their faith in technological progress, and had accelerated the wind energy budget from $1.5 million in fiscal year 1974, and almost $8 million in fiscal year 1975, to approximately $25 million in fiscal year 1977. The CEC expected to qualify for a part of the federal budget. In its effort to get support from DOE, however, the agency "received enough Catch-22 responses to write a novel about it." (Johnson 1979: 48) Underlying all this was President Jimmy Carter and Governor Brown's longstanding disagreement over nuclear power. The intergovernmental conflict between the wind and solar alternatives, and the dream of centralized unlimited nuclear energy, reached its height in November 1978, when the Carter Administration published a national energy study, which told that California, in effect, needed thirty-two nuclear power plants to meet its energy needs by the year 2000. The Brown Administration resisted federal pressures to go nuclear (Rubin 1979: 204).[24] The lack of support from Washington placed California in a vulnerable front-runner position, treading new paths with no guidance from the national level.

A year before, the proposed CEC wind program had already been a clear statement from the Brown Administration that its views on energy policy significantly differed from that of DOE. Although the intention of the CEC plan resembled that of the federal program, its sense of urgency and level of commitment essentially differed. The overall year 2000 program goal was to generate at least ten percent of California's electricity from wind resources – meaning ten thousand megawatts of installed wind power. It was estimated that this would result in tremendous oil and monetary savings.[25] These rewards were believed to justify a sense of urgency in the development of wind electric power, not only in California, but also in the United States. It was argued that the potential benefits of saving time and the immense costs associated with delay dictated a 'parallel development' philosophy. A 60 million-dollar start-up program was proposed to develop three different large scale wind turbines. The machines would be specifically designed to operate reliably and economically in the California wind regime.[26] A severe criticism of the DOE program could be heard in the proposal.

First, it was claimed that an adequately funded federal program, aimed at developing several designs for various wind and environmental conditions, would make the California program redundant.

Second, it was held that the federal program suffered from an 'R&D syndrome,' which meant that it was overly technology development oriented. According to the CEC, an adequate technology base already existed, and it was badly needed to gain operational experience as soon as possible. DOE politely disregarded this appraisal. In March 1978, George P. Tennyson (1978: 156) acknowledged that "In fact we have probably produced more paper than energy."[27] Still, he immediately added "But this is the year when that is changing." The FWEP manager noticed that the coming years would be critical for the FWEP, but reassured the scientific community by revealing that the first Mod-0A machine was already operating, and that the first-generation Mod-1 would be up late that year. Furthermore, he announced that the Boeing Mod-2, designed for quantity production, was almost leaving the drawing boards, and that the second-generation machine's first turn was expected in late 1979 or early 1980.

Third, the federal program was scolded for its minor interest in wind site prospecting. By the way, it must be added to the latter two points that the CEC argued that both involving utilities and locating and confirming sites for wind power development, were tasks ill-suited to the federal program. Unlike research, development and demonstration (RD&D) of wind technology, these responsibilities could best be coordinated at the state level.

Finally, the DOE program was criticized for the fact that its principal involvement was with the aerospace and electrical industry, with minimal interfacing with electric utility companies (Lerner 1978: 13).

2.2.3 The Achilles Heel of the FWEP

The federal wind program's contractors shared this point of critique. During the past two years, they had learnt that it was impossible to "finance a production line out of a development contract or out of a bank until you have got data that says you are practically bound to get a production order soon ... [or] ... a signed order." The phenomenon that utilities were not inclined to buy wind power projects on an experimental basis was not an unexpected event. Several industry and utility representatives had pointed out during the first WECS workshop that the utilities "are not used to this. They are used to an industry coming to them with a working model with the cost available." (NSF/NASA 1973: 230) But now the contractors had experienced it the hard way. The utility companies' inertia had become a serious and real-life threat to their plans of becoming multimegawatt size wind turbine manufacturers. However, expecting some fine test results in the near future that would have direct appeal to the utilities, the FWEP manager reassured the contractors by saying "We are working on studies to give you data, and toward efforts that will lead to signed orders." (Tennyson 1978: 156)

The utility industry was also handled with velvet gloves with respect to its policy toward small power producers. Prior to PURPA, most utilities charged discriminatory rates for backup service, and purchased excess power from home wind systems for only a fraction of the retail rate (Gipe 1983: 56). Indicating that these practices put up serious barriers for the widespread use of small wind systems by home owners or farmers, the federal administrator acknowledged in guarded terms that the existing regulatory structure "was not set up to handle or favor the recent developments in alternative energy power sources." (Tennyson 1978: 159) This was no reason, however, for the FWEP and the small wind turbine manufacturers to take an adversary stance. Instead, he impressed on the members of the American Wind Energy Association (AWEA) – which represented a growing party of small home owner windmill manufacturers – "We must do it by stating that we have something to offer that the consumers, utility organizations, and the state utility commissions all need, that we want to cooperate ... We want to work with them and arrive at acceptable and profitable solutions to problems, acceptable and profitable for everyone involved." (*ibid.* 160)

Why did the FWEP manager go so gently? Tennyson was convinced that the wind machine was 'a better mousetrap,' and that, in the end, wind energy would be the least expensive form of alternative energy. Its economics would prove attractive to taxpayers and politicians, and would put the utilities on the spot. But in the meantime, the utilities' involvement in wind energy was based on the companies' own free will, and at odds with the existing regulatory system and technological conservatism of the utilities. Consequently, their voluntary and kind commitment formed the Achilles heel of the FWEP, and explains why its supervisor did not want to put the utilities' back up. Moreover, the intense debate between the utilities and the soft energy proponents had multiplied the risk of rubbing the utilities the wrong way. This theoretical dichotomy surrounded the FWEP with tension in several ways.

First of all, the ideological tone given to the energy debate jeopardized the external support for the program; both the bipartisan political support and the anticipated voluntary utility support. From the beginning, the engineers and administrators involved had described the energy problem in terms of economic and ecological criteria, and had supposed an R&D program to try out technological solutions to solve it. Solving the energy problem through technical fixes was broadly supported by most of the politicians. They wanted the rapid development of new, cleaner and cheaper supply technologies to meet the expanding energy demands and secure the future stability of the American society. Furthermore, the program harmonized with the utilities' supply-oriented perspective. Politicizing either the FWEP or alternative technologies would likely decrease its present political support and minimize its chances for support from the utility companies in the near future.

Secondly, the technological issue posed a threat to the internal structure of the program. The nuclear controversy had not only stirred up the debate about whether energy technologies should in principle be hard or soft, but also whether they should be large or small, complex or simple. The FWEP goal was to achieve the wide-scale use of wind energy, in principle regardless of size. The program distinguished large multimegawattt scale systems, hundred kilowatt scale systems, and farm and rural use systems. The technological issue added a different political flavor to each part of the program. The taste of the intermediate systems component was bland.

The 'small is beautiful' proponents associated the large wind systems portion of the program with "multinational construction and manufacturing firms, large private utilities, and the military-industrial complex, all of which had a vested interest in perpetuating and elaborating the large technological systems already in place." (Pursell 1993: 633) This argument was valid. In fact the 'suspected' involvement of key purveyors of generation equipment, was a resolute attempt to make wind energy become a normal option for the energy industry. For example, the first two contractors within the Mod-program, General Electric Company and Westinghouse Electric Company, stood out in the turbine-generator business and dominated the U.S. nuclear industry. Wind generation technology would complete the set of generation technologies their utility customers could choose from. The utility managers also had a different perspective on what was big or small. The soft energy advocates held that the draughtsmen of the wind power program envisioned wind turbines of giant proportions. Their yardstick was the electric power individuals used. For the utilities who served millions of customers, the proposal of a 3-MW machine was peanuts, and it reminded them of the small steam turbines that Westinghouse and General Electric had offered them at the beginning of the century.[28]

The small wind system element was encouraged by the soft energy advocates because of their theoretical interest in small-scale, decentralized energy production.[29] Small machines would increase social choices and their use would improve energy consciousness. Most of the utilities did not take the megawatt size turbines serious, not to mention the small wind machines. Nevertheless, as a result of the soft versus hard energy controversy, the budgeting of the federal program received explicit political meaning. The soft energy set criticized the FWEP's emphasis on the development of large turbines and argued that the program supported the present electricity power structure and hegemonic technological culture. Vice versa, the utilities felt that an increase in the funding of the development and testing of small wind turbines was threatening their monopoly position, and damaging their image as a public servant.

2.2.4 The Mello Wind Bill Taken to the Cleaner's

Assemblyman Henry Mello (D) was prepared to sponsor a bill that would require the Energy Commission to implement a state wind energy program, which would be "oriented towards testing and demonstrating the most cost-effective wind technology."[30] The Mello Wind Bill was referred to the Committee on Resources, Land Use, and Energy, which denounced the idea of a state-planned program. The Committee abandoned the initial R&D objective. Instead, it wanted a program "to speed commercialization of wind electricity systems."[31] This thwarted the aggressive wind power initiative that CEC had prepared. The Committee amendments made it entirely clear, that California lawmakers – despite their commitment to alternative energy – were not keen on funding large-scale prototype development and testing. They abhorred state involvement in developing technology, and found this should be left to the market. Of the anticipated 80 million dollars for the six-year development phase, the Senate voted in favor of only about 800,000 dollars over a three year period. This amount was to be allocated chiefly to wind resource development, which was seen as a proper state task.[32]

It can be concluded that the California wind program proposal was CEC's indictment of the half-hearted and failure-prone federal DOE program. It was a strategic move in the 'cold war' between the Carter and Brown Administrations. The political fencing was part of "a major federal-state confrontation over the control and direction of energy policy, the kinds of energy alternatives that will receive financial priority and regulatory approval, and ultimately, the kind of society we want in the 21st Century" (CEC 1979: 7). The California Legislature did not support the Administration in this intergovernmental conflict. The reverse was true. It strongly opposed aggressive state action to speed up the development of alternative energy technology. Therefore, the legislators took the sting out of the original CEC proposal, which imposed centralized state planning. Nevertheless, California lawmakers did promote the decentralized use and commercialization of renewable energy, in particular solar energy.

2.2.5 The Orphan Windmill

With its abundant sunshine, the Golden State had a long history of successful solar use. From the turn of the century until the 1930s, several thousand solar water heaters were sold in California, mostly in the Los Angeles area (CEC 1981: 30). At the end of 1960s, backyard inventors, tinkerers and do-it-yourselfers breathed new life into the California solar thermal business. In the mid-1970s, the infant industry provided more than two thousand jobs and started to lobby for the maximum feasible solar implementation in the state.[33] Many politicians started to see switching to solar as a way to create jobs and to build a strong decentralized energy industry. The government effort, therefore, championed "actions by private businesses and local groups working toward a 'Solar California' – such as local solar utilities, community-run solar enterprises, local solar codes, small solar businesses, community planning efforts, and community-based solar education programs." (SolarCal Council 1979: 10)

In 1976, the Private Energy Producer Act (PEPA) was enacted, which was designed to encourage private producers – including consumers, companies, cities, counties, districts, and public agencies – "to competitively develop independent sources of natural gas and electric energy ... solely for [their] own use."[34] A year later, Assemblyman Hart proposed a measure to promote the private use of solar energy: a 55 percent income tax credit to individuals who purchase solar systems.[35] Interestingly, wind technology was overlooked. Several reasons can be discerned. First, just like the utilities had placed their hope on the development of nuclear breeder reactors to sustain the centralized high-energy society, the soft-energy advocates promoted solar technology as a *backstop technology*. In an intellectual and speculative sense, the absolute solar solution was more attractive than the modest contribution wind energy could offer. Furthermore, at the time the extent of the wind resource was unclear. Finally, whereas the wind industry was a marginal business, the solar industry had a certain magic to it, and was politically very active.

The arguments for the solar energy tax credit revolved around multiple issues – lessening the state's dependence on imported oil, the labor intensiveness of renewable energy, and the obscene subsidies being dished out yearly to support conventional fuels (Mirviss 1983). At the time, the fact that a five-billion-dollar surplus existed in the state General Fund facilitated political experiments, and the proposal was easily accepted. This mood of financial generosity deteriorated rapidly as a result of a 'taxpayers' revolt,' that started in June 1978 with the approval of Proposition 13. This ballot measure severely restricted local property taxes, which immediately reduced local-government income by almost 60 percent, an amount exceeding seven billion-dollar. Governor Brown and the legislature decided to underwrite the local governments through the state surplus. In fiscal year 1978-1979, the state distributed close to five billion dollars to counties, cities, and special districts, keeping local budgets at 90 percent of their previous levels. In short, Proposition 13 implied budget cuts and program reductions on all levels of government. Politically, it signaled a new conservatism in the state, a movement toward economy in government (Harvey 1985: 217-219).

A program to accelerate solar space and water heating escaped the 'spirit of the new austerity.'[36] There was a strong belief that the technology was expensive and still 'just around the corner' from being really cost-effective (Drager 1979: 393). Nevertheless – spurred by business people and active citizen lobbying – the legislature enacted a comprehensive set of legislation for its promotion. The September 1978 solar legislation comprised issues like financing, job-training, utility-involvement, solar rights, and a measure protecting consumers from fraudulent businesses (CEC 1981: 11-12). Furthermore, since it was realized that California possessed some excellent wind sites, the coverage of the solar tax credit was expanded to wind,[37] to stimulate a market for small scale (1-15 kW) and medium scale (15-100 kW) wind turbines. These machines were believed to operate in a variety of applications including rural residences, water pumping, remote small communities, farms, ski resorts, and certain industries.

In contrast to the 1977 solar tax credit law, which had mainly stemmed from the infant solar business community that aspired to grow, the need for the 1978 solar/wind tax credit was defined within the CEC, and the state Office of Appropriate Technology. Despite Proposition 13, the solar tax credit expansion was 'easy to sell,' because there was no noticeable wind power development at the time. Hamrin, who wrote the law when she was at the Energy Commission, vividly recalls that Legislators thought "You want one, you can have one. There is no one building anything. No big deal!"[38] Experiencing no noticeable activity in the field of wind energy, the Legislature refrained from adjusting the rest of the solar legislation to wind.

2.2.6 The Office of Appropriate Technology

Since the mid-1960s, the Farallons Institute in California had developed a critique of American domestic technology. When *Small is Beautiful* became a rallying cry, attention to the problem of technology choice rose quickly. In his similarly titled book, subtitled *Economics as if People Mattered*, the British economist Fritz Schumacher argued that the sophisticated, highly capital technology of modern industry was "inherently violent, ecologically damaging, self-defeating in terms of non-renewable resources and stultifying for the human person" (1973: 163). Contrary to this, "intermediate" or "appropriate" technologies were defined as those that are cheap enough to be accessible to nearly everyone, simple enough to be easily maintained and repaired, suitable for small-scale application, compatible with man's needs for creativity, and self-educative in environmental awareness (Pursell 1993: 632).

The rise of Appropriate Technology was part of a broad counter cultural movement that pointed to a gap between the ideals and the reality of post-war liberalism. The technocracy of the American power elite, its slow response to racial injustice and the humiliating military defeat in Vietnam, the corporate economy, the stalemated labor movement, the energy crisis, and Watergate – these all led to a loss of establishment credibility evident throughout America during the 1960s and 70s. As part of the broader crisis in the power structure of American society, the motives and integrity of leaders in the electric power industry were maligned. The oil shock and regular black-outs, combined with the perils surrounding nuclear power and rapidly rising utility rates, turned a well-established trait in the national character – the distrust of authority – against the power industry (cf. IEEE 1974). According to the Appropriate Technology movement the energy problem was "but one manifestation of a far more fundamental crisis involving nothing less than the future course and/or survival of industrialized society." (Robinson 1982: 33) It criticized the American growth machine for causing a rapid depletion of natural resources, and scolded industrial society's narrow focus on economic efficiency. It claimed that, "the roots of many of our problems ... are in concentrated power by a few over natural, technological, political and economic resources." (OAT 1981: 6) It therefore concluded that "an essential function of U.S.

energy policy should be to break down the political power base of the major energy companies, either through the strict application of antitrust laws, or the reduction of special privileges to the industry, such as oil depletion allowances and specially tailored tax benefits." (Clark 1975: 128)

The movement passionately argued that 'local action' and 'community self-reliance' would do an even better job in decentralizing control over production and resources. They envisioned a democratic, cooperative society in which local producers would use local resources to satisfy local needs, the response of individuals on a community level to the energy and material shortages of the future" was regarded to be "crucial to survival" (*ibid.* 581). In short, Lovins and company believed that onsite solar collectors and wind generators were appropriate and resource efficient technologies, that could be tailored to the needs of the community. The use of small scale and decentralized energy technologies was a means to free the community from utility dominance, and to achieve energy self-reliance. Moreover, the sharing of these energy resources was thought to promote solidarity between community groups. Renewable energy technologies were seen as liberating technologies, which would guide community development and lead to the preservation of community values.

Governor Brown established the Office of Appropriate Technology (OAT) in 1976. The OAT comprised his most outspoken political experiment, as it was a serious effort to mobilize the ideas of the Appropriate Technology movement and develop appropriate technology. The OAT was responsible for small wind energy conversion systems, which was regarded as an ingredient in the 'quiet movement toward community self-reliance,' that was perceived to be taking place in some cities and towns in California (cf. OAT 1981). The Governor's Office worked very closely together with the CEC's Solar Office on this issue, and its approach was very practical. The OAT wanted to take "the mystery out of what needs to be done to get a wind system up and working." (Park and Obermeier 1982: 1) For this reason the state Office managed a few demonstrations. Its main emphasis, however, was on the dissemination of knowledge "to simplify the process of planning the construction of a wind energy machine" (*ibid.* 5).

Besides technical data, the state Office provided information on dealing with economic, legal, and social issues. For example, to encourage local governments to bring wind technologies in, the OAT and CEC developed a Model Wind Ordinance for small wind systems. Their 1979 publication *Common Sense Wind Energy* (Park and Obermeier 1982) was widely distributed to Californians interested in wind energy. This guide was primarily intended for the individual user of wind energy. Small scale wind systems permitted "individuals to be more closely in touch with their energy needs," and fostered "an ethic of energy conservation." (Lerner and Ginosar 1979: 140) But most readers were impressed neither by this ethic of self-constraint, nor even by the notion of 'energy self-sufficiency producing savings.' Instead, they were intrigued by the opportunity of becoming a commercial energy producer. This possibility was created due to congressional adoption of the National Energy Act in October 1978.

2.3 Independent Power

> PURPA is downright revolutionary. When pondering its potential, one cannot help but wonder how a law such as this passed the congressional gauntlet of special interests. It is enough to revive one's flagging faith in democracy. *Paul Gipe, 1983: 54*

In the spring of 1977, Jimmy Carter appeared on television, and told the American people that the era of cheap and abundant energy was over. He warned that the transition to a period of costly and scarce fossil fuel would bring hardship. Declaring this effort as 'the moral equivalent of war,' the President spurred Congress into enacting the nation's first attempt at a comprehensive energy policy. After nearly a year and a half of deliberation, the National Energy Act (NEA) passed. The NEA aimed at decreasing America's oil imports.[39] It was composed of five bills covering every aspect of the energy sector from conservation to production and distribution. Of those statutes, the Public Utilities Regulatory Policies Act (PURPA), and the Energy Tax Act (ETA) played a crucial role in the development of the renewable energy industry in California. In addition to the California solar/wind tax credit, ETA's energy tax credits created an additional financial incentive for private investment in alternative generating technologies.

Addressing the problem of rising utility rates, PURPA attempted to reform utility pricing and policy. It became most widely known for section 210 that enabled independent power producers to break up the utility monopoly on electric power generation. Funnily enough, this revolutionary provision received very little attention from either the public or private sector at the time the NEA was passed.[40] In essence, section 210 required electric utilities to buy power from qualified cogenerators and small power producers up to 30 megawatts, and to sell power to these 'qualifying facilities' (QFs) at just and reasonable rates. In one step, PURPA removed three major institutional barriers.

First, it assured small system users of backup power at fair rates. Second, the law mandated that the rates for purchases from QFs were not to exceed utility 'avoided costs.' In other words, the rates paid to small power producers for the electric energy they feed into the power lines, were not to surpass the costs the utility 'avoided' by not having to generate the power itself. Finally, it exempted QFs from laws, which regulated large-scale power producers. PURPA's implementation on the state level caused a political and analytical battlefield. Congress had left the interpretation of PURPA largely in the hands of the state regulators. The California Public Utilities Commission's favorable interpretation of PURPA was a major contribution to the wind power development in California.[41]

2.3.1 Third-Party Wind Turbine Enterprise

It is still a mystery how PURPA's section 210 came about, and the fact that it barely received attention from either coalition at the time the NEA was passed adds to the puzzle (Starrs 1988: 115). Some have argued that the public clamor over high rates had forced utilities to permit interconnections, e.g. with home wind systems, and saw PURPA as a 'goodwill gesture.' (cf. Gipe 1983: 54) Others believe that Congress passed PURPA to promote the local use of renewable energy sources, and saw section 210 as the federal equivalent of the Private Energy Producer Act in California.[42] Section 210, however, went beyond the California law. The latter was indeed aimed at self-production and only required the utilities to transmit power on the behalf of the *private energy producer*. In addition, section 210 obligated utilities to purchase power from qualifying *small power producers* at the avoided costs. The private energy producer was thought to be a producer and a consumer at the same time, a so-called prosumer. In the provisions of PURPA the connection between energy production and consumption was lost.

The probably unintended consequence of PURPA was that it opened up the utility's monopoly in the electricity generation market, and created chances for independent power producers. Some people seized the opportunity and said, "this looks like this could be good business! I am going out there and I am going to build something."[43] On the new small power producer market, several options were offered. Businesspeople and engineers alike proposed everything: cogeneration plants, geothermal power plants (cf. Willett 1980), small hydro plants (cf. Rios 1981), *et cetera*. Wind farming was another niche market that was anticipated to evolve. It was an imaginative market, a speculative market of expectations about new technological developments. It was built on the prospect that the still-untested "wind turbine technology would be developed in plenty of time."[44]

In short, the ideal of local development of renewable resources implicit in PURPA contested the ground with another, even stronger ideal – the notion of opportunity, of career, of the self-made man. These forces unleashed in the field of wind energy an entrepreneurial zeal, a rage for business, for profits, and for opportunity. PURPA created a new type of player within the field of alternative energy: the QF project developer. Contrary to the utilities who had taken a 'wait and see' attitude, the third-party wind power developers believed that an adequate technology base existed to start using wind machines on a large scale, and were strongly motivated to make wind energy 'happen.' They acted as central nodes in the wind power market of early promises, by attracting the attention of investors who were willing to assume the financial and technical risks necessary to make wind farming a reality. They negotiated for a power purchase agreement with the utility companies. Furthermore, they selected manufacturers who produced wind turbines. Simultaneously, their ideas and planning efforts defined a future market for wind technology. Trading on it, many small and new firms entered the wind turbine manufacturing business. Last but not least, appreciating the fact that the wind power market is a political market, they built up a network of valuable contacts within the Legislator and the Administration. They lobbied to protect their interests and funneled money into campaign donations.

Basically, there were four strategies for making things happen. Wind farm developers could either invest in large wind turbines – already under construction by the federal program for five years – or in small wind turbines. Moreover, they could either buy these wind machines from manufacturers or choose to build their own. All these approaches were practiced to a certain degree, and shifts from one strategy to another occasionally occurred.

2.3.2 Pursuing Large-Scale Wind Farms

In 1978, Windfarms Limited (WFL) started to pursue large-scale wind projects, anticipating the following scenario (cf. Flavin 1981: 24-25; Slichter 1982). Firms just like itself would enter the wind energy business, raise money and set up projects. This would create a growing market for wind energy machines. Wind turbine manufacturers, eager to capture their share of the market, would be spurred on to reduce their turbine costs. Cheaper turbines would allow an even more rapid expansion of the market. And at a certain point in time, the scale of investments would justify manufacturers' use of large-scale production and assembly techniques in the wind turbine industry. Mass production of wind machines would bring the costs of wind electric power down to a level competitive with conventional sources of energy.

This scenario had not appeared out of the blue. Huge corporations like Boeing and Lockheed that were involved in the federal wind energy program anticipated the same type of market development, and published studies that predicted even stronger cost reductions (cf. Lerner 1978: 3). They, however, had experience that as long as performance and reliability data were scanty or nonexistent, the utilities chose not to speculate on future developments and imminent low cost. Instead, the electric power industry preferred to "wait for the kinks to get worked out."[45] The utilities' refusal to commit themselves to the development of wind turbines ran counter to the above commercialization theory, which implied both user and manufacturer motivation. Taking advantage of the need for motivated users, WFL declared that they would fill the "gap between the manufacturers and the utility companies." (Flavin 1981: 32) The announcement put an end to the innovation deadlock that had existed since the beginning of the federal wind program, and gave the potential manufacturers of large scale wind turbines new hope.

In mid-1980, WFL contracted with the Hawaiian Electric Company to construct an 80-MW wind farm made up of 4-megawatt turbines built by Hamilton Standard, which represented "the first large wind energy project undertaken jointly by an investor-owned utility and private-sector financing." (March *et al.* 1982: 18) Showing faith in its own product, the wind turbine manufacturer both took an equity position in the project, and provided guarantees on the performance of its WTS-4 turbines (*ibid.* 19). Nevertheless, the privately funded prototype WTS-4 proved a lemon, and the firm left the wind business in 1981 (Serchuk 1995: 231). WFL was not deterred by Hamilton Standard's retreat. Furthermore, together with the Denver-based Aero Turbine Energy Company, the firm was not put off by the meager test results of the Mod-1 machine.[46] Both led by the opinion of the FWEP administrators that the Boeing machine was "a complete design with very careful cost estimation worked out" (Tennyson 1979b: 2) as well as by the trustworthiness and technological reputation of its manufacturer, the two wind farmers determinedly shifted their ambition and expectations toward the federal program's Mod-2 turbine.

In the meantime, the investor-owned utilities in California seemed to have lost their serious skepticism about the technological potential and economical feasibility of renewable energy technologies. In 1980, Southern California Edison (SCE) stunned the industry by committing itself to a corporate strategy based on renewable resource development. In a letter to all company employees, SCE chairman William R. Gould wrote "As a result of some significant successes in a number of research and development areas, we now believe that some forms of power generation [wind, geothermal, solar, fuel cells, and small hydroelectric] which a few years ago were speculative or unproven have progressed to the point that they can be aggressively developed and relied upon to provide a significant part – perhaps about 30 percent – of the electricity to supply the additional needs of our customers later this decade." (quoted in Fenn 1984: 89) In 1981, PG&E announced that it too would concentrate on the development of renewables. SCE planned to add 120 megawatts of wind power to its grid by 1990, but PG&E considered an even more ambitious goal. In 1981, the PG&E signed a letter of intent with WFL to undertake the joint development of a 320-MW wind farm to be located in a windy pass in Solano County (March *et al.* 1982: 20), which was thought to be sufficient to give the Boeing Company enough incentive to set up an assembly line (cf. Harding 1982).

Apart from supplying the location, PG&E did not take any chances. First, by contracting with Windfarm Limited, the investor-owned utility shifted the economic risk to the third-party developer. Michael Russo of PG&E's planning department explained "We have benefits from this in the power which we will be getting and in the elimination of capital costs we would otherwise have to pay if we wanted to handle such a project ourselves." (quoted in Fenn 1984: 92) Second, the power produced would mainly substitute for oil-fired electricity. PG&E intended to pay the wind farm developer the same price it would normally pay for oil-based generation of electricity. In addition, there was a guaranteed market for the wind power. The state's Department of Water Resources agreed to buy the electricity generated during off-peak times for pumping purposes. In fact, the wind farm project was at the risk of WFL. The electric utilities confined themselves to investigating the wind energy potential in their service territories, to wind turbine design testing,[47] and to keeping an eye on private developers erecting wind turbines.

Although, in practice, the utilities' actual involvement in wind energy was not particularly impressive, the utilities' apparent shift toward renewables suggested that they were making a major break with traditional utility planning concepts (Fenn 1984: 89). The dissolvement of the CPUC's objection to utility involvement in the field of soft energy technologies towards the end of the 1970s, serves as an important explanatory factor. The Environmental Defense Fund (EDF), a national environmental organization, played a key role in weakening CPUC's opposition against utility involvement and changing utility's attitude toward small-scale power plants.[48]

2.3.3 EDF's Businessman's Approach Toward the Use of Renewable Resources

For most environmental groups, conservation was a virtue. Since virtues do not seek profit, it was perceived and accepted that conservation would be necessarily expensive. The EDF contested the truism, 'It doesn't pay to save the world.' Led by economist Zach Willey, EDF argued that conservation-mindedness would profit both consumers and utility's shareholders (Roe 1984: 59). David Roe argued that Willey's perspective had "the soulless texture that is the economist's trade mark ... If it had any ideology at all, it was the ordinary ideology of capitalist industry." (*ibid.* 12) In industry's terms, EDF combated utility-plant-building as economically disastrous in repeated rate hearings before the CPUC. Its position gained much sympathy in the upper reaches of the CPUC after its claim had been strengthened by computer calculations. In 1978, EDF first used its so-called ELFIN model in a PG&E rate proceeding.[49] The investor-owned utility was fined for 14 million dollars (*ibid.* 165). The CPUC criticized PG&E's management for its "minimal level of support" for the development of cogeneration.[50] EDF's computer model showed that the principles that the electricity industry preached for its own parish did not lead to giant power plants, but instead to the very things that the utility industry was vigorously trying to fend off. It revealed that a 'renewable and alternative energy supply' strategy might be more profitable, for both the public and the utility, than the conventional grow-and-build strategy. Small was no longer merely beautiful. EDF's argument was that it was also cheap (*ibid.* 119).

The rate of return penalty imposed on PG&E was the first indicator that the CPUC was losing its objections to utility involvement. Severe changes in the economic conditions of the electricity sector caused the Brown Administration to stop its policy of keeping back the 'soft energy solution' from the utilities. Throughout the decade, rising fuel costs and inflation had weakened the financial health of America's electric industry. By the end of the 1970s, the financial position of the California utilities was so precarious, that state regulators could no longer afford the luxury to deny EDF's claim. Tremendous pressure to balance the financial health of utilities and protect consumers from unreasonable cost increases, pulled Jerry Brown from the ideological barricades. The CPUC announced that it would use its authority to direct utilities into large-scale conservation. According to its President, this effort to redirect the focus of energy regulation toward conservation "is a very fundamental change, reflecting the economies of the energy supply."[51] In addition to its venture into energy conservation, the CPUC declared itself in favor of both dispersed producer, and utility investment in alternative sources of energy. These moves deprived communities and small entrepreneurs of their privilege to political support on conservation and solar energy. The unexpectedly low growth of solar energy in the last years made insiders around Governor Brown realize that time was too short for a surge of popular interest to pay off.

In retrospect, the soft energy intellectuals had anticipated that the masses would share their convictions as a matter of logic and principle. They had assumed that a whole number of communities and small entrepreneurs would get involved to make conservation and solar work. As a result, their expectations on the development of conservation and solar energy systems had been high. On top of that, the idea that hardware would follow ideology had led the energy specialists in the Governor's office to believe that extensive political action was superfluous.[52] In due course, however, their common climate of 'absolutist enthusiasm' became disturbed.

First, Brown's advisers discovered that public awareness of solar development was low. This violated their Jeffersonian faith in participatory democracy as the best means to solve the energy production problem. Moreover, they saw an explosion of corporate interest in solar energy. This was contrary to their expectation that the widespread use of understandable alternative technology would restore honest competition in the energy business. The fact that, along with small manufacturers, big oil companies had been moving into the solar business refuted the notion that appropriate technology 'runs up against the brick wall of vested interests'. The very complexity of the historical process itself had proven the orthodox soft energy doctrine wrong. The soft energy advocates in power were among the first to admit this. Time had played a dirty trick on them. At one moment in history, the rise of the Appropriate Technology movement had seemed to coincide with the rise of the utility industry's problems. More than just offering a solution, the movement's ideal program of reform had been thought to make the existing utility monopoly obsolete.[53] But history had not moved in a straight line. The quandaries of the industry had kept on growing, but the support the reform movement found had not. This had broken the assumed symmetry between the industry's problems and the movement's solution. The Environmental Defense Fund had used this stalemate to promote the 'soft energy path' as a technical solution to the problems of the electric utilities, rather than an ideological one.

In short, the orthodox soft energy advocates had championed appropriate technology as a means to guide community development and minimize resource depletion. They had criticized industrial society's focus on economic efficiency, and expected that common sense and action at the local level would compensate for the inability of its institutions to adapt to changing conditions. The pragmatic 'soft' energy proponents shared the 'loss-of-flexibility' criticism, but didn't blame capitalism for it. They held that economic self-interest was a powerful argument. These 'hard' 'soft' energy advocates[54] argued that the 'hard-headed principles' based on 'the ordinary ideology of the capitalist industry' would impose the alternative energy solution to the utilities (cf. Roe 1984: 12, 79).

The investor-owned utilities reluctantly adopted the businessman's approach of EDF. Their construction plans were substantially off track, and caused serious cash flow problems. Four nuclear power generating stations were behind schedule, with no assurance they would ever begin operating. For example, PG&E's Diablo Canyon nuclear project at San Luis Obispo was supposed to have been completed in 1973 at a cost of 350 million dollars. But, by the turn of the decade, its cost had quadrupled, and the two 1,100-megawatt reactors were still unlicensed. Moreover, the near meltdown at Three Mile Island on the East Coast in Pennsylvania in March 1979, had triggered a nationwide chain reaction, complete with protest rallies, confrontations, and political clashes (Rubin 1979). The construction delays led to serious financial problems, and spread concern about capacity shortages.[55]

With the Legislature, the CPUC and the CEC on their back, the California utilities reluctantly admitted that, "The age of the dinosaur – the large, central power plant with a ten-to-fifteen-year lead time – may have been passed." (Fenn 1984: 92) The grow-and-build strategy was no longer perceived as the cheapest and most reliable way to meet the growth of customer demand. Aside from the regulatory environment, the major reason for the growing interest in renewables was that their small scale and modularity were thought to give utilities greater flexibility and reduce financial risks (*ibid.* 88-94). Photovoltaic plants, geothermal and biomass plants, and wind farms, could all be built in two or three years. Both trying to please the regulators, and seeing renewables as a potential business opportunity, the utilities started to set up research and development programs. Maybe more important, these efforts were attempts to restore their public image. Christopher Flavin (1981: 30) of the Worldwatch Institute commented that "PG&E and SCE seem to be playing a game of leapfrog as each attempts to one-up the other in a fight for leadership and public recognition in wind energy development."

2.3.4 Oil Prices and Investor Confidence

For each wind farmer, raising capital for development work was of vital importance. Encouraged by the state government[56] and charmed by the potential for financial success, California's financial community showed itself willing to invest in commercial wind development. There were several approaches to raising capital (cf. March *et al.* 1982: 20). WFL – founded in 1978 by Wayne Van Dyke, a venture finance specialist – arranged for debt and equity financing from such conventional sources as insurance companies and investment banks. The San Francisco-based firm expected to use large multimegawatt turbines purchased from manufacturers such as Boeing and Hamilton Standard (see above). The company anticipated that once the turbines were manufactured in quantity, wind farms would generate electricity for 4-6 ¢/kWh, compared to the at the time average utility oil cost of 6-7 ¢/kWh (Slichter 1982). Thus, the profit potential of large-scale wind farms seemed formidable. Furthermore, their rate of return would not be subject to state regulatory authority (March *et al.* 1982: 81-82). On the widely held assumption that oil prices would keep on rising, WFL's method proved successful. By 1982, the company had contracts with three oil-dependent utilities for more than 450 megawatts.

The wind farms were expected to begin generating power around 1985, but, for various reasons, have never materialised. The most straightforward reason was that the technology was lacking. The Mod-2 technology soon turned out to be troublesome.[57] The chances that an aggressive federal R&D program would correct its serious faults had disappeared with the arrival of the Reagan Administration. Arguing that "the technology was proven and economic, and deserved no further support," the Administration reduced the budget from 89 million to 19 million dollars for fiscal year 1982 (Slichter 1982: 49). Boeing engineers had already started to modify the Mod-2, but the Boeing management felt hesitant about the quality of its own product. Experiencing a strong decrease of political support for alternative technologies, Boeing Aerospace Company judged that it was too risky to further its plans of moving into the business of manufacturing megawatt scale wind turbines. WFL proved to be a go-getter. The firm decided to set up its own R&D program, and succeeded to arrange 105 million dollars for it. In 1984, however, the prices of oil started to fall. Eventually, the disastrous effect of falling oil prices on the cost-competitiveness of wind technology caused the impressive deal to collapse (Serchuk 1995: 231).

2.3.5 Tax Credits and Investor Confidence

> The Department of Water Resources has associated itself with private industry in ways that set precedents and open new and exciting possibilities for future cooperation between the private sector and government for meeting energy needs.
> *Huey D. Johnson (California Secretary for Resources), 1979: 51*

The vast majority of the wind developers opted for smaller sized wind turbines, and chose a different method to raise capital for building wind farms. Their strategy was to attract individual investors interested in sheltering taxable income. In this approach the individual investor purchased a wind turbine, which the wind farm developer installed in one of its wind farms. The investor next entered into a management contract with the developer to service the equipment and to arrange sale of the power to the utility. In this way, the investor's income is derived from income produced by the sale of electricity and from the benefits of the tax shelter.

This approach was pioneered by the Massachusetts-based firm U.S. Windpower (USW). The company became known as the lone wolf of the wind power business, because of its clear-cut, and 'stubbornly self-sufficient' course of vertical integration: operating its own turbines in its own development projects (Davidson 1989a).[58] In 1979, the firm developed the nation's first PURPA-protected wind farm. It was located in New Hampshire and had twenty 30 kW turbines. The project turned into a big flop, but its failure did not upset the company. Paul Gipe remembers that "When they put up the first 40-foot diameter machine the thing broke, but they were already designing the next one, the 56-50, and they were already raising money for that."[59]

The money actually was raised in the Golden State, where the Department of Water Resources (DWR) was playing an important pioneering role in advancing private renewable energy projects. Headed by Ron Robie, DWR was already engaged in several geothermal energy projects, and became also inspired by USW's artist concepts of the Altamont Pass covered with thousands of small wind turbines, like flowers on a meadow.[60] Excited by the futuristic pictures, the largest single user of electrical power in California decided to enter into an association with USW "for power from what may develop into the first large-scale commercially feasible wind project in the country." (Johnson 1979: 51) This venture caused USW to move westwards in 1980, where the private company started to erect its new 56-foot diameter, 56-50 kW machines for the 100-MW project. The technology was disappointing and very unreliable, but after solving many technological troubles and more than seven hundred turbines and sixth versions later, the company eventually developed a dependable and high performance machine before the end of 1983: the 56-100 (56 foot diameter, 100 kW).

2.4 The Wind Farm Rush

> America has a tradition of individual and corporate independence in times when our government fails to provide real leadership and seems impotent in dealing with crisis situations which affect all parts of our society. ... In California alternative energy options are being converted from dreams to realities and from drawing board to operating hardware in a number of significant and exciting instances.
> *Huey D. Johnson (California Secretary for Resources), 1979: 50*

The possibility of putting up wind farms using small wind turbines proved to be an inspiring and attractive idea. In 1982, some fifteen companies were already developing projects, and only two years later this number had doubled. The entrance of wind turbine manufactures on the California market kept pace with this increase.[61] Farm windmills had been optimized with respect to the consumption of one family or one small factory. This effectively limited the capacity of these little mills to approximately 20 kW. Wind farm developers were not bound by size. In contrast, economies of scale favored larger turbines,[62] and caused a strongly growing demand for medium-sized turbines. In 1982, more than a thousand wind turbines were installed and during the 'gold rush' period almost five thousand machines were installed annually.

Governor Brown's Administration judged favorably on the efforts of the private industry. State administrators had tried to convince the Legislature, the Congress, and the Department of Energy that renewable technologies could innovate faster. Their attempts, however, had not been successful. The Californian lawmakers had rejected the idea of an ambitious centrally planned wind power program. Secondly, the federal government was spending only five percent of its R&D budget to various forms of solar energy. Finally, they had not succeeded to get aid for wind energy from DOE. Disappointed by the lack of political support, the Brown Administration was "excited to see that the alternative energy options neglected by politicians and federal administrators" were "picked up by private industry, concerned individuals and organizations." (Johnson 1979: 50) Dying to suit hard action to their soft energy words, the Brown administrators took advantage of the new feverish activity induced by PURPA. They offered services and information to the QF developers, and strongly stimulated wind developers to capitalize California's ideal sites for wind energy production. In short, the Brown Administration became a mother to the burgeoning solar and wind industry, and made it feel at home in the Golden State.

2.4.1 The Burgeoning Wind Industry's Technological Problems

The vast majority of the wind farm developers had a background in financing. They were interested in making money, and setting up wind farms was a way to do that. The tax benefits attracted a lot of investment money from rich individuals to wind energy projects. Therefore, the demand-side existed for both individual investors who were interested in a short-term tax shelter and who were not concerned about wind turbine technology, as well as wind farm developers who were unsophisticated with respect to the technology.[63] On the supply-side there was only a handful of American manufacturers in the early eighties who could deliver wind turbines larger than 20 kW.[64] The technology these companies sold was weak and underdeveloped.[65] The influx of capital to inexperienced and unknowledgeable wind farm developers created a demand for untested and immature technology. The regulatory system did not do much to prevent the commercial development of low performance, uneconomic technologies with poor longevity.

First, although the solar tax credit had been expanded to wind in 1978, the Brown Administration had not adjusted the rest of the solar legislation to wind. In particular, measures for protecting consumers from fraudulent businesses and bad products, which had been introduced with regard to solar energy technologies, were lacking in the field of wind energy. The main reason for this was that it was anticipated that solar technologies would come in at a rapid speed, while wind energy was expected to show a slow development. In retrospect, the administrators overestimated the development of solar, and underestimated the development of wind energy.[66]

Secondly, the tax credit system tied the level of the subsidy to capital investment rather than to energy production. The fact that proponents, for fear of its abolishment, were kept from raising this matter, prevented improvement of the tax credit program.

Thirdly, until 1985 the Administration made little effort to provide up-to-date factual and reliable data on the quality of the technology. The whole wind business suffered from a serious lack of trustworthy technological information.

While the murkiness of the market enabled competition on an unreal basis, the fierce competition between American manufacturers seemed to justify it. The technological naivete of most of the wind developers, the lack of authoritative product information, combined with the cut-throat mentality that characterized the American business culture, created a market where manufacturers were competing on numbers that did not mean anything. Or, as Paul Gipe expressed it, "They were competing on dollars per kilowatt of installed capacity, and on fancy color pictures."[67] The absence of institutional mechanisms that would have protected consumers from buying bad technology, combined with very favorable public subsidies, caused a form of unbridled competition and prevented a style of competition based on technological merits. At the root of all this was the fact that all actors in favor of wind energy – policy makers, politicians, wind turbine manufacturers, and wind project developers – failed or feared to acknowledge that an appropriate and sufficient technology base was lacking to start up rapid commercialization.

During the 1970s, most manufacturers had lacked the means to achieve their goals of reducing the cost and improving the reliability of their small turbines. The sales had been so low that producers "could not even afford to introduce improvements they knew could greatly boost performance." (Flavin 1981: 22) Once wind farm developers started to express their need for intermediate systems, many manufacturers of residential wind machines had not even been able to match the performance of the 1930s era windchargers (*ibid.*). This unforeseen demand for medium-sized turbines made the inadequacy of the manufacturer's technology even more succinct. The technology not only lacked reliability and longevity, and was not only too expensive, all of a sudden it also was too small. On the positive side, money poured into the capital-starved industry, which enabled manufacturers to advance the changes they had wanted all those years, and incorporate them in the design of their medium sized turbines. They were confident that by pushing "the state of the art as far as was practical in the direction of high speed, low solidity wind turbine design," they would kill two birds with one stone. The use of 'slender lightweight, fiber glass blades,' 'high-speed turbine design,' and 'flexible towers' were believed to lead to bigger, less complex and more cost-effective wind turbines (cf. Flavin 1981: 21-22; Heymann 1992: 21).

Thus, although in the early eighties, the manufacturers had barely been able to bridge the technological gap that existed between their products and the machines of their predecessors from the 1930s, they were already getting ready to use new technological concepts for building wind systems at three or more times the size than they were used to. In other words, they wanted to start dancing before they had learned to walk. Moreover, attracted by the potential for financial success of such a risky and ambitious technological approach, investors and wind developers were willing to assume the financial and technical risks necessary to make wind farming a reality. This cash enabled manufacturers to wipe the slate clean with regard to the technology, and to start fulfilling their dreams in all respects; from a technological, economical, and sometimes political point of view. In this way, technology needs and technology promises began to reinforce each other and filled the California air with high hopes.

The initial wind farms were nearly all experiencing severe blade fatigue problems. The perception that these first failures were start-up difficulties and just normal growing pains of a young industry (cf. Flavin 1981: 22; Tyler 1982: 107) prevented a timely contest of the belief that applying 'known materials and known technological concepts' to wind machines would result in fast technological progress. Consequently, several U.S. companies were already backlogged two years on machine orders before all kinds of mechanical, electrical, and aerodynamical problems showed up on all sides, and revealed that the nascent American turbine industry was undergoing structural problems (Tyler 1982: 107). In theory, the richly filled order books provided the means for rapid technological innovation, but, in reality, had a stranglehold on the 'successful' American companies. To meet the demand, manufacturers put most of their effort and capital into putting up new production lines, as their first priority was to deliver their turbines to their sponsors, namely the wind farm developers, on time. Because the latter agreed that the optimum size would be in the 100-500 kW range, the producers' second priority was to scale up their turbines for the near-future market.

Strangely enough, technology development to a certain extent was of no direct concern to the manufacturers. It was only third rank, because it was felt that technology was related to the future market. Although urgent technological problems were solved on an incidental basis, more structural technological problems were often shelved, because the process of analyzing problems and redesigning wind turbines was very time-consuming. Since ignorant wind farm developers did not have the expertise to value proper solutions, structural solutions did not directly lead to more profit. The full order files of the 'better' American companies also gave lower-performance and even non-performing turbines an effortless market share, and provided American and foreign newcomers a relatively easy access to the California wind turbine market.

2.4.2 Tax Credits under Fire

> Assembly GOP leader Robert Naylor, and other Republican lawmakers, have been the greatest surprise allies. Although renewable energy advocates have argued for years that the issues (decentralization, small business development, national security, etc.) are all 'conservative' issues, this is one of the first times in California that Republicans are taking a leadership role. *Alan H. Mirviss (Director SolarCal Council), 1983: 3*

The arrival of Republican Administrations in Sacramento and Washington brought fresh topics to the political debate, which placed the tax credits under fire. In 1980, the newly elected President Ronald Reagan made his first attack on the federal tax credits, but his effort was turned back by bipartisan support of the emerging industry.[68] Three years later, George Deukmejian replaced Governor Brown and surprised the industry and the Legislature with a proposal to retroactively eliminate California's tax credit. Pointing at the one billion-dollar state budget deficit, Deukmejian had promised the voters to undo the 'damage' brought to California by Brown. The new Republican Governor gave an early indication of the new political climate both by abolishing the Office of Appropriate Technology and by proposing to eliminate the solar tax credit that had been two proud symbols of his predecessor's soft energy policy. However, as occurred at the federal level, a bipartisan coalition of legislators turned back the Administration proposal.

The surprising support from the Republican lawmakers reflected the material and political change the renewable community had gone through during the last years. At a time of severe unemployment, the wind (solar) industry employed more than 750 (20,000) Californians and, in the meantime, rang up about a hundred million (one billion) dollars in annual investments (Mirviss 1983). By promoting themselves as "a new breed of capitalists, using the market to propagate an environmentally sound and virtuous energy option – and to enrich themselves in the process" (Hamrin quoted in Serchuk 1995: 245), the California QF-developers broadened the moral space of the renewable energy community. The renewable energy field was no longer the sole terrain of activists opposing the established electric system. It became invaded by Republican businesspeople who set hard-nosed short-term goals and whose greatest wish was to increase their companies' credibility within the financial and utility communities.

Lost in anti-Jerry Brown rhetoric, Deukmejian had overlooked these changes. Criticizing the ex-Governor for being anti-business, the renewable energy businesspeople confronted him with their view that it was "ironic that Deukmejian would say he supports business in California and then as one of the first acts sends these businesses packing to other states." (Mirviss 1983) Considering his extreme vulnerability to such charges, the new Governor felt he could not ignore the bipartisan support of the credit. His power politics, however, impeded a constructive political debate and limited the options to either extending or abolishing the tax credit program.

It was clearly a missed opportunity, since the two fundamental problems with the structure of the credits were already widely acknowledged at the time, by both opponents and proponents (see Starrs 1988: 140-145). The first problem was that the credits rewarded capital investment (in wind turbines and associated equipment) rather than production of energy. The second problem was the tax sheltering qualities of investments in wind energy. Investment opportunities were tailored to individuals who could invest perhaps $150,000 so as to benefit from tax credits totaling $75,000 in the first year alone.[69] As a result, wind farm developers particularly solicited funds from people in the upper-income brackets, who were seeking a write-off on their federal and state income taxes. This practice nourished the idea that wind farms were no more than tax scams, caused the conservative financial periodical *Forbes* to sneer about "the great windmill tax dodge" (Paris 1984), and made the state's top income tax collector call the solar credit "welfare for the wealthy." (Walters 1983: 256)[70]

Proponents were well aware of the tax credit program's flaws. Especially the CEC looked on the credits sadly, because investors obtained tax benefits whether the equipment performed adequately or not. The CEC principally favored an immediate legislative solution that tied the level of the subsidy to energy production of the wind turbines rather than to their cost. But recognizing that "reopening the legislative issue before the credit's new 1986 sunset" was "like opening Pandora's Box," the Commission chose a moderate administrative response, and developed a program that called for voluntary industry reports on the performance of their installed machines (Edson 1983: 227).[71] When the program started in 1985, the damage was already done. It is doubtful whether the program would have prevented abuse of the tax credit. Because the expected revenue from the sale of electricity was a small fraction of the initial return from the investment, investors had little incentive to ensure that the wind farms were designed and constructed properly.

2.4.3 The Interim Standard Offer 4 Contract

> We were supposed to deliver in five weeks. For four weeks of it we argued about the basis of the price and we got nowhere, just absolutely anywhere. The utilities used this gas-forecast. I said 'You should be using a coal-plant proximity, because that is what you have in your resource plans.' We got no agreement. Then finally, somebody put a screen of numbers on the table in front of me, and said 'No, we want this based on the gas-forecast!' I looked at the numbers and I said 'Timeout, a minute.' I talked to the people I represented, and said 'I don't think it matters what the basis is, I think we've got a number we can do business with.' *Jan Hamrin (Former Executive Director IEP), Interview Oct 25, 1995*

Besides the tax credit programs, PURPA played a key role in the development of the wind energy industry in California. Section 210 had required the Federal Energy Regulatory Commission (FERC) to establish rules forcing utilities to purchase power from qualified facilities, at the utilities' avoided cost. After FERC had promulgated its final rules in the beginning of 1980, the California Public Utilities Commission carefully monitored the quarterly figures for energy payments and the utilities' latest proposed offers for cogenerators and small power producers (Bryson 1980: 70). Despite PURPA's mandate, and while they were paying lip-service to a corporate strategy based on renewable resource development, the utilities in California were slow to comply with the FERC rules, and failed to negotiate power purchase agreements with third-party power producers. Driven by common frustration, several independent power producers recognized the need for a trade organization. In April 1982, the Independent Energy Producers (IEP) association – a *contradiction in terminus* – was founded and started to intervene in the CPUC process, with notable success.

The CPUC's response to the utilities' laxity was twofold. At the end of 1982, it penalized SCE for misusing their monopoly position; in other words for abusing their negotiating advantage as the sole buyer of electricity.[72] Secondly, the CPUC had required the utilities to develop QF contracts, or Standard Offers, based on both short and long-run avoided cost (CPUC 1993: 65-67). The utilities submitted three short-run standard offers, which were approved by the Commission in December 1982.[73] Both the IEP and the CPUC were "somewhat surprised that so few independent power projects signed up." (IEP 1993: 9) Because the technologies had no track record and there was little experience with putting up projects under the short-term contract provisions, these projects were extremely difficult to finance due to "leery lenders concerned about uncertain revenue flows." (*ibid.*)

The CPUC was easily convinced by the IEP about the fact that a long-term contract was essential to bolster the fledgling industry. To speed up the development of contracts for long-run avoided costs, the CPUC abandoned its customary litigation process and instead ordered the utilities and IEP to negotiate an 'interim' long-term contract option in the Spring of 1983 that would at least allow the independent power producers to do business. The Brown appointed commissioners planned to have a regular proceeding after the settlement "to look at all the methodologies, and to figure out the correct way to do all this stuff."[74] Furthermore, the utilities were given a deadline. Jan Hamrin, who represented IEP at the negotiations, recalls that "The CPUC told all three California investor-owned utilities, 'You come back in five weeks and you have a long-term contract or else we will give you one, and you probably like what you would do better than what we would do'."[75]

Helped by this big stick, the negotiations – though difficult – bore fruit. From them emerged the standard contracts based on the utilities' own forecast of future gas prices.[76] In September 1983, the Commission approved these contracts, and called them Interim Standard Offer Number 4 (ISO4). It was expected that no more than one thousand megawatts of QF projects would sign ISO4 contracts, but for a variety of reasons, conventional wisdom underestimated the market response to ISO4.[77] The availability of the interim contract in combination with the energy tax credits caused a real investment boom. By 1984, the state had more than ten thousand megawatts of capacity on contract; by 1985 the number exceeded fifteen thousand. In 1986 these small power projects, particularly gas cogeneration projects, were producing approximately 6 percent of California's electrical supply (Starrs 1988: 135).

This resounding success became a subject of concern for the utilities. Facing a capacity shortfall in the early 1980s, the "utilities had desperately wanted that QF power and it was worth every penny that they were paying for it. ... They were signing up contracts and being fairly cooperative."[78] But once the utilities finally brought their nuclear units on line, and therefore the risk of being caught short of generating capacity faded away, the "conflict and war between the utility and the independent power producers" started.[79] In the utilities' eyes, the independent power producers changed from 'ministering angels' to serious and dangerous political and business competitors. Firstly, the successful and rapid development of small power projects made the utilities really nervous about the possibility that the regulators might decide to disallow their current billion dollar investments in the nuclear plants. In addition, the exponential growth of QF contracts represented a serious threat to their future market share, i.e. the quantity of capacity the utilities can make money on.

However, things were not as bad as they looked for the utilities. Unexpectedly, their call to suspend the ISO4 subscription became supported by the IEP. The explosion of contracts had made the trade association worry that the attractive contract would incite its members to sign more contracts than they would be able to deliver, and thought that it was in the industry's best interest to get it suspended.[80] Justified by both sides, the Commission suspended ISO4 in April 1985.[81]

2.4.4 The Danish Dominance

Although the financial incentive programs were effective in attracting investment funding for the wind industry, they were not efficient and permitted "fly-by-night operators to compile fraudulent investments, that were clearly at odds with the policy goals of the credit." (Starrs 1988: 140-141) The media's often negative image of wind energy as a tax shelter, and fraudulent and even criminal cases, combined with the industry's poor performance record, had thwarted the high expectations of the early eighties, and had squandered the trust of the private investor in the American wind energy business.[82] The Danish wind turbine industry escaped the credibility crisis in a miraculous and inventive way, and actually managed to change the situation for the better.

The import of Danish turbines during the second half of 1982 caused the market conditions in California to change. This was initiated by the availability of the superior technology of the Danes. More important in the long run, however, the business climate in California altered because of the gradual introduction of the Danish style of doing business and the Danish way of managing wind technology. On the Danish home market, only certified turbines qualified for subsidies. Moreover, the buyers of wind turbines were committed to wind energy, and were primarily interested in the price per kilowatt-hour performance of the technology. In fact, the production data of almost every single wind turbine was registered and published quarterly by the windmill owners association. The transparency of the Danish wind-turbine home market had forced manufacturers to compete on performance figures. On the obscure California market their certifications and production projections came in handy to provoke some first minor delivery contracts. Experiencing poorly performing wind farms, and realizing that the American manufacturers were not able to live up to their promises, some serious wind developers were looking for alternatives, and were willing to try some 'exotic' European wind turbines.

Although, initially, the Danish turbines suffered from many of the same problems as the American machines, the acquaintance with the Danish technology was a relief for many wind farm developers.[83] The main reason for this was the higher endurance of the more robust Danish machines that stood the test of time in the rough wind climate several times better than their American competitors. Paul Gipe comments, "Sure, maybe the Danish machine would only work for three years before it needed some major fix, but the U.S. turbines would work for only one year or even shorter, before it needed a big fix. To me it means three times better."[84] While repairing American turbines was like swimming against the tide, the Danish machines gave both the developer and their manufacturer just enough space to take two steps forward before they had to take one step down.

In short, while the American producers slowly faded away in the technological swamp they had created for themselves during the days when the one-eyed man was still king in the country of the blind, their Danish competitors succeeded in keeping their heads above water and managed, by the skin of their teeth, both to gradually improve on their apparently just sufficient technology basis, and to build up a steady relationship based on trust with several American wind farm developers.[85] As a consequence, the Danes gradually came to dominate the wind turbine market in California. In 1983 and 1984, their market share was 11 and 33 percent, respectively. In 1985, at the peak of the California wind boom, the Danes supplied three thousand of the approximately five thousand turbines that were installed.

The dominance of the Danish industry did not result only from their relative technological supremacy. The availability of the ISO4 contract and their confidence in the quality of the national wind industry, made Danish banks and institutional lenders interested in putting their money in commercial wind farms in California. After 1984, the major part of the Danish export was financed by Danish investment money. The support and faith of the Danish financial world enabled the Danes to use so-called package deals to sell their wind turbines. They offered the wind farm developer not only their machines, but also proposed "project financing, insurance, a 5-years product guarantee, a 3-year guarantee on the main components, and service contracts." (Karnøe 1991: 218) In short, the Danish industry not only shipped their turbine technology overseas, much more it exported its whole so-called 'accumulation system,' including investment and insurance structures, to California.

2.4.5 The Expiration of the Federal and State Tax Credits

During the 'wind boom', the view that the government had made wind investment too good a deal gradually got the upper hand, and soft energy proponents were not able to convince Congress to keep intact or at least to gradually phase out the federal tax credit. This caused the abrupt termination of the support at the end of 1985. The renewable energy business community, however, succeeded on the federal level in getting transitional rules written, which allowed developers in case they both had a signed power purchase contract and a wind farm site in stock to still get the federal tax credit after its expiration date. They heaved a sigh of relief since this meant that most of the projects based on the favorable ISO4 contract could be put in place in due time, and still qualify for the 'expired' federal tax incentive.[86] The federal pressure was off, but pushed on by the realization that according to the schedule the state tax credit would be abolished at the end of 1986, the unsteady wind industry put on a laborious finishing spurt. The turbine sales nevertheless plummeted to 2300, of which no less than 2000 were Danish machines (see Kjær 1988: 25; Karnøe 1991: 217; and, Serchuk 1995: 252).

2.5 Entrepreneurial Modernization[87]

The CPUC was proud their concept of standard power purchase agreements had worked so well in aiding the development of the nascent independent power industry. Singing the praises of their initiative, its President Stanley Hulett stated that "Under PURPA, cogeneration and renewable resources have helped defer construction of some new utility plants. ... Renewables are displacing oil and gas consumption units, and cogeneration units are burning gas more efficiently than utility plants. ... PURPA has spurred the development of new, alternative energy sources, improving the diversity of California's resource mix and reducing the risks of over-dependence on any single technology."[88] Despite its success on several key policy issues, the CPUC recognized that – in its enthusiasm – it had made two 'classic economic blunders'.

Thompson explains "It was volume and prices. We did not adjust the prices and we did not restrict the volume that could sign up. So, we got all the supply at the same price. Maybe the prices were right for that first block of megawatts, but we should have recalculated the fixed prices and the fixed capacity value in let say 200, 300 megawatt blocks."[89] The Commission's economical naiveté and political opportunism had caused two major problems. First of all, the large number of PURPA contracts using ISO4 combined with the utilities' new plant capacity created the problem of over-capacity.

Second, at the time of the negotiating conference, the utilities – like most experts – predicted continuing price increases over time. In contrast to the utilities' own forecasts, however, beginning in 1984, the oil and gas prices started to fall. The widening gap between the fixed ISO4 contract prices and the utilities' actual avoided cost caused the problem of overpricing (CPUC 1993: 69; Starrs 1988: 135). Admitting it had made a 'handful of mistakes,' the Commission chose to shift toward a more market-based policy. The CPUC decided to complete the development of Final Standard Offer 4 (FSO4), and to develop a two-year competitive bidding process to allocate these long-term power purchase contracts. Starting in 1986, it took no less than eight years before the Commission finally approved the rules under which competition should occur.

2.5.1 The Long and Winding Road from ISO4 to FSO4

For several reasons, setting up the bidding procedure turned out to be a complex and time-consuming process. First of all, the CPUC was in no hurry because of the situation of over-capacity. Secondly, their monopoly position made investor-owned utilities very reluctant players. Disliking competition, they used any kind of procedural means to slow down the decision making process. Thirdly, and related to this, Thompson explains that "Because the regulators recognized that the utilities would have every motivation to manipulate the way they evaluate bids, so that less independent power would be developed, it was the CPUC's desire to make an entirely transparent bidding and evaluation process; an entirely rigid contract process."[90] In particular, the Commission spent a lot of time on setting up policies governing pricing and access to utility transmission services by non-utility generators (cf. Halligan 1993: 7-9).

The allocation of FSO4 contracts to QFs was supposed to take place in a three-step process. The first step was for the CEC to determine how much new generation capacity each of the three investor-owned electric utilities in California would need to add in the next eight years. Every two years, the CEC would come up with such a plan, the so-called *Biennial Report*. Next, the CPUC would utilize that document in terms of telling the utilities whether they needed or not go out to bid for new resources. The utilities would decide what facilities they would build, and would determine the cost of power from these new facilities. The third step was for independent producers to offer electricity from alternative facilities, and to try and beat the benchmark bid by the utilities.

The first time this planning process – named the *Biennial Resource Plan Update* (BRPU) – went into play, the CEC concluded that no new electrical capacity was needed. Two years later that changed. In 1990, the CEC identified a need for approximately 1350 megawatts of new resources. As a result, the CPUC initiated an administrative proceeding to determine the BRPU bidding rules. Representing renewables businesses (e.g. CAL SEIA, USW, and Zond) and non-profit organizations (e.g. EDF, NRDC, Sierra Club, and UCS), the *Center for Energy Efficiency and Renewable Technologies* (CEERT) was much involved in this, both directly through filings and testifying at CPUC hearings, as well as indirectly by legislative lobbying.

In 1990, CEERT played a key role in developing and passing Assembly Bill 3995,[91] adding Public Utilities Code Section 701.1, which required the CPUC to place values on the costs and benefits of environmental impacts, and resource diversity, and explicitly consider them in its resource planning decisions. A year later, the lobbying arm of CEERT was successful in getting the Legislature to support the so-called 'renewable set-aside' law,[92] which demanded that a specific portion of the future electrical generating capacity would be reserved for renewable resources until a procurement methodology that values environmental and diversity costs and benefits associated with various generation technologies would be completed. On June 23, 1993, the CPUC gave its final approval for the state's three major electric utilities (PG&E, SCE, and SDG&E) to issue requests for bids for new power plants to be supplied by competing independent energy producers. To ensure diversity in the BRPU results, the CPUC had set aside nearly 300 megawatts of the new power for bidding by renewable technologies only.

2.5.2 America's Need for a 5-Cent Cigar[93]

USW concluded that the political situation in the United States is so practically hopeless that the only way to make a really big business out of this – a General Electric or Westinghouse size industry – was to have a windmill so cheap, that you would be stupid to buy anything else. They began this great publicity campaign about the 5-cent per kilowatt-hour windmill. In my book I mention then a vice president of the United States, who said many years ago, "What this country really needs is a 5-

cent cigar." I paraphrased that and said "We need a good 5-cent windmill." That is what Gerry Alderson's [former CEO & President of USW] campaign essentially was all about. I understand the necessity for doing that. We here in the United States have to take the cost down, lower, and lower, and lower. Because it's the country we live in. We want everything cheap here. We don't care about quality. If it is cheap, it is good, quality doesn't matter. Just give it to us cheap. *Paul Gipe, Interview Nov 2, 1995*

As oil and natural gas prices decreased in the mid-1980s, many of the legislative, tax and regulatory supports for alternative energy facilities had been repealed or allowed to expire. The backlog of ISO4 power purchase agreements enabled surviving wind farm developers – like USW, Zond, SeaWest, and FloWind – to continue to develop and operate wind power plants. Linked to the expiration of the federal tax credits in 1985, and state credits in 1986, a decline in new capacity appeared from 1986 to 1989, followed by a development rush during 1990 and 1991 to install projects before the expiration of ISO4 contracts. Although the developers were all running around building new projects, there existed a widespread doubt about the future market potential of wind power. Whereas political support was absolutely needed to make wind turbines cost competitive with fossil fuel generating facilities, it was also rather inconceivable during the Reagan era.

Since the late-1970s, the goal of USW had been "to reach that point in time where wind power is considered a generally accepted utility technology." (Kenetech 1993: 2) While the political and economical environment had altered, the company's goal remained unchanged. In 1986, the management of America's only surviving manufacturer of medium-sized turbines, perceived that the above goal could only be reached by developing "a technology of utility stature." USW managed to get support from the utility-funded Electric Power Research Institute (EPRI) in Palo Alto to find out the requirements of such a technology. In 1987 a group of experts, including representatives from California's PG&E, and several power corporations from other states, was established to conduct a feasibility study. Carl Weinberg explains that "PG&E had an interest in being involved in alternative energy, because that was part of the direction in which the State of California was going."[94]

Moreover, PG&E's management saw wind power as a potential business opportunity. In fact, access to USW's next generation technology, was expected to give the utility the possibility to out compete the independent wind power developers in its territory – including USW – by undertaking utility-scale wind projects itself. EPRI thought that "the cost of energy from a wind power plant owned by an independent power producer could be 30 to 40 percent greater than that of energy from the same plant if owned by a utility."[95] Such a scenario would restore the utility's monopoly position in the electricity generating business and force USW to abandon their old business strategy of operating its own wind turbines in its own development projects, and concentrate on wind turbine manufacturing, building projects for third parties, constancy, and operation and maintenance (O&M) activities. Indeed, two years later, USW adopted such a new "open door strategy." (Davidson 1989a)

The utility partners gave USW an appreciation for the necessity of an advanced electronic system that would not send objectionable current distortions back onto the utility lines. In addition, in order to minimize the O&M costs, they wanted the turbine to be easily erectable in the field, and that components would be easily accessible.[96] For USW, the main aspiration was to show that a next generation wind turbine could be cost effective without subsidies. Eric Miller explains "Our entire focus was actually in convincing [utility] people that wind power was competitive. Any kind of message that you need to be subsidized – at least in the United States – completely undermines that message. Our message had to be 'We are competitive. We are competitive. We don't need subsidies. We don't need assistance,' because of the bigger picture of people taking us seriously."[97] The wind company thought to achieve this by developing a variable speed machine. Weinberg[98] recalls that the variable speed aspect was not a demand from the utilities, but was picked because "Theoretically it was expected that the variable speed would provide you with a lot more energy capture and better control of the machine. Furthermore, it was believed to lessen the requirements on the geartrains and all of the mechanical components." The latter would make lighter designs possible, and thus lower the cost of wind electricity.[99]

Guided by these constraints, and the R&D lessons from the early 1980s, which had shown that it was wise "to stay away from making quantum leaps," it was decided to develop a 33 meters, 300-kW variable speed turbine. This so-called 33M-VS machine, which – except for the variable speed part – was a sort extension of their very successful 100-kW machine, was expected to produce electricity for a record low cost of 5 ¢/kWh, given an average wind speed of seven meters per second. Pleased with this outcome, PG&E engineer Bill Steeley commented "This is the first time everybody – developers, utilities, and industry researchers – has sat down together, and come up with something everybody wants."[100] As a result of the feasibility study a consortium – formally called the *EPRI-Utility-USW Advanced Wind Turbine Program* – was formed in Autumn 1989, to cover the initial R&D costs.[101] The first prototype of the 33M-VS machine was erected in Spring 1991, and two years later, the turbine was launched on the commercial market.

Carl Weinberg notes that "The development of the 33M-VS turbine stimulated the rest of the wind industry in the United States to look at the next generation machine." It created a new atmosphere of technological optimism, which revived the interest of Congress in restarting the federal R&D program. The National Renewable Energy Laboratory (NREL),[102] an arm of DOE located in Golden, Colorado, received funding to start the *Advanced Wind Turbine* (AWT) program. USW wondered why the Government should give their competitors money to compete against them, and fought very hard to stop the program. In early 1990, NREL selected turbine designs, which according to some industry observers reflected "an emphasis on lightweight, high-tech designs typical of the 'American approach.'" (Davidson 1990a) One of the winning proposals was a redesign of the ESI-80.[103] This idea came from Bob Lynette, a former Boeing engineer, who had worked on the Mod-2 turbine. In 1993, R. Lynette & Associates, together with FloWind, started to market the resulting AWT-26; a 275-kW turbine, which employed blades that were designed at NREL (cf. Gantenbein 1994).

Zond's Z-40 wind turbine later also became equipped with NREL-designed airfoils. Spurred on by the development activities in the rest of the industry, Zond's management decided in 1992 to develop their own turbine. Up till then, the firm had always used Vestas machines for their wind farm projects. The Danish firm, however, was focusing its R&D effort on developing larger turbines, which were expected to be more attractive for their European utility customers. One result of this strategy was "a surprisingly steady state in cost of energy ... between the years 1988 and 1995."[104] The market conditions in America differed from those in Europe. In the U.S. it was believed that reducing the cost of electricity was the key to success. As a consequence, the products offered by Vestas no longer suited Zond's demand. Although somewhat disturbed by USW's 'unrealistic' publicity campaign about the 5 ¢/kWh wind turbine (cf. Gipe 1995: 226-227), Zond nevertheless felt forced to develop such a machine themselves for the American market. Starting off with the Vestas design,[105] the company developed the Z-40 turbine, with a 40-m rotor and 550-kW generator, in less than two years. This was done in collaboration with NREL as part of their cost-shared *Value Engineered Turbine* program, which was established under political pressure to fund 'practical' designs (*ibid.* 82). NREL provided some 25 percent of the development costs, and reviewed and tested the design.[106]

2.5.3 FERC's Opposition against the BRPU Competition

Meanwhile, the wind farm developers, had been busy leasing land, measuring winds, doing conceptual designs, and entering into a certain amount of permitting to put the bids together for the BRPU auction, which was finally issued by the CPUC in the Fall of 1993. Just before, USW had gone public. The company was expecting to do well on the BRPU and needed money to finance the production of the new 33M-VS turbine.[107]

The CPUC had made the bidding process transparent in order to reduce the amount of manipulation the utility could exert in the process. The rigidness of the bidding protocol enabled the IEP to put software together that reflected the rules that had come out of the nearly four-year negotiation process. Eric Miller recalls "We had a total of 48 different bid-packages. We bid a little over 1,200 megawatts. ... We were modeling all these different projects. We had software that would simulate the bid scoring right into the model. So, we were constantly playing with pricing, and figuring out what would give us the best bid score."[108] The result of the bidding was that the independent power industry beat utility prices by an average of 33 percent (CEERT 1994). More surprisingly, the wind project developers did extremely well. In ranked order, the ones that were winning most of the contracts were USW, SeaWest, Zond, and the next one in the line was FloWind.[109] The result brought the California wind developers in the first flush of victory. DiVirgilio recalls "We were happy at Zond. We had some good land, and good projects. We were going to go for it, and develop the projects."[110] Miller remembers "USW's stock went up very high, and everyone was very excited."[111]

The delight, however, was short-lived. SCE and SDG&E refused to sign contracts with the winning bidders, and pursued administrative appeals to halt the bid process. Hamrin explains that the way in which – in particular SeaWest, and USW – had put together their bids "definitely allowed the utilities to come in, and legitimately question what was going on." She continues "My personal opinion is that the industry sabotaged itself. Some of the bids that were done and submitted were so inconsistent with the spirit of how it was supposed to be done. Even if the rules didn't exactly say you couldn't do that, everybody understood how the game was supposed to be played."[112] The CPUC realized that the transparency of the BRPU had caused more problems than that it had prevented. Nevertheless, the Commission chose to go on with the bidding, and decided in June 1994 to open hearings to look whether or not people were actually cheated.

Not satisfied with this ruling, SCE and SDG&E went to Washington for relief. Both filed petitions before the Federal Energy Regulatory Commission (FERC) in January 1995, arguing that the bidding process violated PURPA. The Southern California utilities criticized the fact that the BRPU was a QF-only bidding, and claimed that the State of California had no authority to require electric utilities to account for air quality impacts or accommodate clean technologies (Jablonski 1995). The FERC bought both of these arguments, and declared the bid illegal on February 22, 1995.

Although the FERC ruling was just an advisory opinion, it was hard for the CPUC to disregard it, because the utilities could take the CPUC to federal court and have that FERC ruling applied. As a result, the CPUC recognized the invalidity of the contracts, but also encouraged the utilities and QF-developers to negotiate, and to modify the power purchase agreements.[113] At first, the wind developers resisted this idea, but gradually they realized that they were fighting a losing battle, and entered into the slow, and difficult settlement discussions with the utilities, unsure about whether this would result in some new wind turbines being installed in California this decade.

2.5.4 The Electric Utility Restructuring Debate

The main reason why the developers gave in was that they started to appreciate the change in political climate that had taken place. Since 1993, the *electric utility restructuring debate* had gained momentum, and had changed the BRPU into an anachronistic regulation. This debate was triggered in the early 1990s, by the high price of electricity in California during a period of deep recession.[114] Within the CPUC, some Commissioners were getting frustrated with the ever-increasing complexity of the regulatory process, and placed the problem of price in a wider context. Their analysis was that "The regulatory process did not give good outcomes, because it was grounded in a monopoly legal framework from 70 years ago, when in fact the industry had just changed dramatically, and was becoming increasingly competitive."[115] In September 1992, the Republican Commissioners directed the Division of Strategic Planning to do a study about the historical development of the regulatory process, and its future. A month later, Congress passed the Energy Policy Act (EPA), which allows for wholesale competition in electricity generation, with the wholesale price of electricity determined not by cost (as it was until then) but through a competitive auction (Fuchs 1995: 6).

The Division of Strategic Planning regarded EPA as *the* key development, and expected that, eventually, competition would be extended to the energy retail market. In that situation, energy consumers would be able to enter into bilateral contracts with producers, third-party providers, or utilities of their choice. The transition to such a competitive electricity market would require a redefinition of the regulatory compact, since "utilities no longer enjoy an exclusive franchise, nor do they retain the duty to serve customers, who elect to procure services independent from the utility." (CPUC 1993: 5). Together with three less radical scenarios, the strategy promoting a competitive market for electric services, was described in a staff document, known as the *Yellow Book* (CPUC 1993). After the report was released in February 1993, the CPUC had hearings and comments, which did not attract much attention from the Legislature. The Commission draw back with all the information gathered, and after a lot of *internal* discussions came out in April 1994 with the outline of a plan to begin restructuring in the electricity industry.

The so-called *Blue Book* (CPUC 1994a) proposal received enormous (political) interest throughout the State and the nation, and triggered a heated debate over the market structure. It was realized that some fundamental political changes were in the wind. These *winds of change* woke up the normally 'very technical, sleepy industry.' Suddenly, the IEP and investor-owned utilities – the two archenemies within the energy debate over the last two decades – understood that they could no longer ignore or condemn each other and left their trenches to seek for compromise. The Governor's office provided a forum for the two very reluctant parties. The difficult negotiations resulted in a *Memorandum of Understanding* (MOU) that shows the intent to rapidly establish a vigorous competitive market in the supply of generation services.[116]

In response to the MOU, a diverse set of public, environmental, and consumer groups came up with a document called *Framework for Restructuring in the Public Interest.*[117] On the basis of these two proposals and the resulting hearings, the CPUC issued a final restructuring order on December 20, 1995. The decision established a wholesale power pool, and a five-year phase of direct access for all classes of retail electricity customers, who wish to negotiate their own power arrangements. For renewable resources, the CPUC seemed to support a *Renewable Portfolio Standard* (RPS) which would secure that a minimum percentage of all electricity sold each year in California be generated from renewable resources.[118] In an intense battle over renewables policy in California, the renewable energy advocates lost the RPS policy, but secured some funds to maintain the existing base of renewables in California.[119]

2.5.5 And There Is Wind's Old Bugbear: Reliability

> And there is wind's old bugbear: reliability. Although the three main U.S. turbines hitting the market – the 33M-VS, the AWT-26, and the Z-40 – all have substantially better engineering than turbines built 10 years ago, they are nonetheless unproven in long-term commercial applications. *Douglas Gantenbein, 1994: 84*

The wind companies had been focused on the BRPU, and on developing breakthrough wind turbine designs that were expected to redefine the market (Karnøe 1993: 73). In the past, the wind developers had built projects, owned them, and just sold power to the utilities primarily under ISO4 contracts. This had been adversarial, because the utilities felt that those contracts had been forced on them. Now it was felt, however, that the time was ripe to develop a domestic utility wind turbine market, since it was expected that "the U.S. utilities would be interested in investing themselves in owning and operating wind facilities, as a sensible part of their resource mix."[120] USW's 50-MW deal with the Sacramento Municipal Utility District (SMUD) at the end of 1993, was the first time a U.S. utility had committed itself to own a commercial scale wind project, and gave strength to the new market strategy (Asmus 1993). Moreover, utilities throughout the United States followed SMUD's lead, and started to express interest in the latest developments in the field of wind technology.

The sudden arrival of the restructuring debate, however, abruptly changed the mindset of the utilities, and stopped them from looking at any long-term purchases. The deadlock on the American market forced the wind companies to explore the international market, basically without having a clear strategy, without much experience on foreign markets, and lacking the organizational capacity to do so. Moreover, the apparently restored confidence in American wind energy technology again took a terrible knock when the 33M-VS machines in USW's newest wind projects in Minnesota and Texas encountered widespread blade and generator failures. Because of "an inability to meet maintenance and warranty obligations, as well as a lack of capital, repair costs, deregulation of the electricity market, sales declines, and a difficulty in re-scheduling contract obligations," (Davidson 1996) the company filed for bankruptcy in May 1996. Meanwhile, the rest of the American wind industry is anxiously looking for ways to survive within the forthcoming competitive electric energy market.

3 Denmark: Living Wind Energy

"[T]he living word" is ultimately ... the program for a type of politics that strongly favors both tolerance of ideological diversity and peaceful conflict resolution.
Steven M. Borish 1991: 261

Unlike other European countries that rapidly recovered from the war, Denmark only started to benefit from the favorable global economic climate at the end of the 1950s. The sudden jump toward prosperity began with the Danish Great Leap Forward (1958-1962), a four-year period in which the national product at fixed prices increased by a total of 26 percent, a rate never before or since equaled (Borish 1991: 216).

Before, several factors had slowed down the interrelated and mutually reinforcing processes of agricultural rationalization, urban industrialization, international trade, and state intervention. First, it was not until the 1960s that agricultural legislation allowed the amalgamation of farms.

Secondly, Denmark had a tradition of medium-sized and small industrial and craft-based enterprises, whose production had been traditionally oriented to the domestic market. In comparison with countries whose industrial sector was dominated by large (multinational) corporations, it took the Danish industry more effort to become integrated into the international trade system.

Thirdly, the country was relatively slow in adopting the welfare state model. During the 1950s, the political battle was between two totally different societal models: the liberal society with a minor role of the state and the idea of the welfare society (cf. Fonsmark 1990: 206-215). The latter was strongly promoted by the social democrats, but eventually also other parties lost their skepticism on state planning. The consensus on the 'good nature' of state intervention resolved around the issue of growth. On the one hand, the Conservative Party and the liberal *Venstre* party agreed that some state regulation was needed to compensate for market failures and became interested in stimulating economic growth via public infrastructure investments. On the other hand, the Social Democratic Party believed that economic growth was necessary to finance the extension of the welfare system.

After 1958, the above political and economical transformations led to fifteen years of high growth. In this period, the drop in the number of farms and the increasing mechanization of farming caused agricultural workers to move into urban areas. The state initiated extensive house-building and major construction projects to create housing for these farmers as well as for the 'baby boomers'. This new labor force was absorbed by the public sector and industry, which overtook agriculture as the most important source of export income. The country experienced a massive wave of industrialization, particularly in the construction and chemical industry, electronics and pharmaceutical industries as well as oil refineries and power stations.

The construction of large central power stations after the war had formed the basis for the establishment of two electricity networks (Rasmussen 1982: 226); one grid covers the Jutland and Funen area and runs synchronous with the European continental system, and the other supplies the industry and households in Sealand and Lolland-Falster with electricity and runs synchronous with the Scandinavian grid. In the mid-1950s, the regional power companies in each of the two areas divided by the Great Belt established the power pools Elsam and Kraftimport (since 1978 called Elkraft), respectively.[1] These two associations became responsible for planning, design and construction of new generating plants, load dispatching, fuel purchases, operation of the transmission grid, as well as international trade (DEF 1994b). In 1970 the Danish electricity consumption was generated by twelve power stations that primarily used oil.[2]

3.1 Breaking the Environmental Silence

> There was little public debate in the 1960s about environmental deterioration. As such, the pre-phase in [the development of environmentalism in Denmark] can be characterized in terms of an increasing 'cognitive dissonance.' *Andrew Jamison and Jeppe Læssøe, 1990: 67*

The generation that had experienced the depression of the 1930s, and the repression of the German occupation, saw the 'happy sixties' (*de glade tressere*) as the 'golden age' that they had hoped and striven for. As such, there was a 'benign collective neglect' of problems associated with Denmark's rapid modernization and international integration. At the end of the decade, however, several groups began to point at the contradiction between material growth and social welfare. The industrially constructed blocks of flats, for instance, began to be criticized for their lack of refinement, which was believed to cause vandalism and social disintegration (Himmelstrup *et al.* 1992: 52-53). Pollution became a hot political issue in 1969.[3] In the early 1970s, public debate focused on visible and tangible local pollution. Chemical and electric industries were depicted as the chief causes of local air pollution.[4]

The government soon admitted that measures had to be taken, and institutionalized environmental concern. In 1971, the new social democratic government established a Ministry for Pollution Prevention (*Ministerium for Forureningsbekæmpelse*). This started to prepare a comprehensive legal package that was passed in June 1973 by the social democrats and the Conservative Party (Jamison and Læssøe 1990: 78). It included a law to limit local sulfur pollution by power plants by placing restrictions on the sulfur content in oil. The strongly increased environmental protection activities of the state more or less forced the environmental movement to leave the basic pollution themes to the established political culture; i.e. the political parties, industrial organizations, and labor unions (*ibid.* 94-95). OPEC's rise in oil prices was even more significant in "the epochal shift in the development of environmentalism from 1974 onwards," (*ibid.* 90), since it brought the energy issue to the center of interest. In particular, the fight against nuclear power pushed the political ecological debate into the background.

From March 1969, this discussion was promoted by NOAH, a group of students at the natural science faculty at the University of Copenhagen, which at that time was the center of the ongoing student revolt. By picking up the pollution debate they participated in the struggle for a more democratic and modern university system. The social criticism of most students at the social and humanities faculties were inspired by Marxism. The NOAH activists were not grounded in social theory, and instead gradually developed their criticism through protest actions and intensive debates. In NOAH's 'bible' – *Nogle Oplysninger om den Jord, Vi Sammer Lever på* (Some Information about the Earth, on which We are Living Together) from November 1970 – common sense ideals are confronted with social reality, and the need for political activism is connected with popular social values, such as social responsibility, an active local democracy, tolerance towards minorities, and demands for equal rights among the sexes. Jamison and Læssøe (1990: 85) conclude that, "In relation to Danish political culture, [NOAH's ideological position] is ... more connected to the humanist and culture-radical traditions than to the revolutionary new left." In the 1960s, this cultural radicalism[5] took political expression in the form of the Socialist People's Party (*Socialistik Folkeparti*, SF). This party was a crucial factor behind the relative strength of the alternative political culture in Denmark, and "established a 'third way' in the Danish left political spectrum between, and in opposition to, the social democrats and the pro-Soviet Communist Party." (*ibid.* 69)

3.1.1 The Political Earthquake of 1973

The youth revolt was one of the three political fields of strong tension that existed between 1968 and 1973. The two other domains that contributed were the welfare system with its growing tax burden and the European Community (EC) membership (cf. Fonsmark 1990: ch. 5). Successive socialist and liberal governments proved unable to cut down on public expenditure and reverse the deficit in the balance of payments.[6] In the early 1970s, both the social democrats and the conservative parties came to see membership in the EC as a solution to Denmark's economic problems. The important agricultural sector was particularly eager to follow the lead of Ireland and Great Britain – Denmark's main agricultural export market – and join the Common Market (Himmelstrup *et al.* 1992: 47-48). Opponents coalesced into the 'people's movement against the EC,' which covered the entire political spectrum. The movement feared increasing political and economic concentration and the segmentation of the capitalistic system. But, perhaps most important, the opponents were afraid "to relinquish Danish cultural identity, to give up the social welfare democracy, the Nordic legal tradition and the Grundtvigian notion of *folkelighed*." (Fonsmark 1990: 323)[7] Rather than fostering a narrow nationalism, "*Folkelighed* ... must be seen as promoting ... a means of defending a small country such as Denmark from being culturally crushed by more powerful nations." (Lawson 1992: 118) In short, the youth revolt, the anti-EC struggle, the growing deficit in the balance of payments, the increase in state expenditures and the corresponding tax burden, all helped to undermine confidence in the established political culture and to revitalize and renew alternative political cultures. Political and economical stability was further undermined by the oil embargo.

This all helps to explain the dissolution of voters' loyalty that was expressed in the outcome of the election of December 4, 1973. This 'landslide election' (*jordskredsvalg*) was a disaster for all the parties represented in the previous parliament. The 'four old parties' – the Social Democratic Party, the *Radikale Venstre*, the Conservative Party, and *Venstre* – and the Socialist People's Party, which had established Danish welfare democracy, now lost support to five new political streams. The *Kristeligt Folkeparti* (Christian People's Party) was a reaction to free abortion and pornography, and won seven mandates. The *Centrumdemokraterne* had splintered off from the Social Democratic Party and stood for an ideological attack on the youth revolt and the rising 'alternative political culture' – both the new left, the culture radicals, and anyone else who wanted to change traditional values (Jamison and Læssøe 1990: 92). The established political culture (represented by the four old parties) was relatively tolerant toward the three left parties: Denmark's Communist Party (DKP), Social People's Party, and the Left-Wing Socialist Party (*Venstre Socialister*). In contrast, "The Progress Party's ideological stance was not considered as truly *folkelig*, at least not in the sense that people would characterize it as a grassroot movement." (Fonsmark 1990: 340) With its demand for lower taxes and cuts in government expenditure, the just founded radical right-wing Progress Party (*Fremskridtspartiet*) won no less than 28 seats of the 179 seats in the Danish parliament (*Folketinget*).[8]

The entrance of five new political parties caused a further diversification of the political spectrum, and a corresponding ideologically more refined distribution of mandates. The resulting dispersion of political power caused a break with the parliamentary tradition of mostly majority governments (Christensen and Christiansen 1992: 40-43). Ever since the so-called 'voters' revolt' (*vælgeroprøret*) of 1973, all Danish governments have been minority governments,[9] which had to survive by making agreements with other parties within the *Folketing*. This frequently forced a socialist or conservative government to look for support from parties on the other side of the political fence, and made it possible for small parties to play a more than only marginal role. Especially the party that is sitting on this fence, the *Radikale Venstre*,[10] represents a crucial swing party, which has often been able to tip the scales in parliamentary decisions (Jamison and Læssøe 1990: 68).

3.1.2 The Nuclear Power Issue

> NOAH had shaped visions and hopes for many people about a society that was less polluted and resource demanding and more decentralized, democratic and open. 'Small is beautiful' was the slogan that the economist E. F. Schumacher had coined in 1973, and as opposed to it, the plans of the Danish electricity companies in 1974 to build centralized, life-threatening, mammoth nuclear plants was seen as a provocation that could not go unanswered. *T. W. Nielsen (OOA pioneer), 1984: 4*

> Atomkraft? Nej tak!
> Vi kan sige nej tak – uden at komme til at mangle noget. *OOA slogans*[11]

OPEC's production strategy tripled oil prices within a year. The supply crisis led to myriad events. First, the Ministry of Trade (*Handelsministeriet*, HM) charged the electric utilities to reduce electricity consumption with respect to 1972 by 10 percent. Helped by an effective information campaign and economizing consumers, the utilities succeeded on a 'voluntary basis.' (DEF 1973-1974a: 5)[12] Secondly, guided by the changed price-relationship between oil and coal, and helped by the fuel flexibility of its plants, the Danish power sector reacted by reducing the use of oil to about 40 percent in two years.[13] Besides these short-term policies that were aimed at a direct reduction of Denmark's dependence on oil, the utilities announced that they wanted to make extensive use of nuclear power in the near future. The Association of Danish Electric Utilities (*Danske Elværkers Forening*, DEF) held that, within the electricity production sector, "nuclear power was the only realistic alternative to fossil fuel," and emphasized that its use was in line with the goals of "supply security and the wish to produce electricity at a reasonable price." (DEF 1974-1975a: 2)

Already for a couple of years, Elsam had been looking for favorable sites (DEF 1971-1974a). Being aware of the nuclear debate abroad, the Jutland company had been waiting for the right time to make their plans public. Ironically, the oil embargo was just what Elsam needed, and already at the end of 1973 the Jutland company published a list of ten potential sites for nuclear plants. The day after, a small group of students, guided by the student priest, Jens Brøndum,[14] NOAH, and the folk high schools founded the Organization for Information about Nuclear Power (*Organizationen til Oplysning om Atomkraft*, OOA). Both the extensive discussions about growth, ecology, and decentralization that had been inspired by NOAH in the previous years, as well as OOA's strict neutrality in relation to political parties, enabled the anti-nuclear movement to grow very quickly (Jamison and Læssøe 1990: 91-92, 97).[15]

OOA was an integral part of the alternative political culture, but dissociated itself from the neo-Marxist pole of that culture. OOA took over NOAH's populist dialogue, but "reversed the positions of aim and means present in NOAH's vision." (*ibid.* 97) NOAH carried out a broad public information effort to raise environmental consciousness and to attack established values for ideological reasons. In contrast, OOA just wanted to prevent nuclear power and used information as a campaign instrument. As a result, the anti-nuclear organization did not insist on a direct cessation of nuclear power, but instead demanded a moratorium for three years in order to have a public debate about nuclear energy. Its efforts led the Ministry of Trade in 1974 to set up an Energy Information Committee (*Energioplysningsudvalg*), which was intended to carry out "a broad information effort, with the aim to orient people about the energy policy problems, which the Danish society will confront in the coming years." (HM 1976: 8)

3.2 The People's Democracy

In May 1976, when the information campaign was already going for a year and a half, the Ministry of Trade finally came up with a national energy plan (HM-plan) that stressed the need to reduce Denmark's oil dependency by saving energy and by speeding up "the development and use of other energy sources than imported oil." (HM

1976: 7) In the short-term, this implied a fuel shift from oil to coal that had already taken place in practice. With respect to long-term planning, the Social Democratic government took over the utilities' proposal concerning nuclear power. It was assumed that, in the future, old plants would exclusively be replaced by nuclear plants, which implied that, in 1995, almost two-third of the electricity production would be based on nuclear power.[16] Close to 4 percent of the total production in that year was reserved for renewable energy sources, in particular wind energy produced by large wind turbines (*ibid.* 39, 51).

The public, OOA, and the parliamentary Energy Policy Committee (EPC), all criticized the Ministry of Trade for coming up with only one energy scenario. The EPC asked the Minister to work out an alternative energy plan without the use of nuclear power. On July 13, 1976, Per Hækkerup answered that there was no need for such a plan (cf. Blegaa *et al.*: 7). This led to reactions within and outside the parliament. Within the *Folketing* the Left-Wing Socialist Party (*Venstre Socialister*, VS) responded by proposing a package of four bills to foster an alternative energy policy. This left-wing party aimed at an electricity supply system without atomic power, to safeguard Denmark's "independence from the multinational concentration of capital."[17] This goal was thought to be achieved by energy savings, self-supply, and by spreading the necessary fuel import among several suppliers. The four bills were meant as a "concretization and a way to speed up some parts of the government's energy plan," and were, among other things, concerned with the development of wind power. VS argued that – except for the really big turbines – the technology was already developed. The neo-Marxist party was in favor of a forced development that would cause wind turbines to produce no less than 14 percent of the Danish electricity consumption by 1980.

The Ministry of Trade's unwillingness to work out a nuclear-free energy scenario also inspired a group of eight engaged scientists to put together an alternative energy plan (AE-plan).[18] Within a month the alternative plan was a fact, and in October it was published with the support of both OOA, and the Organization for Renewable Energy (*Organizationen for Vedvarende Energi*, OVE), which concentrated on the development of practical alternatives to nuclear power. In contrast to the government, which espoused that nuclear power was "commercially available" (HM-plan 1976: 32), and its "security problems could be solved with reasonable certainty before the end of the century," (*ibid.* 16) the authors of the counter plan questioned the economics of atomic power plants, and emphasized "the long range of unsolved problems connected with the storage of radioactive waste, military protection of nuclear plants and waste transports, the misuse of radioactive material, ... *et cetera*." (Blegaa *et al.* 1976: 8)

The scientists mentioned that the principal choice was between the use of uranium and thorium in fast breeder reactors, and the use of solar energy; between the 'plutonium society,' in which democratic rights would be challenged, and the 'solar society,' in which decentralized energy systems would keep the future open to human intervention (*ibid.* 112-114). Nevertheless, the ideology behind the alternative was de-emphasized. Not because it was unimportant but because it was not relevant in reaching the interim goal they had set: namely, to broaden the information basis for making parliamentary decisions. By presenting their ideas in an accommodating manner, the researchers ensured that the results of their analysis were both connected to the dominant value of supply reliability within the electricity industry, as well as the main contemporary political issues such as unemployment and foreign exchange problems.

To facilitate a direct comparison between the AE-plan and the HM-plan, the alternative plan took over both the assumptions made in the HM-plan, e.g. about economic growth, as well as the primary aim of the governmental energy policy to reduce oil dependency. On the supply-side, no nuclear power was used. In return, the AE-plan aimed at a stronger development of both decentralized cogeneration plants running on natural gas, and renewable energy sources. Gathered in fifteen parks, fifteen hundred 1.25-megawatt wind turbines were envisioned, which would cover 12 percent of electricity production in 1995 (*ibid.* 55). On the demand-side, a continuous strong effort to save energy was proposed.[19] The scientists argued that, with respect to the HM-plan, their plan would lead to "a more robust and secure energy system, fewer limitations on physical planning, higher savings of foreign currency, and more Danish jobs." (*ibid.* 24)

Both the VS law proposals and the AE-plan served as input for the November debates in the *Folketing* about the energy plan of the Ministry of Trade. The Social Democratic Party, RV, and SF were sympathetic to the VS plan but did not fancy the forced development that the left-wing socialists strove for. The researchers' alternative energy plan basically contained the same elements, but was taken more seriously because it was inclusive, and proposed change at a lower pace. Moreover, the alternative depicted by these 'limits-to-growth' avant-gardists was adaptive to both the growth ideology of the social democrats and the zero or selective-growth ideology of RV and SF. The growing popularity and political influence of the anti-nuclear movement, combined with the range of new energy policy options that had entered into the energy debate through the comprehensive AE-plan, caused the postponement of the nuclear plans for one year.

3.2.1 The ATV Wind Energy Committee

During the same November debates, MPs discussed the state of the art of wind turbine technology, and the share and implementation speed of wind power within the electricity system. A large majority, that crossed the left-right divide within the parliament, found that wind power was a good idea but emphasized that first the technology had to be developed. As a result of the debate, the liberal *Venstre*, the social liberal *Radikale Venstre*, the Social Democratic Party, and the Socialist People's Party, all gave their support to the joint Wind Power Program (WPP) of the Ministry of Trade and the utilities. The WPP was mainly based on the work during the previous years of the Wind Energy Committee (WEC), which was set up in October 1974 by the Danish Academy of Technical Sciences (ATV), under the umbrella of technology transfer from the universities to the industry. Inspired by the report for the *Club of Rome*, ATV President Niels I. Meyer, had taken this initiative in order to interest the government, the utilities, and the industry in wind energy.[20] Considering Denmark's rich tradition in wind energy, the formation of a study group seemed a natural thing to do.[21] In May 1975, the carefully selected committee members[22] concluded that "effective research and development work" can make it "possible to establish electricity and heat producing wind power plants profitably on places where the wind is strong." (ATV 1975: 4) Against this background, a 'new' WEC was formed to formulate concrete R&D projects, and to establish contacts with wind energy researchers abroad (ATV 1976: 3).

In January 1976, the ATV Wind Energy Committee had completed its second report, an action program (ATV 1976). In line with the conclusions of the first ATV-study, the program concentrated on two categories of wind power plants: large grid-connected utility wind turbines (rotor diameter larger than 30 m), and small turbines (rotor diameter smaller than 15 m) for heat production. The Danish program was clearly inspired by the Federal Wind Energy Program in America.[23] As in the American program, about 80 percent of the budget would be spent on the development of large wind power plants. Half of this money would be used to construct three demonstration machines. The terminology, planning schemes, and R&D strategy of the Danish program were also copied. Just like their American colleagues, the Danes talked about so-called '0-series' turbines, whose design would be based on "well-known calculation methods and familiar technology." (*ibid.* 34) After three years of studying and testing, another demonstration series ('1-series') would be set up. This time, the optimal size and construction would be in accordance with the projected studies (*ibid.* 22).

According to the WEC, part of the results from these studies would be directly applicable for small plants. But because it was anticipated that small wind turbines would be used for producing heat instead of electricity, some special projects for small plants were defined (*ibid.* 35). Moreover, it was suggested that a test station for testing both in and outside constructions be constructed.[24] The *Folketing* largely accepted the proposed R&D program, and authorized the Ministry of Trade to start the Wind Power Program in 1977. The utilities had no great esteem for wind energy, and reluctantly cooperated in the program. They wanted a hand on the wheel,[25] and realized "that in Denmark's difficult energy circumstances, no energy contributions [could] be rejected." (DEF 1978a: 5)

3.2.2 Grassroots Activity and Amateurism

Besides the activity of the ATV Wind Energy Commission, a lot more action was going on in the field of wind energy. The do-it-yourself activities of Christian Riisager can serve as a good starting point (cf. Petersen *et al.* 1993: 13-15). Riisager was a carpenter, and during his training in the mid-1940s, building windmills happened to be a part of his apprenticeship. In the early 1970s, he started to build a water mill for fun. In 1972, he took his next step, and built a windmill in his backyard that delivered both direct current and hot water. By using both Vogt's and Juul's research, Riisager constructed a new windmill with stall regulated blades. The rotor drove an asynchronous motor that worked like an AC generator. This meant that by using a simple and inexpensive generator it became possible to maintain the proper grid frequency.

First in 1975, the wind turbine was connected without permission to the public grid. A bit nervous about the result, Riisager checked with the neighbors whether they had problems with electricity. They obviously had none, and so Riisager went to his local electricity company in Herning, and asked for a permit to run his mill. But, since the utilities had not anticipated that a private person would come up with a way of delivering electricity to the public grid, regulations were nonexistent. Riisager was invited to several meetings to establish technical requirements. Permission, however, only came after an inquiry from the company's management. At the time, the electric utilities started to get a common interest in working out rules for grid-connected wind turbines. Still, the problem was not very urgent, as only a few folk high schools had installed wind turbines that could be connected to the grid.

In the same year that Riisager put the plug of his windmill into the socket, a subgroup within OOA decided to establish the Organization for Renewable Energy (OVE). They believed that it was not sufficient to rely on public research and energy policy. Instead, they held that people themselves should get involved in the development of alternatives to nuclear power. Together with a couple of progressive folk high schools, and OOA, OVE wanted to stimulate this process, and set up a network of Energy Offices (*Energikontor*), for the dissemination of information and advice about how to use renewable energy systems. OVE also enabled amateurs and professionals who were engaged in the construction of wind turbines, to meet each other and discuss ideas and experiences. In particular, the Wind Meetings (*Vind-Træf*) – which OVE organized up to four times a year – were of central importance for the exchange and production of technological knowledge (cf. Jørgensen and Karnøe 1992: 9).

3.2.3 The Tvindmill

> Let a hundred windmills flourish. *Tvind Højskole, 1976*

> Energien den koster mange penge
> Svineriet det øges dag for dag
> Ned i lommen på de store oliedrenge
> Noget vil vi til at gøre ved den sag. *Tvind Højskole, 1976*

This practical approach represented another form of argumentation within the antinuclear struggle. In contrast to the intellectual argumentation of OOA, and NOAH, the so-called 'ecological practitioners' created counter images, and emotional symbols of alternative visions. The most prominent example was at the Tvind folk high school in Jutland, where, in the Spring of 1975, construction began on a 53 m high, 2 MW wind turbine. The recent oil price hike, that had drastically increased the costs of warming up the school buildings, had triggered this tremendous effort. As illustrated by their 'Windmill Song,' the Maoist folk high school put this event into a political perspective. The first strophe went like this (Tvind 1976):[26]

> *Energy costs a lot of money*
> *This dirty business gets worse day by day*
> *Down in the pockets of the big oil boys*
> *We will do something about that.*

The teachers at the Tvind schools were sure Denmark would get nuclear power plants. To escape both the economic stranglehold of the oil monopolists, as well as the undemocratic centralization policy of the utilities, they decided to build a wind turbine. This was seen as a direct and practical way of changing the centralization tendency within society. "The power of the people (*de folkelige kræfter*) would be the main power within this development." (Tvind 1976)

The idea of cooperating in this huge project appealed to the imagination of many (folk-high-school) people, and volunteers from all over Denmark (and even Holland) helped to build the so-called *Tvindmill.* Besides lay people, several engineers were glad to join this technologically challenging large-scale grassroots experiment, and willingly accepted the invitation to design the basic components, the blades, and the electrical system.[27] In this way, Tvind became both a place of pilgrimage for ecological practitioners who wanted to study its impressive wind turbine, as well as a common experience for experts and practical-minded activists. And, although the utilities as well as the government considered the leftist-oriented Tvind people as "a bunch of freaks", the project gained "great sympathy within the Danish population," and "won the respect of many professionals."[28]

3.2.4 DEF's Provisional 1976-Guidelines[29]

Herning's power station supplies wind power. *Herning Folkeblad, March 25, 1976*

Meanwhile, the publicity that his private windmill had attracted had stimulated Riisager to take his chances and move into the wind turbine business. He constructed two 30-kilowatt prototypes. One of the test turbines was put up at the place of the journalist Torgny Møller, an advocate for wind power, who expressed his point of views in a column in the national newspaper *Information* (Petersen *et al.* 1993: 14).[30] Hunted by the press, and forced by the "growing private interest in wind energy," that confronted their members, the Association of Danish Electric Utilities (DEF) asked the Price and Tariff Committee to come up with definite rules for connecting wind turbines to the electricity grid (DEF 1976b). In August 1976, DEF published its provisional guidelines, and recommended its members "to support the private initiatives to promote wind power," and "to buy the surplus electricity production for the costs avoided by the utilities." (Hansen 1993: 7) Despite this official policy, however, most wind turbine owners did not get paid for the electricity they delivered to the grid.

3.2.5 DV and FDV

Dissatisfaction with this situation and the technological state of affairs eventually incited hobbyist and windmill owners – like Torgny Møller, Flemming Tranæs and others who had got to know each other during the previous Wind Meetings – to form a pressure group. In May 1978 they established the Association of Danish Wind Turbine Owners (*Danmarks Vindmølleforening – Danske Vindkraftværker*, DV). These, primarily idealistic, but also critical buyers created a small market of some hundred turbines, which made it possible for a small wind turbine (component) industry to come into existence (Jørgensen and Karnøe 1995: 67). About ten small firms tried their luck in the wind energy business, and in 1978 formed their own trading group: the Association of Danish Wind Turbine Manufacturers (*Foreningen af Danske Vindmøllefabrikanter*, FDV), basically to promote their interests within the *Folketing*.

The early turbines did not perform as hoped and incidents involving runaways, damaged gearboxes, burned generators and lost blades were numerous. These troubled both the entrepreneurs as well as the windmill owners and, at a Wind Meeting in 1978, they reached the conclusion that a 'safe' wind turbine needed to have a double brake system (Karnøe 1993: 79). Paul Gipe (1995: 59), one of America's leading wind energy experts, maintains that the requirement for a fail-safe, redundant braking system "probably did more than any other [provision] to further Danish wind technology, because it attempted to ensure the survival of the wind turbine when something went wrong."

3.3 The Wind Turbine Export

Until that moment, Danish MPs seemed indifferent about the private development of renewable energy sources. But appearances are deceptive. Since the end of 1977, the *Radikale Venstre* had made one of its famous political masterstrokes that would have a great influence on the development of the market for wind turbines. Starting in August 1977, the social democrats – who formed a minority government – entered into negotiations with the Conservative People's Party, *Venstre*, and the RV to work out an employment plan. As part of the so-called second August compromise (*augustforliget II*), RV managed to get support for an investment subsidy on renewable energy technology.[31] The agreement compelled the Minister of Housing to draw up a market subsidy law. The Ministry of Trade was to decide about the right moment in the development of the alternative technologies to introduce the subsidy program.

3.3.1 The Renewable Energy Investment Subsidy

In the Spring of 1979, the Ministry of Trade told their colleagues at the Department of Housing, "Please go ahead, now it is O.K. for wind and solar." The bill was discussed in the *Folketing* in June 1979. During the debates, the Minister of Housing, Erling Olsen, clarified that the 30 percent subsidy was "not meant to support the basic development work ... [nor] pioneers," but about "creating production opportunities for the Danish industry, in such a way that series production could be achieved".[32]

At the time, the situation with balance of payments was almost intolerable, and, to ensure that government money would be well spent, it was decided that a type approval was obligatory to qualify for the subsidy.[33] Except for the ultra-right Progress Party, all parties agreed with the renewable energy subsidy. Even the Conservative Party accepted the proposal. Its commitment to the August compromise forced the KF to keep silent about its antipathy toward wind energy. The strong increase in world oil prices in the first few months of the year made it easy for the conservatives to hide their real motives, and, like the utilities, they argued that, in this situation, it was wise "to use as many energy sources as possible."[34]

3.3.2 DEF's Provisional 1979-Rules

The market subsidy law came into force in the autumn of 1979.[35] Meanwhile, the joint complaints of DV and FDV about the utilities' negative attitude had found response within some of the smaller political parties. In particular, DR, RV, SF and VS regretted the fact that most electricity companies refused to comply with the rules set by DEF in 1976, and asked the Minister of Trade, Arne Christiansen, whether "the state should not interfere with a law."[36] They criticized the utilities for discouraging so-called cooperatives (*fællesskaber*, or more specific *vindmøllelaugene*) in setting up wind turbines for electricity production.[37]

Rushed by Christiansen, and watched by the Energy Agency (*energistyrelsen*), the utilities' umbrella organization established a fast-working committee to revise the old regulations, and, by the end of the year, new rules were published that applied to both individual as well as cooperatively owned wind turbines (*fællesmøller*). DEF accepted the private initiatives "by way of experiment," and settled an 'experimental period' of three years (DEF 1979a: 6) As a result, from 1980 onwards the vast majority of wind turbines sold in Denmark were bought by small local cooperatives (typically consisting of some ten to thirty families). There was widespread political support, which was regarded as *folkelig*, in that it was connected to community-based self-reliance, and to the ideal of social cooperation. It was seen as coming from the people, and (thus) as being beneficial to the people.

3.3.3 Energiplan 81[38]

The adverse effect of the second oil crisis on the balance of payments again put the energy issue at the top of the political agenda, and incited the new Social Democratic government in the Autumn of 1979 to set up a separate department to deal with energy-related matters. The Ministry of Energy (*Energiministeriet*, EM) was given the task of preparing a new national energy plan since the decision of the *Folketing* in May 1979 to introduce natural gas in Denmark from 1984 onwards and the repeated postponement of a governmental decision around nuclear power[39] had rendered the 1976 HM-plan a useless document. In contrast, the AE-plan gained political relevance at the end of the decade,[40] as it became central to OOA's campaign, 'Make Plans Without Nuclear Power' (*Planlæg Uden Atomkraft*), which received massive support.[41] The draftsmen of the AE-plan became indirectly involved in the preparations of the new policy document, as they were asked to comment on preliminary versions of the plan.[42] At the end of 1981, the final version was presented (EM 1981).

The so-called Energy Plan 81 (*Energiplan 81*, EM-plan) contained "all kinds of measures and proposals from the AE-plan from 1976."[43] First, two long-term energy scenarios – one with and one without nuclear – were presented. Second, where the HM-plan aimed at a temporary curtailment of the increase in energy consumption, the EM-plan emphasized the need for permanent energy savings, and contained the explicit goal of reducing the coherence between increasing energy consumption and national economic growth (IRP 1993: 18).[44] Third, the plan wanted "a maximum development with decentralized cogeneration plants," and "a stronger accent on renewable energy sources." (EM 1981: 8)

With respect to wind energy, the government aimed at "60,000 small wind turbines," which would cover about 8.5 percent of electricity consumption in 2000 (*ibid.* 166-167). The use of large electricity producing wind turbines was regarded as an 'alternative' to the small wind turbine development. In several ways, the new target represented a major shift within the government's perspective on wind energy. First, small turbines were originally defined as heat producing machines, now wind machines of all sizes were seen as electricity generating plants, and small turbines were presented as the most important option. Connected to this, there was a change from a policy related to R&D funding directed to the development of utility wind turbines, to a policy of market and export stimulation devoted to the development of a wind turbine industry. In more abstract terms, this shift implied a change from an energy-technology development perspective to a strategy that tried to create a symbiotic relationship between energy and industrial policy. Third, and almost needless to say, this policy change indicated that the two instruments that were most relevant in creating a new industry – the investment subsidy that stemmed from an employment policy and the establishment of the Test Station for Small Wind Turbines at Risø – became institutionalized within Denmark's energy policy. Fourth, the shift exemplified that the concept of decentralized energy production systems was in fashion again. Finally, it represented an official recognition of the steady development the wind turbine technology, market and industry had undergone during the past few years.

3.3.4 The Test Station's Central Role

> It is not a matter of building the best turbine technology, instead it is more important to develop an industry. *Helge Petersen (First head of the Test Station), 1988*

As illustrated by the above judgement, the people that worked at the Test Station for Smaller Wind Turbines identified themselves more with the wind turbine industry than with the research community. There were at least three obvious reasons for this (cf. Dannemand Andersen 1993: 205-207). First, the five people who started the Test Station at the atomic research institute at Risø in 1978 shared a strong interest in and commitment to wind energy. Moreover, three of them – young engineers from the Technical University – were members of OVE, and adhered the goal of that organization: "to introduce wind power in Denmark, and ... also to avoid nuclear power." (*ibid.* 205) Related to this, the meaning of wind turbines as a symbol of protest against atomic power provoked other members within the Risø research community to perceive the

Test Station as a counterculture.[45] Finally, the fact that the Test Station was financed on the basis of a short-term contract of three years, and was ultimately dependent on the results of the industry, both stimulated and required a market oriented attitude to justify its institutional existence.

The first tasks of the Test Station were to test small wind turbines in- and outside Risø, and to be a consultant for the manufacturers (DEA 1979: 91-92). Initially, the wind industry and Risø cooperated on a voluntary basis. This changed fundamentally in 1979, when the above tasks came to include the new approval procedure for commercial wind turbines. This forced companies to interact with the Risø Test Station, and made it become "the common test, consultancy, and R&D department for the whole industry." (Jørgensen and Karnøe 1995: 67) Acknowledging the lack of a proven and reliable scientific knowledge base, the Test Station made use of normal engineering sense and applied aero- and structural dynamics for testing, interpreting measurements, and developing a proper licensing procedure (see Karnøe 1993: 79-82). In the first years, the approval focused on the reliability of the wind turbine, and was based on on-site inspection of components and a talk with the designer of the system. This focus related to the staff's sensibility to the negative publicity about wind energy. Like one of the early engineers at Risø aptly formulated "It was not only the wind turbines that we tried to save, but it was the whole wind turbine development, and the image wind turbines had." (quoted in Dannemand Andersen 1993: 205)

In this period, several product types were tested, ranging from Darrieus and gyro mills over multivanes (so-called windroses) to propeller types. From all these attempts one particular concept emerged: "the horizontal axis wind turbines, equipped with a three-bladed upwind [stall-controlled] rotor, a grid-connected induction generator, and an active yawing system." (Madsen, Rasmussen, and Madsen 1990: 15) Although the approval system most likely speeded up the selection process, Risø did not select the so-called 'Danish design.' Instead, the selection was made in the emerging market (Karnøe 1993: 82). Around 1979, the three firms – Bonus, Nordtank, and Vestas – that would dominate the Danish wind industry in the 1980s entered the market, and went for the Juul/Riisager design. This design was chosen for its simplicity, the availability of some crucial standard components (i.e. blades, electronic control system, gear, and generator), and because there were already some operating on the market – in particular the fifty 30 kilowatt turbines that the pioneer Riisager had produced between 1976 and 1978 (*ibid.*)

After some failed experiments with a vertical-axis Darrieus wind turbine, the biggest firm, Vestas, started in 1978 to cooperate with the smith, Herborg, who was the first pioneer to use a three-blade upwind rotor with two generators and an active yawing system (Karnøe 1991: 195). The two other companies started on their own, and bought components on the market to 'piece together' wind turbines about 10 m in diameter, which powered 30-kW generators. Nordtank originally produced oil and water tanks for vehicle tankers, but due to stagnating markets decided in 1979 to diversify into the wind energy field, and manufactured its first wind turbine in 1980 (Nordtank Energy Group 1995: 13). Bonus stemmed from a company that produced irrigation systems, and started to build wind turbines both due to bad market conditions, and out of sympathy for wind energy (Karnøe 1991: 195). Lacking any specialized knowledge on wind technology or aerodynamics and standard components, these new firms concentrated on the kind of work they were used to doing, and were good at; namely, designing constructions that are strong enough to last for many years in operation (cf. Karnøe 1993: 78).

3.3.5 The Gedser Wind Turbine

> Only now I know how wind turbines should not be constructed. *Johannes Juul (Designer of the Gedser wind turbine), 1962*

The technological concept that evolved in the realm of the small wind turbine industry was also at the center of the Danish Wind Power Program for the development of large-scale turbines. Both had their roots in the 1950s experiments of Johannes Juul (1961). Ever since 1904, when he had completed the course for rural electricians at Askov Højskole under Poul la Cour, Juul had desired to get involved with wind turbines again (cf. Petersen *et al.* 1993: 11). At the age of sixty, his long-life employer – the power company SEAS – made his dream come true.

After the Second World War, SEAS took over the power supply on the island Bogø, and introduced an AC supply system, which made the local DC power stations superfluous. Among these stations was a F. L. Smidth propeller-type wind turbine that Bogø had bought during the occupation to make up for the cutting off of oil supplies. SEAS was interested in wind energy as "a supplement to the expensive foreign fuel," (SEAS 1987: 71) and offered Juul the opportunity to experiment with the turbine. Somewhat later, SEAS authorized Juul's proposal for a research program that aimed "to construct an AC wind turbine that could directly deliver electricity to the power grid." (*ibid.* 72) This enabled Juul to build his first experimental wind turbine: a two-blade downwind rotor of 8 meter in diameter, and an asynchronous 10-kW generator. After a few simple experiments, he found a fatigue-resistant rotor design: a three-blade, stall regulated upwind rotor, with movable blade tips as airbrakes, mounted on the hub by a system of stays and struts.

In the 1950s Juul's experiments became part of an international program. The year before, it had been clearly stated at the International Electricity Meeting in Lake Success in the United States, that the known fuel deposits were rather limited, and that especially Western Europe would run into serious problems in the long term. In the light of this conclusion, the OEEC (Organization for European Economic Cooperation) established a working group on wind energy. Denmark was asked to participate. The Ministry of Public Affairs (*Ministeriet for offentlige arbedjer*) asked DEF to set up a committee that could represent Denmark (Petersen *et al.* 1993: 11). Hence, the Wind Power Committee (*Vindkraftudvalget*) was set up in 1950 to "explore the feasibility of utilizing wind power for making the national electricity supply less dependent on fuel imports" (DEF 1973b: 20) and "to study the possibilities of using the construction of wind power plants as a way to relieve unemployment." (*ibid.* 2)

A year later, Juul reconstructed the old F. L. Smidth *Aeromotor* with a new model rotor of 13 meter in diameter, and a larger 45-kW AC generator, to test both the upscaling of the design, and the switching on of the turbine to SEAS's high-voltage grid (Karnøe 1991: 168). With two satisfying experimental turbines in operation and producing data, the Wind Power Committee decided to proceed with a design that would be representative for a modern, electricity producing power plant. The Ministry of Public Affairs agreed, and, with the help of a Marshall loan, the development and construction of the 200-kW Gedser wind turbine was financed. On July 26, 1957 the now famous turbine, nicknamed *Oliemøllen* (oil mill), was inaugurated.

On this occasion, DEF president, Robert Henriksen, noted that the peaceful use of atomic energy in nuclear plants had come to the fore in Europe.[46] He argued that, with respect to the rapid development that took place within this field, it was "justified to ask whether the time of interest for the use of wind power has not passed." (Thorndal 1993: 11) Still, Juul's R&D program continued for five years. The Wind Power Committee valued the technology of the Gedser turbine, and stated that "if the industry would be interested ... it [could] be used as a model for a more industrial produced wind turbine." (DEF 1973b: 6) Nevertheless, the committee decided to stop the research effort because "Under the current price conditions wind power will not be able to be competitive with steam power." (*ibid.* 21). Juul disagreed, and found it unfair to compare the mature steam-engine technology with a prototype-phase technology like wind turbines, of which he had just found out how *not* to construct them.[47]

3.3.6 The Danish Wind Power Program

> The utility involvement in the Wind Power Program right from the beginning was founded on the expectation that the utilities would most likely be the owners and operators of future large-scale wind turbines connected to the Danish electricity grid system. *Henning Grastrup (Elsam Power Station Engineering), and Poul Nielsen (DEFU), 1990: 24*

Although the program had stopped, the Gedser turbine stayed in operation until a gear wheel broke in 1967. In view of the low oil prices in these years, SEAS decided to dismantle it. Ten years later, the old machine was refurbished and put to the test as part of the new Wind Power Program, in order "to reconnect to the past and study the basic concept of the Gedser wind turbine." (Grastrup and Nielsen 1990: 25).[48] No less attention was given, however, 'to connect to the future,' and by the Spring of 1977 a Danish delegation visited the United States – the place where the future was being shaped. The objectives of the study-trip were fourfold: to keep informed about America's future research plans, to maintain and extend the contact between the two national wind power programs, and to discuss the Danish-American cooperation with respect to the measurement program of the Gedser turbine (Grastrup *et al.* 1977: 1). The main practical aim of the journey, however, was "to buy a ready-made rotor, or otherwise to buy a calculation method to construct a Danish rotor." (*ibid.*) With respect to this latter aim, the commission returned empty-handed because the American firms (for example, Hamilton-Standard) were not interested in just exporting technological know-how. Moreover, the glass fiber reinforced polyester (GRP) blades these firms gladly exported, turned out to be too expensive (*ibid.* 22).

This threw the Danes upon their own technological resources, which meant that Risø took over the development of the blades. Also, with respect to other design features, the Americans provided little insight. The Danish delegation became somewhat surprised about the fact that – in contrast to a visit the year before – the American researchers had become far less sure about the construction costs, the ideal size of wind power plants, and especially about whether the rotor should have three or two blades. (*ibid.* 37) With little guidance from the American experience, and without the test results from the Gedser turbine, it was decided in 1977 to build two horizontal axis 630-kW wind turbines, equipped with a three-bladed upwind rotor of 40 m in diameter at Nibe in northern Jutland (see Grastrup and Nielsen: 25-26). Because of limited financial resources, a hybrid structure of the rotor blades was chosen, comprising both steel and GRP.[49] The main difference between the two machines was related to the system output control: Nibe A had a stall-controlled rotor, and Nibe B had blade pitch control. The first grid connection of Nibe A took place in September 1979, followed by Nibe B in August 1980. The tests and trials revealed a lot of problems, which caused delays for days to several months. The most serious trouble was metal fatigue in the steel parts of the rotor blades.

In contrast to the growing activity in the field of small wind turbines, industry's interest in developing large machines was low (Maribo Pedersen 1990: 6). The Social Democratic government showed more faith in the export potential of clean energy supply technology in general (cf. SEAS 1987: 99), as well as in the technology developments with regard to large-scale wind turbines within the WPP in particular. This was proven in December 1981, when – despite the absence of a tradition for public ownership of industrial enterprises in Denmark (Himmelstrup *et al.* 1992: 23) – the Ministry of Energy, together with SEAS, established the wind turbine company Danish Wind Technology (*Dansk Vindteknik A/S*, DWT). "Out of the research results from the Wind Power Program" this company "would develop and market large wind turbines".[50]

The Ministry of Energy's drive to promote the commercialization of large-scale wind turbines had wider repercussions. First, the authorities asked the utilities to proceed with the development of large machines in order to promote industrial innovation. Elkraft proposed to build a wind farm on the Masnedø island consisting of five units close to the Nibe B machine in size and technology. Logically, DWT was chosen to manufacture the 750-kW wind turbines, with rotor blades entirely made of GRP to avoid the fatigue problems encountered with the Nibe prototypes. The Jutland power company, Elsam, chose to design an upscaled version of the Nibe B machine, the 2-MW Tjæreborg wind turbine with a rotor of 60 m in diameter. In contrast with the Masnedø project, no manufacturer in Denmark was willing to construct a prototype of this size, and, as with the Nibe machines, the Tjæreborg turbine was procured on a multi-contract basis. The wind farm and the Tjæreborg turbine were only officially inaugurated in January 1987, and June 1988, respectively.

Secondly, the offensive state intervention, combined with growing industrial activity and private development, gave the utilities a fright, and incited DEF to set up an ad-hoc committee[51] "to examine the judicial conditions surrounding the establishment of production wind turbines, and wind turbine parks". (DEF 1983a: 26) The committee concluded that to make wind turbines fit into the existing electricity generation system, "Potential wind turbine parks or production wind turbines ought to pay for themselves, and ought to be owned and driven by the electric utilities." (*ibid.*) DEF advocated that subsidized private wind energy development was at odds with the principle of lowest possible consumer price, and maintained that private development could not easily be incorporated in the planning system of the electricity sector, and therefore threatened supply security. Moreover, private ownership would make it hard to accommodate wind parks and big wind turbines within the running production system. DEF doubted whether wind turbine owners would accept the utilities to disconnect their turbines in situations where this would be cheaper (e.g., when cheap hydro power from Norway and Sweden was available). Finally, DEF estimated that in the event of private ownership, "among others, grid amplification and turbine's capacity value [would] be clear subjects for almost irresolvable disagreement". (*ibid.*)

3.4 Home, Sweet Home Market

In the meantime, the utilities tolerated the development of small private wind turbines 'by way of experiment.' In contrast to the anticipated private owners of *production* wind turbines, who were conceived as potential electricity producers, the utilities regarded the individual and cooperative owners of small wind turbines as self-suppliers (cf. DEF 1981a: 15). This perspective had been incorporated in the 1979-rules, which forced cooperatives to adjust the turbine production to their joint consumption, and prescribed that the owner(s) should live within a distance of three kilometers from the turbine. Among other things, DV wanted to get rid of the '3-km rule,' and thought it was fair if the utilities would contribute to the connection costs. Consequently, DV started to negotiate with the Wind Turbine Group[52] within DEF's Price and Tariff Committee to both change and elongate the provisional rules that were set up two years before.

3.4.1 The 1984-Agreement[53]

The negotiations were tough going. To get an idea of the capacity value of wind turbines, DEF insisted on first carrying out measurements on twenty private wind turbines across the country (DEF 1982a: 15). DV agreed, in exchange for prolonging the 1979-rules. Frustrated by more disputable delaying tactics used by the utilities in 1983, DV wrote a polemic in the wind power magazine *Narturlig Energi*, which triggered an open conflict between the two organizations (DEF 1983a: 23). This was rapidly settled, but the negotiations only gained momentum at the start of 1984, via a bill from the *Radikale Venstre* (DEF 1984a: 17). RV argued that the uncertainty surrounding the setting of rules had led to the renunciation of many projects, and proposed to lengthen the provisional guidelines with ten years by law. Moreover, the party suggested that: (a) the utilities should pay a part of the connection costs, (b) the 3-km rule should be replaced by the demand that wind turbines should be sited within the supply area of the electric utility company that serves their owner(s), and (c) there should be no limit to the maximum installed capacity.[54]

FDV, DEFU and the Test Station at Risø became involved in the negotiations, and by May 1984 a 10-year agreement was reached, which came into force in October (see DEF 1984b). In its effort to avoid legal intrusion (cf. Hansen 1993: 7), DEF had been forced to seriously moderate its demands, and to accept most of the propositions made by RV. The electric utilities agreed: (a) to pay 35 percent of the connection costs, under the condition that the government would maintain the investment subsidy program, (b) and (c) to buy *all* surplus power at a rate of 85 percent of the consumer price from cooperative wind turbines, of which all owners live in the area of one and the same electricity supply company (cf. DEF 1984b: 2-3).

3.4.2 The Danish Home Market

Between 1976 and 1979, about 170 small wind turbines with rotors of 4 to 10 m in diameter were erected. Most of them were standalone windroses for heating, which were considerably less expensive than the three-bladed grid-connected wind turbines (Karnøe 1991: 204). After the introduction of the investment subsidy and the addition of provisions for cooperatively owned wind turbines to the settling rules in 1979, the latter product type started to dominate the domestic market. In the two years that followed some two hundred wind turbines were sold per year.

At the end of 1981, two terrible November storms swept Denmark and crashed two percent of the wind turbines. Whereas wind energy advocates saw the small number of accidents as a proof of the technology, the event was grist to the mill of its opponents. In particular, the total loss of a Windmatic 55-kW turbine – whose blade debris was scattered over the main road from Viborg to Aalborg and kept it blocked for hours – caused a lot of bad publicity in some conservative newspapers (Van Beek 1981: 19). Because of this and stagnated oil prices the Danish home market almost halved at the beginning of 1982, and, in March, the wind turbine manufacturers started to think about ways to increase their turnover. The Technology Council of the Ministry of Industry subsidized FDV to examine the American market. The report was ready in August, and, almost the day after, sales representatives from Bonus, Vestas, Windmatic, and Nordtank packed their bags and flew to the unknown market.[55] Exports to California took off, with some 40 wind-turbine deliveries before the end of 1982 (Karnøe 1991: 216).

Meanwhile, the burden of foreign debt had become so heavy that Anker Jørgensen's Social Democratic government resigned and surrendered power, without an election, to a new conservative government. On 10 September 1982, Poul Schlüter formed the Four-Leaf Clover coalition (*Firkløveret*) consisting of his own Conservative People's Party, and three other parties: *Venstre*, the Center Democrats, and the Christian People's Party. The new Energy Minister, Knud Enggaard (V), announced that the government would continue to work with Energy Plan 81, but that price regulation by means of taxes would soon be abolished (DEF 1982a: 1-3). A small electricity tax of two øre per kilowatt-hour had been introduced in 1977 to control electricity consumption.[56]

In response to the second oil crisis, the Social Democratic government had boosted the tax to about 13 øre per kilowatt-hour. The utilities and the conservatives had felt aversion to this growth, and argued that the tax had been raised "primarily to cover the costs of other governmental expenditures." (DEF 1980a: 2) But although Knud Enggaard paid lip service to repealing the electricity tax, he was not able to fulfil his promise. Also, under the new political regime that introduced notions like de-bureaucratization, privatization, modernization and technological renewal (cf. Sidenius 1989: 72), the tax kept on growing more or less in pace with decreasing world fuel prices. This neutralized the effect of fuel prices, and kept electricity prices in Denmark constant for consumers during the 1980s.

Enggaard also hinted that he was planning "to reassess the contribution that renewable energy sources could make." (DEF 1983a: 5) Concerned by this attitude, the Socialist People's Party pointed out to both the Minister of Energy and Industry that the export of wind turbines to California had taken off. SF argued that "a stable home market is the alpha and omega, and that the savings under reconciliation would lay poison under the home market," and asked the Ministers "which initiatives [they were] going to take to keep up Denmark's technological lead, and to use the very strong export possibilities".[57] Not impressed by the modest scale of export, but facing a 'green majority' (*det grønne flertal*) – consisting of SD, SF, RV, and VS, the Ministers made the best of a bad job by pretending they were pleased with the new development. They assured the *Folketing* that the subsidy program would be continued in 1983 to maintain the size of the home market. This came true, and some 8 MW of wind turbines were installed. The export, however, surpassed the expectations of the conservative government. It was almost three times bigger than domestic turnover. Once again, SF claimed that a considerable domestic market would be decisive to maintain and extend the export potential. Through its energy-policy spokesman, Ole Kalnæs, the party argued that one way to achieve this was by exempting wind turbine owners from taxes on the electricity they bought from the public grid during windless periods to the extent of their turbine's electricity production.[58] The government agreed.

In the following three years, exports to California boomed, completely outpacing the home market. At its peak in 1985, almost 3,500 wind turbines were shipped overseas, representing an export value of more than two billion Danish kroner. This huge export success made the bourgeois Four-Clover government a new ally of the wind industry (Karnøe 1991: 301). Although the industry was under the spell of the California event, the sudden jump in the number of turbines sold on the domestic market from 7 MW of installed capacity in 1984 to 23 MW in 1985 provided enough to talk about on the political level. The 'Danish boom' was caused by the combined effect of the investment subsidy, the exemption of wind power from electricity tax (which had the effect of a production subsidy), and the 1984-agreement between DV, FDV, and DEF. Under these favorable circumstances, wind turbine parks became profitable investment projects for institutions or local municipalities. The utilities complained to the Ministry of Energy that the projects threatened the planning of the electricity system. DEF pointed out that only one-third of the electricity produced by wind turbines was used by the owners themselves, and claimed that the 1984-agreement was not intended for public or private organizations to set up production wind turbines as investment projects (DEF 1985a: 6).[59]

The utilities also criticized the fact that most of the wind farms were placed in windy areas in northern Jutland, and argued that this concentrated development caused financial and technical problems for the local electricity companies.

3.4.3 The First 100-MW Agreement[60]

> The Ministry of Energy finds it advisable that in the future the development of large wind turbines and wind turbine parks – without local attachment – will be an integrated part of the normal development of the electricity supply system, in order to avoid that the development of wind turbines primarily will be determined by individual investment purposes. *Ministry of Energy, December 20, 1985*[61]

Without even informing DV and FDV, the utilities secretly started to negotiate with the conservative government for a new regulatory compact within the wind energy field (Karnøe 1991: 231). On 20 December, 1985, their protests resulted in a new departmental order with more restrictive rules for state subsidies for wind energy (*ibid.* 7). The new rules stated that in order to get subsidized private investors should either live within a 10-km distance from the turbine,[62] or live within the same municipality. The capital investment subsidy was also reduced to 15 percent. A further constraint was that the expected yearly electricity production should not exceed the owner's prospective yearly consumption by more than 35 percent. In return, the Ministry of Energy forced Elsam and Elkraft to commit themselves to install 55 MW, and 45 MW of installed wind power capacity, respectively, between 1986 and 1990.

This so-called *100-MW agreement* was expected to secure a stable home market with a yearly turnover of 20 MW over the next five years. The utilities had noticed that the political interest in renewable energy had increased (both within and without the governmental coalition), and the 100-MW agreement was a dash for safety, an attempt to keep total control of the planning and production system. Like DEF euphemistically expressed it, "to secure an appropriate siting of large wind turbines and wind parks by taking into consideration the wind conditions, the electricity grid, and the landscape". (*ibid.*) The government became triggered by the fact that in 1985 large investors had swallowed most of the investment subsidies. Its motive behind this crucial political initiative was to stop the speculative development of wind farms in order to continue the promotion of wind turbines with *local attachment* by means of investment subsidies. According to Knud Enggaard, "This development [would] take place parallel to the development of the utilities of large wind turbines and wind turbine parks." (*ibid.*)

3.4.4 The Management of Product Development in Danish Firms

> What we, the Danish manufacturers, must still be able to offer is operational experience and project security. Our projects must produce the energy they are supposed to, using tested machines from suppliers who have the wherewithal to stand by their promises. Research should be directed towards making machines more reliable and

> gaining a fuller understanding of the physical phenomena we deal with on a daily basis. *Henrik Stiesdal (Technical Director Bonus), 1990: 28*

At the end of the 1970s, a certificate from the Test Station had become a prerequisite for entering the domestic wind turbine market. This had forced manufacturers to develop a turbine reliable enough to qualify for the investment subsidy. Type-approval, however, rapidly became a normal aspect of a turbine, and economical efficiency became more and more the dominant distinguishing competitive item. Since the performance figures of almost each single wind turbine in Denmark were published by DV in their magazine *Naturlig Energi*, the potential customer had all the information needed to select the cheapest and most reliable turbine.

The transparency of the market increased competition between different producers, which spurred technological progress. It was commonly thought that larger wind turbines would be more cost effective. At the time, the technological challenge was to upscale the 30-kW wind turbine, equipped with a rotor of 10 m in diameter, to a 55-kW one of about 15 m in diameter (Karnøe 1991: 208-209). This specific jump in size was dictated by the next standardized generator-capacity that foreign suppliers had available. The dimensioning of the other parts had to be optimized with respect to this invariable. From 1980/81 until 1983, manufacturers focused on incrementally improving the effectiveness of the 55-kW design.

In this period more than 600 turbines were sold in Denmark. The four largest firms established a combined market share of 70 percent. The firms mostly operated on a regional market. This meant that manufacturers could service their own turbines. According to Gipe (1995: 56) this geographical proximity "enabled companies to learn quickly from their mistakes and to keep their turbines in operation as physical proof for potential buyers that the company's machines were a good investment." The development was characterized by implementing *ad hoc* solutions and trial-and-error learning, and based on the intuition of the designer as well as rules of thumb (cf. Karnøe 1991: 209). By gradually improving on the solid design and selecting out its weak elements, the cost effectiveness of the average installed turbine increased 50 percent (*ibid.* 215).

3.4.5 Danes Competing Danes on the American Market

> Hvem konkurrerer dansk vindmølleindustri med på det amerikanske marked? Sig selv – kort og godt. (Whom did the Danish wind turbine industry compete with on the American market? Itself – in short.) *Tyge Kjær, 1988: 24*

Danish exports to California began in 1983 with the well-tested 55-kW turbine, which was upgraded to a 65-kW machine. In the previous two years, California wind farmers had become frustrated with the reliability and low power output of the American-made wind turbines, and were willing to take a chance on the European-made machines. The Danish turbines looked most promising, because their salesmen could point to independent (see above) and relatively impressive performance statistics. To some extent Danish manufacturers managed to keep to their promise, and rapidly penetrated the

California market. Paul Gipe, who worked at Zond, summarizes this phenomenon by simply stating that "The Danish wind turbines worked, and the American turbines did not," and adds that the only American turbine that did work, the USW 56-100, was not for sale.[63] In 1984, about 1,600 Danish turbines were sold in California. As shown by their profit rate, which was twice as much as the average within the Danish industry, the wind turbine manufacturers did extremely well out of the early export experience.[64] Attracted by these figures, several new Danish firms (e.g. Wincon, Danwin, Vindsyssel, Dencon, and Windworld) tried to seek their fortune on the lucrative wind turbine market. One of the founders of Danwin, Jan Svenson, explains that, with the end of the tax credit programs in sight, "we took a final chance to take part in the California 'Gold Rush'".[65]

As the market share of the Danish firms increased, their market positions changed in a very fundamental manner. In the first two years, they had technologically outdistanced their American competitors. But after this 'battle' between the 'one-eyed Danes' and the 'blind Americans' was won, Danish firms started competing each other on equal terms from a technical perspective (cf. Kjær 1988: 24). What was at stake in the technological rat race that followed was the development of larger and more efficient turbines. Within this competition, the older firms were rushed by the newcomers, whose strategy it was to leapfrog their competitors by combining the best available technology to introduce new and larger turbines. The existence of a 'public', simple, and solid design, as well as the availability of free advice from the Test Station, enabled the new firms to do this.[66]

For several reasons wind farm developers wanted larger turbines. Thousands of small wind turbines had been erected in the three windy passes (Altamont in the north, Tehachapi and San Gorgonio in the south). The large number of machines that would be erected in the coming years made good sites a scarce, expensive resource. Installation (building roads, foundation, control system), operation (size of the crew) and maintenance (number of inspections) costs were more or less independent of turbine size, and thus provided larger turbines the benefits of economies of scale.[67] Finally, the prospect of expiring energy tax credits created a need for more effective wind turbines.

The huge demand for larger turbines started a fierce competition based on technological competence (Kjær 1988). The innovation process went so fast that several turbine types were barely on the market for one year before other producers introduced new, bigger, and more effective types. The rapid techno-economic obsolescence of wind turbine technology meant that the 'old' turbines could no longer be sold for their original price. To fulfil the exponentially increasing demand, manufacturers used the most resources for gearing up their production technology, and rapidly shifted from batch to series production (Karnøe 1991: 218). The cost reduction within the production process, however, was not able to counteract the price reduction due to product innovation (Kjær 1988: 38).

The paradoxical result was that, by 1985, at the height of the California export boom, a quarter of the turnover involved a loss, and seven out of the twelve Danish companies had negative trading results (*ibid.* 20). A year later, the number of turbines shipped overseas almost halved. The reduced marketing possibilities in California made competition even more intense and caused profits to decrease. This was strongly aggravated by the falling dollar exchange rate.[68] As a result, the portion of transactions in 1986 that involved a loss increased to 70 percent and no less than eleven out of the twelve firms suffered from negative running costs (*ibid.*).

The process of forced upscaling, combined with hasty series production led the Danish industry in general to export new, relatively unproven wind turbine designs (Madsen 1988). Like in the early industrialization period on the home market, most of the manufacturers now experienced problems with gearboxes, generators, oil-cooling, and blades. Several factors, however, aggravated these difficulties during this hectic export period. First, the slovenly assembly of turbines worsened things (cf. Karnøe 1991: 220). Second, doing business on the international market removed the advantage of 'geographical proximity,' which made it harder for the Danes to learn from their mistakes. The fact that the California wind farm developers serviced their own turbines slowed down the feedback to the Danish designers about the problems their machines encountered in the field. Things were even worse because the Danish firms took for granted that their machines would perform as well in California as they did in their home country. Initially, the Danes somewhat ignored the problems the American maintenance mechanics experienced in the field, and said to them, "This turbine has been thoroughly tested in Denmark, it should work in your country. It is your fault. You guys can't fix anything."[69] Because of this posture, the Danes had to learn the hard way that the very rough wind and weather regime in California essentially differed from the moderate maritime climate back home, and indeed put extra demands on their technology.

The increased number of Danish wind turbine manufacturers, the falling, but still huge, and thus attractive demand for larger and more effective turbines, the intense techno-economic competition that compelled firms to sell turbines below the production price, the forced shift to series production, the assumption that wind energy technology could be easily transferred to another environment, the disturbance of the innovation process that was based on the quick feedback from the service crew to the designers, all help to explain why the Danish industry ran into trouble on the U.S. market in 1986 and 1987. In combination these factors caused the industry's "inability to deliver wind turbines to wind farm developers that generated the profits promised,[70] ... [and] to honor service and guarantee agreements when technical problems arose." (Stiesdal 1990: 28)

On the one hand, most companies had neither the managerial competence, nor the organizational and technological capability, to maintain the operational experience they had built up on the home market during this frantic period of hasty technological development and rapid economic growth. On the other hand, most manufacturers were small and undercapitalized firms that did not have the wherewithal to stand by their promises and offer project security when severe technical problems arose. The list of companies that went bankrupt started with Vestas and the blade manufacturer Alternegy in 1986, and Nordtank in 1987. The individual bankruptcies had a negative self-perpetuating impact on both the image and earnings of the whole industry (Kjær 1988: 21). When in 1988, the export to California hit bottom, the companies Wincon, Tellus, Windmatic, and Vindsyssel filed for bankruptcy, and many feared the collapse of the total industry. The U.S. subsidiary of Micon completed the sad series of liquidations the year after. According to its management, this decision was made "due to the evaporation of the business investment environment," or more bluntly because "there was no sales activity" (Davidson 1989b).

Only a few companies managed to cope with the export boom and its successive crash. Bonus stood out as an example of prudent management. This company had been careful about stepping up production and expanding, and had secured operational experience by gradually introducing its new turbine models (cf. Stiesdal 1990). In general, however, Danish companies had not been able to convert their leading technological position into hard cash. No wonder Tyge Kjær – who was hired by FDV in 1988 to analyze the strong and weak aspects of the Danish wind turbine industry – described the California experience as both 'a heroic technological success,' as well as 'a devastating financial failure.' (Kjær 1988: 16)

3.4.6 Back to the Cradle to Restore Exports

> One of the main goals of this Bill is precisely that we here in Denmark will get a set of rules for approval of wind turbines, in particular with respect to quality control, that respond to rules that apply internationally. *Jens Bilgrav-Nielsen (Minister of Energy), 1989*

Despite the gradual reduction of the investment subsidy for private wind turbines, which was abolished in August 1989, and the tightening of the consumption and siting criteria at the end of 1985, the home market kept growing from about 20 MW of installed wind power capacity in 1985, and some 30 MW in 1986, to a peak of almost 60 MW in 1991 (FDV 1994: 36-38). This slowly growing private market, combined with the utilities' pledge to develop 100 MW of wind power, provided a stable turnover for the Danish industry to fall back on when exports collapsed after the California 'wind rush.' The existence of a basic home market, which was widely supported by the public[71] and politicians, combined with both the belief that the Danish technological concept was solid and would survive in the future, as well as the notion that the wave of bankruptcies was part of the 'normal' maturation process of an infant industry within a market economy, all helped to maintain the trust of entrepreneurs and financial institutions in the profitability of the wind industry. Some firms, like Vestas and Nordtank, were reorganized, while other bankrupt firms fused into financially stronger wind companies (Karnøe 1991: 232).

Thus the technological expertise in the field of wind energy was conserved (Madsen 1991: 83). More important, this continuity enabled the industry to implement the managerial lessons it had learned from its export experience. The companies had learned to reduce business risks by (1) spreading their marketing efforts over several 'political' markets, (2) hedging against exchange rate risks to minimize any adverse effect on earnings, and (3) ensuring the stable supply of components by having two 'reliable' sub-suppliers of each type of component. More important, they had encountered the need (4) to take on the development of new turbines in a controlled fashion, with all the necessary testing along the way, (5) to certify a uniform turbine quality, and

(6) to guarantee project performance and insurance in order to find investors willing to finance wind projects.

The latter three issues were strongly connected. After 1984, the financing of Danish wind power projects in California had been based mainly on the concept of shared financing, by which shares were sold to Danish institutional investors. The projects relied on long-term contracts (ISO4 contracts) that secured the purchase of the produced electricity, and were organized in such a way that the Danish shareholders were responsible for buying the wind turbines, and for their performance, and service. This method of financing received quite a blow from the bad press concerning projects that could not live up to their expectations because of technical difficulties.

In particular, the fiasco with some of the projects financed by the Danish Investment Fund (*Dansk Investerings Fond*, Difko)[72] became widely known. Troubles arose after the mass closure of nearly 500 Micon turbines in late 1987 because of major blade tip failure (Davidson 1988). The manufacturer of the ill-fated blades, the Danish company, Alternegy, had already walked away from these problems by quitting the business the year before. The insurer of the wind turbines accused their manufacturers of fraudulently misrepresenting the turbines, and of poor service, and asked the court to reassess its obligations. To avoid long drawn-out court battles, Difko – involved as an insured party and as operator of most of the wind farms at issue – agreed on a settlement with Zurich Insurance Co., which extracted the insurer from its obligations, and gave the financial institution enough money to take care of the repairs. Events like these caused a loss of reputation by Danish turbines during the late 1980s, scared off Danish institutional investors, and made insurers unwilling to issue policies for wind developments.

Several attempts were made to revive the reputation of the Danish wind turbine industry as a whole. Difko hired Bonus to complete retrofits on its defunct turbines (Kjær 1988: 47). DV lobbied for more structural actions to be taken by the government, to restore the trust of the Danish investment community in wind turbine technology. This process was speeded up by the appearance of large Japanese companies – like Mitsubishi, Sumito, and Yamaha, which were reputed to be reliable manufacturers – on the international wind turbine market (Karnøe 1991: 229-230). These financially strong multinationals were able to finance the sale of their own turbines, and the Danes feared that the Japanese would soon dominate the export market. As a result, the Ministry of Energy set up a committee to study possible ways to relieve shared financing by Danish investors, which would "secure and forward the technological development of Danish wind turbines, and safeguard their competitiveness on the world market."[73] It was decided to establish a private Wind Turbine Guarantee Company (*Vindmøllebranches Garanti A/S*), which was given a state guarantee of 750 million Danish kroner, to enable it to warrant – on usual commercial conditions – the long term financing of large projects with Danish wind turbines outside Denmark.[74] Except for the Progress Party, this export promotion measure was broadly supported within the *Folketing*. Also the energy spokesman of the opposition – the social democrat and former Energy Minister Poul Nielsen – praised the creativity and good will of the new Minister of Energy Jens Bilgrav-Nielsen in acting within "the very important borderland between energy policy and industrial policy".[75]

One of the conditions that had to be met to qualify for the guarantees was that the turbines would be approved according to a new system. The existing approval system, described in a 50-page document, consisted mainly of a design review and an examination of the basis of calculations. DV and the government believed that a new system – that would set levels for design, manufacturing, transportation, installation, and subsequent service of the wind turbines – was needed, since the range of technical problems with Danish turbines in California had strongly increased insurance prices. As a result of this technological uncertainty, insurance companies started to demand that known classification companies should give approval to a project. Firms like Germanisher Lloyd, and Det Norske Veritas had therefore entered into the wind energy field, and were already offering a comprehensive certification system for wind power generators.

The government established a working group under the Energy Agency to work out a new approval system since it was thought that, by increasing the quality level of Danish wind turbines to international industrial standards, and by lowering the costs of certification, exports would be stimulated.[76] After two years of preparation by the Energy Agency in cooperation with the Test Station, the manufacturers, and the insurance and certification companies, the new approval system, of which the technical basis covered more than 300 pages (Dannemand Andersen 1993: 265), was introduced in the beginning of 1991. Around this period, the guarantee company was also established. In the meantime, Danish exports had already increased from their 1988-low of 20 MW of wind power capacity, to about 90 MW in 1991 (FDV 1994: 38). The United States accounted for more than half of the exports, but the most significant development was on the European market, especially Germany and Spain (Madsen 1991: 84).

The new European market varied a great deal from the large utility-scale central-station wind power projects – ranging in size from 40 to 400 turbines – found in California. Instead, these so-called 'extended home markets' were characterized by dispersed installations of wind turbines and by small wind farms averaging in the tens of machines, and thus strongly resembled the market situation in Denmark after the 100-MW agreement. As part of this political settlement, the utilities had agreed to develop wind projects in a way that would "optimize the conditions for a broad industrial and technological development in the field of wind energy, among which the development of new prototypes." (EM 1988: 5)

Although the manufacturers were happy about the prospect of a steady utility market, they feared that the utilities considered the development with the existing wind turbines to be a waste of time, and thought that 'new' designs would be a better utilization of time (Stiesdal 1990: 27). In this situation, proven turbines would be rejected for utility projects, and large numbers of marginally more profitable new designs, which had not been fully tested, or even series of prototypes, would be erected instead. The technical director of Bonus argued that this so-called '*empty-kilowatts scenario*' would be unfortunate, because it involved the risk of "losing the chance of further consolidating that which has proved to be the Danish wind industry's hallmark in the world at large: operational experience." (*ibid.*) The concern that the utilities would rather wait for larger turbines to appear on the market than install the current 150 to 250-kW turbines grew when two years after the 100-MW agreement only two utility projects had been built. The utilities argued, however, that the "chaotic interplay between central, regional, and local authorities, site owners, and local civilians" caused the delay (DEF 1988a: 11). Spurred on by the utilities, the Physical Planning Agency (under the Ministry of Environment) tried to persuade the counties to take into account the development of wind turbines within their regional plans.

Moreover, the difficulties encountered by the utilities in gaining permission to site wind farms on land gave impetus to the idea of siting them offshore. In November 1987, the Ministry of Energy set up a Committee for Offshore Wind Farms to "ensure the best possible start for the first offshore wind power demonstration project". (Jensen, Falbe-Hansen, and Mortensen 1990: 51) The committee located a suitable site north of Lolland, and, in the summer of 1989, the board of Elkraft expressed its willingness to construct the Vindeby offshore wind farm as part of the 100-MW agreement "in order to gain experience in the field of offshore wind turbine installations". (Jespersen 1992: 1) It was also decided that the turbine chosen for the world's first offshore wind project would be "a commercially available turbine of proven standard". (*ibid.* 5) The order to build a very corrosion-resistant turbine that was specially equipped for easy and cheap maintenance work offshore was put out to tender. In June 1990, Bonus' offer to adapt its largest available wind turbine (a 450-kW machine with a rotor diameter of 35 m) to the tough weather conditions at sea was selected. A year later the 5-MW Vindeby offshore wind farm was erected (Jespersen 1991).

At the time, Elkraft had already embarked upon a new project to design and construct a 1-MW machine with a rotor diameter of 50 m, to be situated at Avedøre Power Station in Hvidovre, near Copenhagen (Grastrup and Nielsen 1990: 29). The project was predicted on the discussion concerning whether stall regulation would still be sufficient for megawatt-size turbines. This debate had already been an important topic since the start of the Danish wind power program for large wind turbines, and had led to setting up two turbines with different power control systems (i.e., Nibe A and B). But now this engineering question had gained practical relevance for those manufacturers who were engaged in the utility wind turbine market.[77] To address the above research question, the turbine at Avedøre was equipped with two control systems – one based on stall regulation and the other on pitch regulation – to enable a direct comparison of the two systems. The technological specifications were made by the utilities, interested manufacturers were asked to produce the components,[78] and, at the end of 1993, the measuring and test program started. One of the main reasons why Eastern Denmark's power companies invested in the technological development of large wind turbines was "the increasing problems in finding suitable sites for these installations". (Elkraft 1993b: 13)

According to Elkraft, large wind turbines were the preferred solution both on shore because they take up less space, as well as offshore, for inspection and maintenance reasons. A technical reason was that the utilities were accustomed to servicing their own machinery, including the wind turbine parks they possessed, and wanted to keep up with the know-how to maintain this practice. At the same time, these innovative wind projects were clear statements about what kind of commercial turbines the utilities preferred to buy. This is not to say that the utilities' involvement in the field of wind energy was freely chosen. The government had commanded such an effort, and the utilities "did it and resented it".[79]

3.5 Ecological Modernization

> Sustainable development, which implies meeting the needs of the present without compromising the ability of future generations to meet their own needs, should become a central guiding principle of the United Nations, governments and private institutions, organizations and enterprises. *United Nations' General Assembly, December 11, 1987*

In 1987, the United Nations' World Commission on Environment and Development – headed by Norway's prime-minister Gro Harlem Brundtland – followed in the footsteps of the Club of Rome by describing the looming global problems within the fields of environment and resources. But besides depicting the problems connected to growth, the Commission emphasized "the need for a new approach to economic growth, as an essential prerequisite for eradication of poverty and for enhancing the resource base on which present and future generations depend," and, connected to such an approach, the concept of sustainable development.

In the course of 1988, the 'Brundtland report' became central to the Danish energy debate and it became clear that its recommendation to establish a very tight coupling between energy and environmental policy would become one of the pillars of Danish energy policy in the coming years (DEF 1988a: 7). The most often cited part of the report involved the need for low-energy development. According to the Brundtland Commission, a criterion for sustainable development was that industrialized countries should save energy.[80] The Commission supported the use of combined heat and power production, higher energy efficiency, and renewable energy.

3.5.1 The Green Plan[81]

> So far Danish energy policy has given good results. What must be done now is to ensure that it be reformulated, to let Denmark remain in the lead at a time when all industrialized countries are being forced to reduce energy consumption and its impact on the environment. We can thus set a good example while at the same time offering Danish industry new opportunities to further develop and sell cleaner technologies. *Jens Bilgrav-Nielsen (Energy Minister) in his foreword to Energy 2000, 1990*

> A classic mistake in energy planning is to assume – often subconsciously – that energy consumption is a measure of welfare. This is particularly unfortunate when planning for the long term. Energy – excluding food – is of no direct value to human beings. Oil, coal, gas, uranium, petrol, electricity, *et cetera* cannot be consumed directly. They are useful only as inputs to technologies, which then provide services as their outputs, such as warm meals, convenient transport, comfortable rooms, illumination at the desk, and clean clothes. These energy services are what directly contribute to human welfare. *Jørgen S. Nørgård and Bente L. Christensen, 1993: 318*

The Danish government had already moved in such a direction as a result of two political compromises in 1986. In reaction to the drastic fall of crude oil prices in 1985, the Four-Leaf Clover government – which had just started its second term of office – and RV, had agreed to raise taxes on energy to keep electricity prices outside the business sector more or less unchanged. The so-called 'Easter Agreement' (*Påskepakken*) had increased the electricity tax from 19 to 29.5 øre per kilowatt-hour. Meanwhile, the subsidy on renewable energy was disconnected from the electricity tax, and was set on 20 øre per kilowatt-hour (DEF 1986a: 1).

Subsequently, the government started to negotiate with the Social Democratic Party about the future power supply and intensified actions in energy efficiency. In June, an agreement was reached that stipulated that the future development of the electricity production system should comprise both decentralized cogeneration plants, as well as new, large fossil-fueled thermal power stations. The new liberal Energy Minister, Svend Erik Hovmand (V), stated that the plan would secure "a broad political consensus concerning the development of the electricity system throughout the 1990s." He added "Only with an adequate and well-functioning power supply system will we be able to assure the successful occupational, commercial, and economic development in our country." (*ibid.* 2) According to corporative practice, the utilities had participated in the negotiations, and had welcomed both these words as well as the compromise. The Socialist People's Party, the *Radikale*, plus a faction within the Social Democratic party, however, had criticized the June agreement for environmental reasons, because they feared that it would serve as a permit for building large coal-fired units.

The publishing of the Brundtland report at the end of 1987 destroyed the political stability as assumed by Hovmand, and rekindled discussion on how to merge economics and environment into energy policy. This political effort got a strong impulse when the Four-Leaf Clover coalition crumbled in the middle of 1988 and the *Radikale Venstre* joined the Conservatives and *Venstre* in a new three-party governing coalition. Within the KVR-government – which only remained in power until the election of December 1990 – the *Radikale* managed to capture the ministerial posts of both environment and energy. Lone Dybkjær became Minister of Environmental Affairs, and Jens Bilgrav-Nielsen became the new Minister of Energy. These two politicians had long-standing experience in the field of energy, and were convinced of the need to take up the challenge formulated by the World Commission on Environment and Development, and "to effect changes in the Danish energy policy positively along more environmentally acceptable lines, yet not at the expense of continued economic growth."[82]

During the 1970s, local sulfur pollution from coal and oil-fired power plants had been the central environmental issue within the electricity sector. When the KVR-government took office, the former fly-ash problem was solved, since it had become possible to filter almost all the fly-ash, and use it as a raw material for building roads. In the early 1980s, the electricity sector was confronted with the acid-rain controversy. The regional acidification problem was due primarily to sulfur dioxide (SO_2) and nitrogen oxide (NO_x) emissions. The government had reacted by introducing a law on SO_2 emissions in 1984. Complaints from DEF about the fact that the law was too drastic and Denmark was taking a leading role had kept the Four-Leaf Clover government from taking further steps.

Dissatisfied with the environmental content of the June 1986 agreement, the *Radikale Venstre* had proposed a combined NO_x and SO_2 law. The 'SNOX-law' had aimed to reduce SO_2 and NO_x emissions by 2005, 60 and 50 percent, respectively, relative to their 1985 levels. At the time, the bill had not received sufficient political support, but it soon revived after RV had joined the KVR-government. The SNOX-law was finally adopted in March 1989, and signaled the end of a political process in which the acid-rain controversy had been translated into clear emission quotas, which could be reached by optimizing existing burner technologies and adding end-of-pipe technologies. The utilities complained about the costs of these environmental technologies (which required about 25 percent of their total investment costs). But the utilities also realized that the negotiated demands resulting from the acid-rain debate did not require a fundamental change of the electricity system, and would keep their concept of large central coal-fired power plants intact.

Meanwhile, global consequences of human behavior – mainly the depletion of the ozone layer, and global warming – had become the focal point of international environmental concern. The contribution of the energy sector to these global environmental problems is linked especially to emission of carbon dioxide (CO_2) from the combustion of fossil fuels. The pollution from CO_2 required more radical changes in the energy system.[83]

As one of the many follow-ups to the Brundtland report, the 1988 U.N. conference on 'The Changing Atmosphere' had been of specific importance. It recommended the stabilization of global CO_2 emissions by the year 2000 at the latest, and a 20 percent reduction by 2005 as the international target. The Danish government took up this goal and it became the overall objective of the new energy action plan, *Energy 2000*, that was published in April 1990. The main political initiator behind *Energy 2000*, popularly known as *The Green Plan*, was Jens Bilgrav-Nielsen. Worried about threatening climatic problems he decided to spend his holidays on the topic. The Minister asked Jørgen Stig Nørgård and Niels I. Meyer, from the Energy Group of the Physics department at the Technical University of Denmark at Lyngby, to provide him with books and materials on efficient and sustainable energy futures.[84]

During his summer retreat, the social-liberal Energy Minister concluded that global environmental problems could only be solved through cooperative global action. He argued that, in order to get this international process going, industrialized countries first had to show both the will and the ability to organize their own energy systems in a manner that would be consistent with sustainable development. He found that Denmark, being one of the nations that emitted the greatest quantities of CO_2 per capita, should take the lead, and commanded his administrators to make 'a plan of action for sustainable development' (cf. EM 1990: 7).

As usual, the preparatory work in the form of exchanging ideas and working out technical reports was done in open cooperation with the utilities. "But when it came to an assessment and conclusions needed to be drawn the authorities totally stopped the openness." (DEF 1989a: 4) Instead, Bilgrav-Nielsen gave Jørgen Nørgård a free hand in establishing an energy plan that would focus on the demand side.[85] Nørgård had been one of the scientists behind Alternative Energy Plan 1983 (AE-plan 83), which represented an elaborate and more radical sequel to AE-plan 1976. AE-plan 83 depicted a sustainable energy system in Denmark after the year 2030 consisting of thousands of small renewable energy plants (Hvelplund *et al.* 1983).

At the time, the plan had been denounced, both by the then liberal Energy Minister, Knud Enggaard, as well as by the utilities, as 'unrealistic.' (DEF 1983a: 8) Severely worried about Nørgård's involvement, and their own exclusion, the utilities felt like cornered rats. They stated that the Minister of Energy was about to weaken the electricity sector and harm the whole Danish society "in favor of some short-sighted considerations concerning other energy suppliers or to obtain an immediate political gain by acting as an environmental advocate." (DEF 1989a: 4) Their cry of despair, however, was a voice in the wilderness, since a broad political consensus existed within the *Folketing* about the new 'green' proposals.

This became apparent when, just before the publication of the energy plan, an agreement between the government and the SD on the future power supply system became public.[86] The agreement of 20th March, 1990 put a "bomb under the development of large coal-fired plants in the first half of the 1990s," (DEF 1989a: 4) and made clear that decentralized cogeneration plants would be the essential element in the development of the electricity system.[87] The plan stressed energy conservation increased energy efficiency in the production sector, and the enhanced utilization of cleaner energy sources such as natural gas, biomass, wind energy and other renewables (EM 1990: 13-14). With respect to the latter, it was decided that the power companies' participation in the development of wind energy needed to be continued, "both with respect to further technological development and with the aim of establishing a further 100 MW of wind turbines before 1 January 1994". (*ibid.* 98) This proved that the first 100-MW agreement had not been a flash in the pan, and finally made the utilities realize that Danish politicians demanded a lasting effort in the field of wind energy technology.

In short, *Energy 2000* was a rigorous attempt to add the goal of sustainability to the two traditional overall objectives within the field of energy policy: ensuring energy at the lowest possible cost and security of supply. Although by this venture, the broad support for the plan from the Danish parliament forced the utilities to surrender. And while they made no secret of their skepticism for the DSM planning method and the loosely described action programs, they notified the *Folketing* and the Minister by letter that, "The utilities [were] prepared to go into a constructive operation with the Ministry of Energy and the Energy Agency to work out an improved basis for making decisions". To show their good will they agreed to build another 100 MW of wind power before the end of 1993 (DEF 1990a: 10).[88]

The new agreement gave comfort to the wind turbine manufacturers, who had, in the past few years, invested a lot of money, expertise, and time, to broaden their strategy and product range towards utility-grade wind turbines. Both the permanency of their involvement in wind energy, as well as the fact that the biggest commercial turbines available had reached the already 'respectable' 500-kW size, lessened the utilities' appetite for buying large, new designs to fulfil their obligation as fast and easily as possible. Accordingly, they changed their strategy to purchasing the cheapest 'large' proven commercial wind turbine available on the market.[89] Moreover, the second '100-MW' agreement stressed that an important part of the electric utilities' role with regard to wind power would be "to contribute to the development of the technology and raise quality levels in general." Bilgrav-Nielsen therefore preferred to talk about a 'development agreement' instead of a MW agreement.[90]

To some extent, the 1 MW turbine at Avedøre Holme represented this change of policy. The utilities set down the technical requirements, and the manufacturers checked whether the project fitted their own development strategy, and then decided whether or not to join in. Bonus, for example, seized the opportunity to test several special novelties on the utility-prototype before implementing them in their own products, and Vestas upscaled some of the components the company expected to use for the megawatt-sized wind turbine it was planning to develop (Østergaard 1993: 14). Although the manufacturers were relatively pleased with this situation, they nevertheless aspired to an even more ideal position, in which their *own* new designs would be "developed and tested as *appendices* to major wind farms." (Stiesdal 1990: 27)

3.5.2 Technological Progress and Organizational Lag

Despite the new 100-MW agreement, the installation of utility-owned wind turbines severely slackened. In 1992, some mere 14 MW was installed, and in 1993 only 5 MW was installed, which brought the accumulated installed utility-owned wind power generating capacity at the end of the year to about 110 MW, instead of the agreed 200 MW (FDV 1994: 37). According to Bilgrav-Nielsen this was "largely because of opposition from the Danish public which, while supporting wind power development by private groups of individuals, does not welcome the idea of sizeable utility projects being forced upon it."[91] The utilities did not feel responsible for this situation, since they had entered the 1990-agreement under the condition that the government would ensure sites (MM 1991: 50-51). Consequently, the Ministry of Environmental Affairs (*Miljøministeriet*, MM) had set down the Wind Turbine Siting Committee.[92] In its first report, the Committee described two ways in which the state could interfere in the planning process. Issuing a national planning directive was the most far-reaching central steering instrument. Instructing counties to integrate the development of wind power plants within regional planning was a less radical second option (MM 1991: 12-13). A second report introduced a third option, and recommended that the MM should ask the municipalities to become involved in making plans for wind turbine development (MM 1992: 61-63). SD argued that the utilities were not to blame, and proposed a Bill at the end of 1992 to work out a national planning directive to secure the development of wind turbines. A clear majority within the *Folketing*, however, opposed this centralist planning vision, and held that the construction of wind turbines should remain a local event.

The government adopted the recommendations of the Siting Committee, and took into account the majority view within the Parliament, and in September 1993 send out a circular which demanded the municipalities to indicate sites where wind turbines could be placed. The clause that such action was not required when no space could be found to put up wind turbines severely weakened this policy statement, and led policy scientist Jørgen Grønnegaard Christensen to conclude that the circular was a classic example of 'symbolic action.' (Vestergaard 1993) However, the good reputation held by the young Energy Minister Jann Sjursen (KrF) among renewable energy proponents, the ambitious way he advocated the agenda set by *Energy 2000* (which the former conservative Minister Anne Birgitte Lundholt (KF) had tried to hush up),[93] the correspondence between this weak policy instrument and the notion of local self-governance which was highly praised within the *Folketing*, and the parallel strong decrease in the number of installed private turbines ('despite' the new Wind Turbine Law), were all signs that the situation was somewhat more complex.

The Wind Turbine Law had been put forward at the end of 1992, after the utilities and the windmill organizations had informed Lundholt that it had been impossible to settle the succession of the 1984-agreement themselves. The negotiations had started in 1989, after Bilgrav-Nielsen had axed the installation subsidies for wind turbines. The utilities had argued that the existence of a state investment subsidy program had been a precondition for them to contribute to the connection costs (see above), and had needled DV and FDV by stating that they wanted to stop this activity. In July 1991 it had evolved into a preliminary minor agreement, in which the utilities had promised to extend their contribution to the investment costs until a final agreement was reached, and to continue the negotiations in 1992 (DEF 1991a: 13).

Similar to ten years before, the utilities had wanted to talk about the possibility of disconnecting private wind turbines. Moreover, to decrease grid-connection costs, they had proposed to put down private wind turbines in groups (*klynger*) or parks. The windmill organizations had again put the extension of the siting criteria on the agenda, and had advocated standardizing costs of grid strengthening. Negotiations had restarted on the 1st of February 1992 and, two days later, the parties had concluded their positions were so far apart that an agreement could not be reached (DEF 1992a: 15). The parties had agreed on the principle of a common pool system (*solidarisk puljeordning*) which would distribute the costs for the required grid strengthening equally among all wind turbine owners, but had not been able to settle its financial side.

This deadlock was broken by a political agreement in March between the conservative KVR-government and the Social Democratic Party, which said that, with respect to wind turbines and decentralized cogeneration plants, the utilities had to pay for the amplification costs. The fundamental argument behind this had been that the electricity grid should no longer be conceived solely as a distribution grid, but also as an effective collection grid (DEF 1992a: 12). This had paved the way for the new Wind Turbine Law that came into force in November 1992. Within the law, the basic assumptions for allowing private wind turbines to connect to the grid were preserved. The principle of local attachment was slightly altered in such a way that owners were required to set up their wind turbine within the community they lived in or within a neighboring community. They were also allowed to own a share corresponding to 150 percent of their own yearly electricity consumption, and the capacity limit of 150 kW was abolished.[94] Unfortunately, these small concessions turned out insufficient to get the development of private wind turbines back on track.

Part Two

The Political Game

The two thick descriptions in Part One have thrown up detailed information on technical, economic, and political aspects of the development of wind energy in Denmark and California. Change within the field of technology, change in the policy making process, and change in management strategy is evident. The two cases illustrate how changes within one institutional dimension induce and are induced by changes within other dimensions. The countless number of interactions among the three institutional dimensions forces upon the reader the metaphor of a 'seamless web' of society, industry and technology. As such, the story of wind energy complies with the image of the innovation game as an event in which several interrelated games – the political game, the technological game and the market game – are played simultaneously.

The case studies also show the role of coincidence within the development of wind energy, in the sense that "the constitutive mechanisms and actors that made it happen are described in a way it might as well not had happened, or at least not in this particular way."[1] It is important to appreciate the importance of chance and to increase our understanding of how different actors deal with these ever-changing circumstances. It is crucial, therefore, to acknowledge the dynamic complexity that is embodied within any innovation process. The recognition of these characteristics should withhold policy makers, company managers and technologists alike from taking facile positions on how to best guide techno-economic innovation. Part One is meant to underline this, and to force the reader to come to appreciate the 'whole' dynamic complexity of wind energy innovation. However, the coincidental and mutual development of wind energy policy, technology and industry seems to offer little direct material to address the question of how policy makers should guide innovation into a democratically legitimate direction. To deal with this essential question, several painstaking analytical steps have to be taken so as to clarify the role policy makers have played within the development of wind energy. Such an attempt is made in the remainder of the book.

In Part Two, an initial analytical step is taken to unravel the seamless web from a policy perspective. The focus will be on the political game. In order to analyze the California and Danish political debates related to wind energy, a theoretical framework is set up in Chapter 4. The policy process is conceptualized in terms of interacting ideological coalitions, each consisting of actors who share a set of basic political beliefs that guide their behavior. In Chapter 5, the political analysis starts. First, the relevant coalitions and related policy belief systems are depicted. Next, we examine how various policy beliefs have changed over the years and to what extent this has led to a change in (wind) energy policy. This communication process between coalitions within a policy subsystem is portrayed as policy-oriented learning.

In Chapter 6, the higher-order beliefs of the major ideological coalitions are traced back in history, and are found to be rooted within the various political traditions that together make up the national political culture. A shift in the overall political culture of a nation thus implies a change in the relative discursive power of the various ideological coalitions within the separate policy subsystems. The way in which a change in national political culture has taken place over the period of study is scrutinized. In addition, we look at the ways these changes have effected belief system dynamics and coalition behavior within the field of wind energy in California and Denmark.

The observant reader might have noticed that in Part Two of the book, policy change is in some way detached from change in industry and technology by treating actions within the industrial and technical arenas as external events. Although this goes against the idea of a seemingly seamless web of society and technology, it offers, for the time being, an appropriate strategy to pull the policy and ideology strings out of the web. A thorough separate political analysis, then, will allow us to examine the interdependence between the political game and technical and market games in a much more accurate and sophisticated way. Accordingly, we expect to return to the seemingly seamless web metaphor of policy, technology and industry at full strength in Part Three 'The Innovation Game.'

4 A Theoretical Framework of Policy Change, Tradition and Learning

> The production of interaction has three fundamental elements: its constitution as 'meaningful'; its constitution as a moral order; and its constitution as the operation of relations of power. *Anthony Giddens, 1976: 104*

Traditionally, policy scientists have concerned themselves largely with examining the logic of political decision making in terms of competing interest. Central to this approach is the idea that actors' interests provide a self-evident starting point from which purposive behavior can be scientifically studied. The question of origin of interest is not faced by the politics of interest model. Recent approaches use frames of meaning, rather than interests, as their focus, since these are more inclusive and more verifiable (Fischer 1980; Sabatier 1987; Schwarz and Thompson 1990; Thompson, Ellis and Wildavsky 1990). Within the politics of meaning model (Sederberg 1984), the behavior of policy makers is guided by their policy belief systems, which include the perception of where their interest lies. The logic of political decision making, then, should be examined in terms of competing, or rather, interacting belief systems.

Of course, belief systems do not interact themselves. It is political actors actively concerned with a certain policy issue who do the fighting. Moreover, the political game is not a man-to-man fight but a group thing. Thinking in terms of interacting beliefs, it is a logical step, therefore, to study policy change over fairly long periods of time by aggregating actors on the basis of the particular belief system they share (cf. Sabatier 1987). It turns out that the number of distinct policy beliefs active within a political debate is rather limited.[1] Besides being compatible with the politics of meaning model, this method of organizing the set of actors who are actively involved with a policy problem or issue such as energy, seems to be a very practical one.

In applying this method, two theoretical issues have to be dealt with. First, there is a need for operationalizing the concept of policy belief system in a clear way. Besides a clear conceptualization of the structure of policy beliefs, we need some basic insights concerning the dynamics of belief systems. Changing circumstances stimulate policy actors to constantly revise their policy beliefs. Accordingly, this type of policy-oriented learning can be regarded as a major driving force behind the dynamics of beliefs. In the next section, these issues will be considered. In section 4.2, several additional concepts are introduced to properly study the dynamics and statics of policy processes over a decade or more. In the final section, the plan of the remainder of Part Two will be unraveled.

4.1 Policy Beliefs and Learning

4.1.1 The Structure of Policy Belief Systems

In his attempt to move evaluation beyond the question of goal attainment by extending it to the political assessment of policy goals, Frank Fischer (1980, 1985; see also 1994) has worked out a precise four-level model that covers two orders of discourse. As a practical method for structuring policy belief systems, Van de Graaf and Hoppe (1989) propose using Fischer's prescriptive four-layer depiction of political policy evaluation as a representation of policy belief systems. This is allowed because both evaluations and belief systems reflect political judgements (Van de Graaf and Hoppe 1989).[2] The structure of the policy belief system that results from this is schematically represented in Figure 4-1.

	Level	Argumentation concerns
Second-order policy discourse	Rational social choice	Discussing *preferences about the social order* that are supposed to be affected by the policy.
	System vindication	*Value systems and world views* are vindicated in terms of their contribution to the preferred social order.
First-order policy discourse	Situational validation	*Defining the problem* in the situation under scrutiny and validating objectives, means and ends as contributing to reducing the problem as perceived through second-order values and perceptions in terms of their impact on social reality. Thus *a meaning is attributed to* policy objectives, ends, and means (which may include a technological artifact), which reflects their expected contribution to solving the problem.
	Technical verification	Evaluating efficiency and efficacy of *means-end chains* in realizing the objective

Figure 4-1
Different levels of policy argument and their objects (according to Fischer 1980, 1985, and Grin and Van de Graaf 1996).

First-order discourse is mainly concerned with the design of a specific policy program. By contrast, second-order discourse organizes the policy approach to many different programs and projects. In first-order discourse, major objectives are set and then translated into causal means-end chains. The choice of means and objectives is thought to be generated by more general perceptions and values that constitute second-order discourse. Within both orders, Fischer demarcates two levels of argument.

The level of *technical verification* concerns an evaluation of the effectiveness and the costs of alternative means to achieve a certain policy goal. The choice, efficiency, and efficacy of specific policy instruments are at stake, as well as their possible unintended consequences. Whereas technical verification comprises the empirical examination of the question of what measure is able to efficiently meet an adopted objective, arguments at the level of *situational validation* concern the question of whether certain objectives should indeed be striven for. The latter assumes that a definition of the situation under scrutiny is present. Meaning is attributed to the situation by confronting the policy maker's broader value orientation with the actual problem situation.

At the level of *rational social choice,* these fundamental values are justified by normative and philosophical arguments. American liberal and conservative value systems may serve as a good example.[3] According to Nagel (1987: 8), "Liberal values are those (1) having the government play a more positive role in dealing with social problems, (2) promoting more equality of opportunity if not income and wealth, and (3) having consumers and workers play a more important role in economic decision making. Conservative values are those oriented toward (1) having the marketplace play a more positive role in dealing with social problems, (2) justifying inequalities based on merit or sometimes ancestry, and (3) having managers and owners play a more important role in economic decision making." Finally, on the *system vindication* level, systems of values and perceptions concerning a specific policy area are evaluated. Reasoning at this level involves investigation of the question of whether values and beliefs generate objectives and strategies that strengthen the preferred social order.

4.1.2 Policy Belief Systems as Public Programs

Following many other scholars who see any belief systems as hierarchically and laterally ordered (Steinbruner 1974: 91-103; Sabatier 1987: 667; Majone 1989: 150ff.), Van der Wouden (1990: 77-84) regards a belief system as being built up out of a normative hard or deep core that is surrounded by a protective belt of secondary beliefs. In his thesis, *De Dynamiek van Beleid* (*The Dynamics of Policy*), Ries van der Wouden (1990) uses the work of Lakatos to formulate a theory of the dynamics of policy belief systems. According to Lakatos (1976), the dynamics of science is conducted by competing *research programs,* which consist of a *hard core* and a *protective belt.* The hard core is a global model of (a part of) the reality and the processes therein. For its followers, the hard core is sacrosanct. The ban to attack the hard core advances, what Lakatos calls, a *negative heuristic.* The negative heuristic serves as an aid to problem setting, and prescribes how variations can be generated that will survive the selection process within the research tradition. In the protective belt the *positive heuristic* of the research program can be found, which serves as an aid to problem solving. It helps the researcher strengthen the empirical quality of ongoing research programs.

By thinking of policy belief systems as research programs, or rather public policies and programs, Van der Wouden clarifies the changeability of the various levels within Fischer's model of a policy belief system by considering policy beliefs at the level of first-order policy discourse as the protective belt of the policy belief system, and the beliefs within the second-order discourse as its hard core. Consequently, policy beliefs at the level of second-order policy discourse are relatively stable and resist change. The hard core consists of strategies and methods for converting abstract ideas into action. Such strategies justify and predict the success of certain courses of action, while constraining or excluding others (Sabatier 1987: 667; Majone 1989: 150).

Peripheral first-order beliefs are far more flexible. Analogous to a research program's positive heuristic, a policy belief system's long term destiny is determined by the flexible periphery's capacity to generate a wide range of policy programs in a wide array of different policy areas. Besides a positive heuristic that is generating practical solutions and new ways of dealing with problems of everyday policy making, the protective belt also harbors a negative heuristic. The negative heuristic has a gatekeeper function. It observes whether proposed solutions are in line with hard core values. Protecting the hard core against foreign elements strengthens the internal consistency of a policy belief system. However, it may also weaken the sense of reality of policy actors and, as a result, imply rigidity and stubbornness. Hoppe and Peterse (1993: 36) comment that, "because of this, adherents of the policy belief system may forego opportunities for creative and innovative fusion."

4.1.3 Competitive and Cooperative Forms of Policy Learning

> Integration involves invention, and the clever thing is to recognize this, and not to let one's thinking stay within the boundaries of two alternatives, which are mutually exclusive. *Mary Parker Follett, 1925*[4]

As we saw, a policy belief system should not be regarded as an impregnable fortress. Moreover, public policies and programs are seldom determined by one particular appreciative system. Instead, they usually mark the simultaneous influence of several distinct clusters of normative and causal assumptions. It is exactly this ability to map beliefs and policies on the same 'canvas' that provides a vehicle for assessing the influence of various groups of actors and their policy beliefs over time (Sabatier 1993: 17). Public policies and programs should be seen, therefore, as the continuously evolving outcome of the interaction of the various relevant policy belief systems.

In the course of their interaction, actors are constantly realizing, and modifying, their own beliefs. While the interplay between policy belief systems may bring on policy change, it may also lead to change within the constitutive policy beliefs. It is worthwhile, therefore, to deal with the interaction of different policy beliefs at a somewhat greater length. Follett distinguishes three main ways of dealing with difference or conflict: *domination*, *compromise* and *integration* (see Graham 1995: 67ff.). This subdivision can be used as a starting point for discussing the interaction of different clusters of normative and causal assumptions adhered to by policy makers.

Domination, obviously, is a victory of one side over the other. In this case, state policy is determined by one ruling belief system. In addition, the hegemonic belief is hostile towards other political ideas. Having a monopoly on governmental policy, dominant actors will not easily be inclined to change their beliefs. To maintain their position of power, small adaptations may be perceived as necessary.

When two or more equally powerful groups of actors try to influence a policy, a second method of dealing with conflict becomes more common. Compromise is often said to be the basis of most political tactics. Compromising implies that each side gives up a little in order to have peace, or, in other words, to have any policy at all. This way of settling controversies does not necessarily require the actors concerned to modify their beliefs. However, policy actors may seek to better understand the world in order to improve their bargaining position within the political debate. Ultimately, therefore, this type of *policy-oriented learning* may lead to a change in beliefs.

Follett argues that domination is the easiest way of dealing with conflict for the moment, but is not usually successful in the long run. Compromise is an accepted and approved method of ending controversy. Yet no one really wants to compromise, because that means a giving up of something. Instead, Follett proposes that the most fruitful way of dealing with conflict is integration. *Integration* means that an innovative solution has been found, "in which both [or all] desires have found a place, that neither side has had to sacrifice anything." (Graham 1995: 69) It follows that, ideally, integration leads to policy change which leaves policy actors from various ideological positions satisfied. Since the resulting policy change is beneficial to all actors concerned, they will be inclined to adapt their belief systems to the new political situation. This means that, in this way of dealing with ideological differences, policy change tends to go hand in glove with policy-oriented learning.

Two basic forms of policy-oriented learning can now be distinguished. First, there is a type of policy learning that aims at elaborating one's own policy belief system or at attacking opponents' views in order to strengthen one's power position within a political debate. This *competitive* form of learning is confined *within* the boundaries of the various belief systems. Both the strategy of achieving or maintaining dominance as well as that of seeking compromise prompt competitive forms of policy learning. These are enhanced by the negative heuristic of policy belief systems. According to Zonneveld (1991: 90-92), too much emphasis on the protection of core values may lead to unproductive forms of policy learning that may eventually become fatal to a policy belief system.

In addition, a *cooperative* form of policy learning that is aimed at incorporating different points of view within a public policy or program can be discerned. This type of learning can prosper in circumstances where different policy actors try to deal with conflict by means of integration. Integrative policy learning entails the search for innovative solutions. It is a truly interactive process that requires joint learning *across* alternative policy beliefs.

4.2 The Dynamics and Statics of Policy

Child (1995) comments that the whole tenor of Follett's argument, stressing the possibilities for a fluid and positive reconstruction of structured positions, anticipates Giddens' concept of structuration, which has been so influential in modern sociology. He asserts that, "Rephrasing [Giddens' concept of structuration] in Follettian terms expresses one of the most powerful and hopeful insights into our social life, namely, that we have the power through open debate and discussion to examine and reconstitute the assumptions and structures within which we are confined". (*ibid.* 91)

4.2.1 The Duality of Structure

These inspiring remarks call for a brief introduction of Giddens' views on the relationship between structure and action. In contrast to the objectivist usage of 'structure' as something that is factual and external to the activities of an actor, Giddens' 'structure' has no reality except in so far as it is internal to actors in the form of memory traces (Cassell 1993: 12). It is actors who bring 'structure' into being, and it is 'structure', which produces the possibility of acting. The *duality of structure* consists in the structure's two-sided existence – as both the medium and the (unintended) outcome of social practices. Structure not only constrains action but also enables it.

According to Giddens, it is *rules* and *resources* that structure social practices. By 'rules' he does not usually refer to 'formal' rules. Instead, most day-to-day interactions are based on taken-for-granted knowledge which actors assume others possess if they are 'competent' members of society (Cassell 1993: 105). This so-called *mutual knowledge* is applied in the form of frames of meaning whereby contexts of communication are created and sustained in interaction.

Rules cannot be conceptualized apart from resources. "Resources (focused via signification and legitimization) are structured properties of social systems, drawn upon and reproduced by knowledgeable agents in the course of interaction." (Giddens 1984: 15) Structural properties thus express forms of *domination* and *power*. Power relations in social systems can be viewed as relations of autonomy and dependence. No matter how unbalanced the balance of power may be, actors in subordinate positions are never totally dependent. In other words, they are often very adept at converting whatever resources they possess into some degree of control over the system's conditions of reproduction. Giddens emphasizes, therefore, that, "In all social systems there is a *dialectic of control*, such that there are normally continually shifting balances of resources, altering the overall distribution of power." (Cassell 1993: 243)

4.2.2 The Duality of Structure and the Policy Process

In what way can we use Giddens' broad concept of structuration in our study of the policy process and our concern with policy change over time spans of decades and more? Or, to put it differently, how, with respect to the political game, should we interpret the claim that the rules and resources drawn upon in the production and reproduction of social action are at the same time the means of system reproduction and change?

The notion of the duality of structure implies that the rules of the political game should not be considered rigid, as in a chess game. The political game is closer to children's play, in which initially negotiated rules do not stay uncontested but constantly evolve in the course of the game. Some rules will be contested less than others. In street football, for instance, it is normal to use only one ball at a time, while the size of the goal is less standard and will often be changed during the game. One could say that the football game, therefore, exhibits structural properties, which should be conceived of as "hierarchically organized in terms of the time-space extension of the practices they recursively organize". (Giddens 1984: 17)

The same is true for the political game. In this book, we take the nation-state as our point of departure for studying the political process. With respect to the national political system, the national political culture represents, in terms of space-time extension of political practices, its most deeply embedded structural properties. In other words, the national political culture is related to the specific long-term history and political evolution of a country's political system. As we saw, structural properties express forms of domination and power. Consequently, the national political culture can be viewed as representing political forms of domination and power. The national political culture represents a certain political distribution of power, or rather, embodies a distribution of rules and resources.

Two theoretical observations will make the notion of a distribution of resources more specific and concrete. First of all, we should view national political culture not as a single culture, but rather as a mix of competing cultures (cf. Thompson, Ellis, and Wildavsky 1990: part 3), which all have their origin within the country's specific political history. Thus we may also speak of ideological traditions. The mix of cultures defines the argumentative power of each ideological tradition at a certain point in time. Hence it is precisely the specific mix of political cultures or traditions that constitutes the political balance of power.

A second comment concerns the relationship between ideological traditions and policy belief systems. As we have seen, arguments on the most generic level of the policy belief system concern preferences about the social order that are supposed to be affected by the policy. Another way to put this is that rational social choice arguments represent ideological points of view, that is, they concern fundamental aspects of political accountability. It follows that, at their most general level, policy belief systems represent different ideological traditions. By studying the policy process in terms of competing belief systems, we, therefore, at the same time, describe it in terms of competing ideological traditions. These ideological traditions can be considered, therefore, the principal *long-term* glue of politics. In this book, the policy actors who share hard core beliefs that are cherished within a certain ideological tradition are seen as members of a so-called *ideological coalition*.[5]

The following picture of the statics and dynamics of the political process emerges. The national political culture embodies a relatively stable political distribution of power between various value systems or ideological traditions. The specific mix of political cultures provides rules and resources to the various ideological coalitions that are active within a certain policy area. National political cultures enable and constrain political action, but do not define it in a deterministic sense. Every moment of political decision making contains a degree of freedom that leaves it open as to whether a certain political action will either reinforce the existing political order or weaken it. The accumulative effect of multiple political actions within various policy areas or subsystems may, therefore, eventually lead to a serious change in the political culture that generated these actions. Thus, when we study policy processes over short periods of time, national political culture can be treated as an independent or explaining variable. However, when studying policy change over a decade or more, national political culture itself becomes a dependent variable.

4.2.3 The Two Dynamics of Policy Belief Systems

> There is a double movement in all reflection, a movement from the given partial and confused data to a suggested comprehensive (or inclusive) entire situation; and back from the suggested whole ... to the particular facts. ... Roughly, the first of these movements is inductive; the second deductive. A complete set of thought involves both. *John Dewey, 1910: 79-80*

With respect to the political game, Giddens' concept of the duality of structure can now be grasped by distinguishing two opposite, but interrelated, types of dynamics within policy belief systems (cf. Van der Wouden 1990). A so-called *reflective dynamic* can be discerned working its way down from the top of the belief system, while, concurrently, a *practical dynamic* works its way up from the bottom.[6]

Through the *reflective dynamic*, changes in second-order beliefs come to influence first-order beliefs. It is easy to imagine changes in political culture influencing the dynamics of policy belief systems. This means that a change in the mix of ideological traditions will affect the rate and nature of long-term policy change within policy subsystems. These fundamental sociocultural values (Sabatier 1993), systemic beliefs, or *structural principles* as Giddens (1984: 17) names them, "work their effects through the socialization process that limits and molds behavior across the entire subsystem by determining the universe of legitimate policy opportunities and constraints." (Munro 1993: 125) More directly, the ideological mix defines the (often-uneven) distribution of discursive power over various policy belief systems. In its turn, the way in which power is divided over several ideological coalitions will influence each coalition's behavior. When argumentative power is truly evenly divided or strongly dispersed, one would expect searching for compromise and/or integrative solutions to be the dominant modes of dealing with political differences.[7]

Through the *practical dynamic*, changes in first-order beliefs diffuse from specific policy areas to the wider cultural and social context, and gradually have an impact on second order beliefs. Policy-oriented learning within and across ideological coalitions that simultaneously takes place within various policy subsystems covers this process. Following Hugh Heclo (1974: 306), policy-oriented learning refers to relatively enduring alterations of thought or behavioral intentions that result from experience and are concerned with the attainment (or revision) of policy objectives. Through policy learning, members of an ideological coalition seek to better understand 'the world' in order to further their policy objectives (cf. Sabatier 1993: 19). However, what 'the world' stands for is in the eye of the beholder, i.e. actors who share different policy beliefs will likely have different interpretations about 'the world.' Moreover, 'the world' is constantly changing. Natural and human disasters and changes in economic conditions and technology may alter the resources of subsystem actors and substantially affect a subsystem by undermining the causal assumptions of existing policies. If skillfully exploited, such events and signaled policy weaknesses or *anomalies* may shift the balance of power of ideological coalitions at the policy subsystem level. The two oil crises of the 1970s, for instance, uncovered several weaknesses within the supply-oriented way of thinking that dominated energy politics of Western countries and offered an opportunity for alternative ideological coalitions to open up the energy debate.

4.2.4 Sectarians and Pragmatists within Ideological Coalitions

The above demonstrates that, subject to the national political culture they are part of, ideological coalitions constantly try to cope with their real world experiences within a specific policy area. To deal with both changing political circumstances, as well as with the changing world, an ideological coalition is compelled to be engaged in a permanent, never-ending process of policy-oriented learning in order to stay politically relevant.

Although, by definition, members of an ideological coalition share deep core values, they do not necessarily have to agree on first-order beliefs. To account for the flexibility of coalition behavior, a distinction is made between *extremists* or *sectarians* and *pragmatists* within an ideological coalition (cf. Munro 1993: 126). Sectarians attempt to protect the core beliefs of their respective belief structures. Their behavior is, thus, guided by the negative heuristic of a belief system. In contrast, pragmatists are predominantly guided by the positive heuristic. They are open to cooperative forms of policy learning across the belief systems of various ideological coalitions.

As we saw, dramatic events provide both *windows of opportunity* (Kingdon 1984) as well as *windows of vulnerability* (Van Muijen 1993: 129) to each coalition, offering chances to improve their position and/or undermine the position of rivaling coalitions. Of course, opportunities must be exploited if the political debate is to open up and change is to be realized. Coalitions may take advantage of the newly evolved situation by contending that they perceive a conflict between existing policies and current circumstances. Perceived policy weaknesses or anomalies can play a crystallizing role within the ongoing political debate, and in the long run, may even evoke political crises. A minority coalition may stress the existence of anomalies in order to open up the debate. A majority coalition may react in several ways. Sectarians within the dominant ideological coalition may try to dampen the effects of such anomalies "by suggesting the need for further research, confining change to small experimental projects, or diverting attention to other issues." (Sabatier and Jenkins-Smith 1993: 222)

While political stalemate may provoke defensive behavior and stimulate competitive forms of learning, it may also lead to cooperative forms of policy-oriented learning. Pragmatists within the dominant coalition might well become sensitive to the criticism articulated by the minority coalition and challenged by the solutions they propose. They may try to deal with the political problem at hand by incorporating ideas of those rivaling coalitions within the lower levels of their own belief system. If the problem can not be resolved by changes within the technical verification level, or powerful new issues constantly turn up, there is a fair chance that problem(s) will eventually be pushed upwards in the direction of the hard core.

4.2.5 Learning Alliances and Story Lines

> A political leader creates a story that helps persons structure their experience. He draws from their stories to make his more perfect, more encompassing, more capable of attracting a wider following and gaining greater allegiance ... The archetypes and experiences the leader emphasizes in his story must be sufficiently rich so that others can see themselves in the story, yet at the same time the story will transform how others see themselves. *Martin H. Krieger, 1981: 75*

In the same way first-order beliefs of members of the same ideological coalition may differ, the first-order discourse of distinct coalitions may well be similar. In that case, coalitions are said to form a *coalition of convenience*. To avoid confusion, this will be called an *alliance*. Full agreement on first-order beliefs, however, does not come naturally, but results from long-term interactive and cooperative forms of learning. A particular mix of political cultures will encourage some ideological coalitions to cooperate, and hinder others in organizing fruitful interaction. Since the mix of political cultures, or political balance of power, within one country tends to be relatively stable, it is readily conceivable that certain kinds of alliances will persist over a decade or even more. In a similar fashion, some types of interaction between coalitions may be regarded as taboo for a relatively long period of time. To denote the fact that some type of alliances embody such long-lasting cooperative forms of learning, we introduce the term *learning alliances*.

So-called *story lines*[8] may fulfil an integrating role between ideological coalitions within a learning alliance. They may be defined as "a generative sort of narrative that allows actors to draw upon various discursive categories to give meaning to specific physical or social phenomena. ... Story lines fulfil an essential role in the clustering of knowledge, the positioning of actors, and ultimately, in the creation of [discourse] coalitions amongst the actors of a given domain. ... [Story lines] are the essential discursive cement that create communicative networks among actors with different or at best overlapping perceptions and understandings." (Hajer 1993: 50, 56)[9] In the terms used here, story lines guide actors from different ideological positions (i.e., ideological coalitions) in their joint search for mutually approved policy solutions. In other words, policy story lines or policy narratives are both cause and effect of long-term policy-oriented learning that is aimed at integration. Within learning alliances, the coalitions somehow disguise their ideological positions so as to be open to new constructive ideas. In such a situation, core beliefs are not seen, therefore, as constitutive or guiding in a deterministic sense.[10]

One might say that policy stories are 'bigger' than ideological stories because they include several ideological positions. At the same time, one might say that policy stories are 'smaller' because they are less fundamental to policy actors than ideological traditions. Because such a discussion is not very fruitful, we might better talk of *nested stories*, that is, grand ideologies containing policy stories, while, in their turn, policy stories contain ideologies.

4.3 The Plan of Part Two

The above theoretical concepts will be used in Chapter 5 and 6 to analyze the political game that played such an essential role in the development of wind energy in Denmark and California. The distinction between the practical and the reflective dynamic contributes the structure of Part Two. Analysis of both dynamics aims to provide a comprehensive view on the wind-energy-related policy process.

Chapter 5 is basically concerned with the practical dynamic which contains the process through which various ideological coalitions, in interaction, try to better understand the constantly changing and ambiguous world in order to further their policy objectives. That Chapter deals, therefore, with the following compound question:

> What were the politically significant ideological coalitions and their associated policy beliefs within the political debate related to wind energy innovation in California and Denmark, and how did the interaction between the various ideological coalitions change their policy beliefs over the last two decades?

The remarks surrounding the reflective dynamic delineate the research topic of Chapter 6. In that Chapter, a brief historical description is given of the national political cultures in both the United States and Denmark. Next, the national political culture's influence on the interaction of ideological coalitions within the energy policy subsystem is studied. Chapter 6, therefore, deals with two questions:

> What political traditions constitute national political culture in the United States and in Denmark? How does the mix of political traditions enable and constrain political learning processes within the energy subsystem?

To avoid getting lost in broad historical cultural explanations, however, it seems wise to first scrutinize political practice.

5 Policy Change and Learning

The analysis of speech shows it is oriented toward the idea of truth. The analysis of 'truth' leads to the notion of a discursively achieved consensus. The analysis of 'consensus' shows this concept to involve a normative dimension. The analysis of the notion of a grounded consensus ties it to a speech situation, which is free from all external and internal constraints, that is, in which the resulting consensus is due simply to the force of the better argument. *T.A. McCarthy, 1973: 153*

With respect to wind energy, a theme that runs through the California renewable energy policy is that it has never explicitly promoted wind energy technology.[1] In contrast, the specific promotion of the usage of wind power definitely has been a direct object of Danish policy. Another more puzzling difference is that while, in California, the decentralized use of alternative energy sources found broad support during the 1970s, and the centralized usage did not, the reverse was true in Denmark. These dissimilarities appear to indicate that there exists a huge contrast in the way in which the California and Danish policy community frame, and have framed, the energy quandary. By analyzing the energy debate in both countries, we hope to clarify this diversity. This is done in the next two sections. In section 5.3, the two debates are compared.

5.1 Policy Change in the Field of Renewable Energy in California

California politics squirm with a complexity and intrigue that defy reasonable analysis. *Theodore White, 1968*[2]

Assessing California's commitment to renewable energy sources is only slightly easier than grabbing a handful of water. *Pat Washburn Rubin, 1981*

In the Californian energy subsystem, we can distinguish four different policy belief systems and associated ideological coalitions: the corporate capitalist coalition, the public interest coalition, the libertarian coalition, and the communitarian coalition (see Figure 5-1). Hence many line-ups of allies and opponents are possible and this increases when ideological coalitions are not viewed as unitary actors, but account is taken of discords between sectarians and pragmatists within coalitions. Three mechanisms exist that limit the number of coalition composition.

First, not all four ideological coalitions are politically relevant all the time. For example, between 1950 and 1965, energy policy was clearly dominated by the corporate capitalist and public interest coalition. In this so-called 'golden age,' libertarian and communitarian beliefs had little impact on the political debate. Thus, during certain periods, some ideological coalitions may be politically nonexistent or insignificant. This phenomenon is depicted in Figure 5-1, by simply not drawing the box that represents a certain coalition during the period it is politically invisible or powerless.

The formation of alliances also regulates coalitional behavior, and, therefore, the number of coalition makeups. For example, between 1965 and 1980, two alliances were

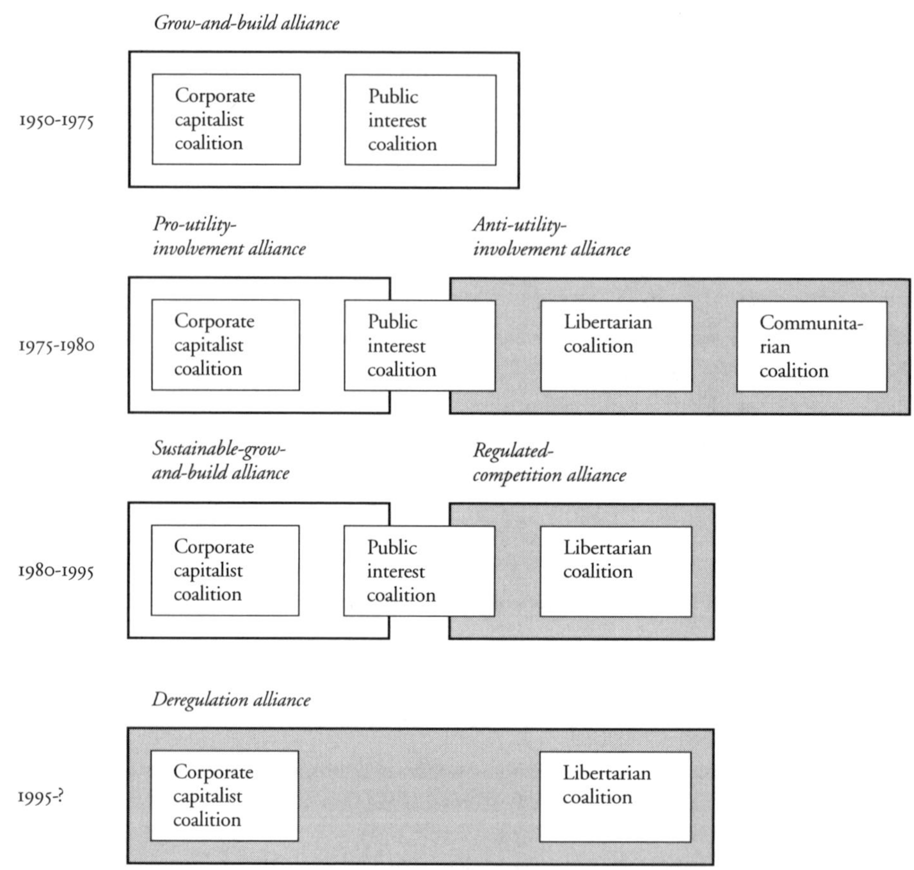

Figure 5-1
Ideological coalition, alliances and story lines within the Californian energy debate. Ideological coalitions are represented by single-lined boxes. When a certain ideological coalition is politically powerless within a certain period, the box that represents such a coalition is omitted. Alliances between coalitions are depicted by thick-lined boxes. The organization of alliances around the story line of ecological and capitalist modernization are depicted by the colors white and black, respectively (see also text).

formed. The pro-utility-involvement alliance, consisting of members of the corporate capital and public interest coalition, saw a major role for the utilities in promoting renewable technologies. In contrast, the anti-utility-involvement alliance, incorporating members of the public interest, the libertarian and the communitarian coalition, seriously doubted whether the utilities could play a positive role. The various alliances that existed over time are illustrated in Figure 5-1, by means of thick-lined boxes. The fact that the thick-lined box divides the public interest coalition must be taken literally. It demonstrates that the question of the proper role of the utilities in stimulating renewables divided the public interest coalition.

A third mechanism concerns the relative stability of alliances. This implies that one can talk of long-term policy learning alliances. It is argued that, in the California energy debate, two learning alliances can be discerned. The corporate capitalist and public interest coalition formed an alliance that was organized around the story line of ecological modernization. This learning alliance gradually tried to change the grow-and-build strategy into a sustainable grow-and-build strategy. The story line of ecological modernization is depicted in Figure 5-1 by means of the color white. The second long-term learning alliance is colored black and is centered on the story of entrepreneurial modernization. Here, introduction of competition in electricity generation was regarded as a panacea for both the environmental and economic problems that the electric industry was facing. These broad story lines are thus conceived as both cause and effect of the relative stability of coalitional behavior. As they structure long term coalitional behavior, they represent the threads of the political debate.

Figure 5-1 schematically summarizes the analysis of the Californian energy debate over the last few decades. In first two subsections, the four ideological coalitions and their policy belief systems will be introduced. Next, the change in policy beliefs that took place over the period under study is described in detail. It reveals how policy change is realized in interaction between the various coalitions. The evolution of various competing alliances will, therefore, be examined.

5.1.1 The Public Interest and Corporate Capitalist Coalition

Until the early 1970s, decisions about energy policy in California, as elsewhere in the United States, were in the hands of the privately owned utilities (cf. Mazmanian 1992). Their grow-and-build strategy found broad political support, and was seen as the proper means to balance demands from consumers, for reasonably priced and reliable electricity, with demands from utilities for a respectable profit margin (CPUC 1993: 25). Ideal market and economic conditions, combined with steady technological innovation, led to a 'golden age' of declining electricity costs and increasing utility earnings, which both satisfied utilities and investors as well as regulators and customers. This stability resulted from an agreement on the situational validation level of two different ideological coalitions: the corporate capital and public interest coalition (Figure 5-2).

In the late 1950s, the perceived efficiency of the utilities, which was demonstrated by their declining cost structure, had made the *public interest coalition* (which included Democratic legislators led by Assemblyman Charles Warren, environmental organizations like the Sierra Club, NRDC, EDF, and consumer groups headed by Ralph Nader) drop their demand for public ownership. This demand had stemmed from its great distrust against the political and economic role of large corporations. At the social choice level, the public interest coalition preferred a social democratic mass society, where the outcomes of a regulated market and the technology it generates reflects the economic and social interest of the public. In contrast, the so-called *corporate capitalist coalition* (which consisted of pronuclear legislators, utility companies, labor unions, and the majority of the CEC and the CPUC) celebrated corporate enterprises as the pillars of prosperity. It believed that, without corporate bureaucratic power and rational planning, a rich and complex society could not exist.

The threat of public power from the public interest coalition had forced the utilities (the most prominent actor within the corporate capitalist coalition) to give their public image an equal place with profits (Roberts and Bluhm 1981: 124). In their attempt to keep their good standing, the utilities went so far as to call themselves 'public service companies,' thus using the core values of the public interest coalition to undermine it and broaden their own appeal.[3] The 'glory days' of the electric utility industry were characterized by an alliance between the public interest coalition and the corporate capitalist coalition concerning the 'traditional regulatory compact' (CPUC 1993: 9), and the utilities' grow-and-build strategy. The public interest coalition's condition for joining such a *grow-and-build alliance* (Figure 5-1) was that the utilities would be regulated in such a way that, in effect, they would behave like a public servant. The corporate capitalists saw regulation as a necessary evil, which conflicted with their second order preferences. The utilities themselves, however, not only bowed to the inevitable, but, instead, actively sought regulation, as they saw it as a means to suppress competition and advance their own economic self interest (*ibid.*).

At the peak of the golden age for electric utilities in the mid-1960s, the public interest coalition again started to conduct a crusade against the corporate capitalist coalition. At the technical verification level, the issue was not price, nor service, but (local) concern about the deterioration of air quality and the immediate aesthetic impact of power plants. The public interest coalition (encouraged by California Tomorrow and the Rand Corporation) began to discern the disastrous effect on the environment of the grow-and-build strategy it was supporting. To counteract the supply-oriented and pro-growth perspective of the corporate capitalist coalition, which at the time seemed steadily institutionalized within the CPUC, the public interest coalition proposed a new state agency. Pro-corporation advocates also lobbied for centralized planning. They envisioned an agency that would help the utilities to realize their power plant building schedules. With the need for comprehensive state planning expressed from two sides, the energy crisis finally breathed life into the CEC. It was plain political opportunism, with no one knowing whether the political coin would fall to the pro-growth or conservationist side.

Argumentation concerns	Corporate capitalist coalition	Public interest coalition
Preferred social order	A complex industrial mass society build through the proper application of corporate power.	A social democratic mass society where the outcomes of a regulated market, reflects the economic, social and ecological interest of the public.
(Empirical and normative) Background theories	Belief in bureaucratic power, rational planning and technological progress to make profit and to solve the social problems within the mass society.	Belief in popular democracy and the public good, fair competition and consumer protection through public policy.
Problem definition and the meanings of technological artifacts in the light of the problem	No problem. The grow-and-build strategy leads to a respectable profit margin for the utilities and an efficient electricity system. Through the use of nuclear power the grow-and-build strategy can be maintained	No problem. The 'traditional regulatory compact' leads to reasonably priced and reliable electricity. The use of nuclear power will lead to low electricity prices and an equitable energy system.
Perceived policy solutions	More of the same	More of the same

Figure 5-2
Policy belief systems of the corporate capitalist and public interest coalition, which dominated the California energy debate from the post-war period to the mid-1960s.

5.1.2 The Communitarian and Libertarian Coalition

Initially, newly elected Governor Brown let the CEC side with the corporate capitalist coalition. But, by the end of 1977, the commission's political stance had shifted in favor of the public interest coalition, and two other marginal coalitions that were gaining political credibility: the *libertarian coalition* and *communitarian coalition* (Figure 5-3).

Two political key events help to explain this shift in policy. First of all, the winter of 1977 was the winter of discontent for a large proportion of the U.S. population. There was frost in Florida, drought in California, and flooding in the Midwest, and a continuing cold wave kept natural gas in short supply in scores of cities east of the Mississippi. The energy crisis had become real, and energy costs increased by several billion dollars. The decade before, declining electricity prices had compelled the public interest coalition to put its antitrust sentiments on ice. But as soon as electricity rates rose and inflation struck again during the 1970s, its argument against the monopoly position of the electric utilities took on new force and received massive media coverage.

A second major event was the coincidental outcome of the anti-nuclear struggle. The political atmosphere created by Proposition 15 caused the corporate capitalist coalition, by way of precaution, to join ranks with the public interest coalition. Three nuclear safety bills effectively curtailed further nuclear development in California. These circumstances compelled Governor Brown to experiment, support conservation and renewable energy supply technologies, appoint 'soft energy' advocates to the CEC and CPUC, and establish the Office of Appropriate Technology (OAT).

Argumentation concerns	Communitarian coalition	Libertarian coalition
Preferred social order	A grassroots democracy in which local producers use local resources to satisfy local needs.	A liberal laissez-faire society, in which the free market acts as the perfect allocator of goods and services.
(Empirical and normative) Background theories	Belief in self-reliance, living a life close to nature and devoted to the community.	Belief in entrepreneurship and equal opportunity.
Problem definition and the meanings of technological artifacts in the light of the problem	The roots of the energy problem are in concentrated control by few over natural, technological, and economic resources.	The energy problem results from a lack of competition. The monopoly position of the electric utilities leads utilities to inertia and technological stagnation.
	The local development and use of renewable technologies creates jobs, maintains a healthy environment and preserves community values.	Cogeneration and renewable technologies enable small entrepreneurs to enter the generating business.
Perceived policy solutions	Stimulating local action and community self-reliance (Private Energy Producer Act, OAT, demonstration projects, energy tax credit)	Restoring competition (PURPA's section 210, OAT, ISO4, BRPU, deregulation, electricity restructuring)

Figure 5-3
Policy belief systems of the communitarian and the libertarian coalition, which came to influence the Californian energy debate during the 1970s.

This new political constellation led to a lot of action in the field of renewable energy technologies that lacked a clear ideological texture. The initial flood of legislation and state programs drew their inspiration from a multifarious mixture of public interest, communitarian, and libertarian beliefs. The three corresponding ideological coalitions supported the renewables option for a diverse set of reasons. At the situational validation level, the public interest coalition appreciated the low environmental impact of renewable energy technologies, and saw them as means to mitigate public exposure to fuel price and supply risks.

The communitarian coalition (around the Appropriate Technology movement and the Campaign for Economic Democracy led by Tom Hayden) saw the energy problem as but one manifestation of the fundamental crisis of industrial society. It situated the roots of problems like the depletion of natural resources, unemployment and pollution in the concentrated control by few over natural, technological, and economic resources. For the actors in this coalition, local development and use of renewable energy supply technologies were regarded as means to create jobs, maintain a healthy environment and preserve community values. The second-order convictions in the policy belief system of the communitarian coalition included preferences for a decentralized democratic, cooperative society in which local producers used local resources to satisfy local needs.

The libertarian coalition (CAL SEIA, SolarCal Office) did not regard the energy problem as a social problem, but as a market problem, and deplored the fact that the electric utilities had a monopoly position in the electricity production business. They argued that lack of competition and business risk had made the utilities rigid and had prevented them from developing and employing cheaper energy sources than oil and uranium. The libertarians saw decentralized energy supply technologies as a business opportunity, whose simplicity, small scale and modularity were thought to enable small entrepreneurs to make cogeneration and solar power cost-competitive energy options. This coalition embraced a vision of a society with a laissez-faire mode of regulation, in which the free market would act as the perfect allocator of goods and services.

As said before, the presence of four relevant ideological coalitions led to a political situation in which many line-ups of allies and opponents were possible, but, in reality, two broad line-ups covered the entire political spectrum. The object of disagreement between the two political alliances concerned the utilities' role in promoting renewables. The line that divided the *pro-utility-involvement alliance* from the *anti-utility-involvement alliance* crossed the public interest coalition (Figure 5-1). Its pragmatic members, i.e., mainstream environmental organizations like EDF, Sierra Club, and NRDC, were strongly in favor of utility intervention for its scale and efficiency. In contrast, its sectarian members seriously questioned whether the utilities could play a positive role with respect to the introduction and commercialization of renewables. This was exemplified by the consumer movement led by Ralph Nader. The so-called *capture theory* presented the central tenet of the consumer groups at the situational validation level. It was held that the agencies that were supposed to be watchdogs over the utility industry were in fact 'captured' by the regulated firms, which manipulated and dominated the regulatory process for profit. Along the same line of reasoning, the sectarian part of the public interest coalition feared that the utilities would frustrate the development of renewables.

5.1.3 The Anti-Utility-Involvement Alliance

Besides 'sectarians' within the public interest coalition, the alliance of political actors that abhorred utility involvement in the development of alternative energy sources consisted of the communitarian and libertarian coalition. Their compact was rooted in the American Appropriate Technology movement and became institutionalized within the OAT. Although the communitarian belief was hegemonic within the movement, proponents of laissez-faire capitalism were also present. Gigi Coe, at the time Deputy Director of the OAT, worked, she says, "in the area where Democrats and Republicans met."[4] Although the core values of pro-small business Democrats and pro-market Republicans significantly differed, they found each other in the struggle against "the continual march of governmental control and the corresponding diminution of individual liberties, corporate concentration, megacities and spiritual emptiness."[5]

In 1976, the anti-utility-involvement alliance supported the Private Energy Producer Act, which was aimed at self-production and required the utilities to transmit power on behalf of the *private energy producer*. The private energy producer was thought to be a producer and a consumer at the same time, a so-called prosumer, which referred to the central ideal of self-sufficiency within the second order discourse of the communitarian coalition. In the two years that followed, no less than twenty pieces of solar legislation to promote the non-utility use of renewable energy technologies were enacted. These included the 55 percent solar energy tax credit, which, in 1978, was expanded to wind. Around the same time, at the federal level, the revolutionary Public Utilities Regulatory Act was passed.

It is still a mystery how PURPA's section 210 came about and barely received attention from either coalition at the time the NEA was passed (Starrs 1988: 15). The corporate capitalist coalition interpreted PURPA as a challenge to the monopoly power of utilities. The anti-utility-involvement alliance welcomed section 210 as a godsend. The sectarian part of the public interest coalition argued that the public clamor over high rates had forced utilities to permit interconnections with, e.g. home wind systems, and saw PURPA as a 'goodwill gesture'. (Cf. Gipe 1983: 54) According to the communitarian coalition, Congress had enacted PURPA to promote the local use of renewable energy sources, and saw section 210 as the federal equivalent of the Private Energy Producer Act in California. In addition to the utilities' obligation to transmit power, section 210 compelled utilities to purchase power from *small power producers* at the avoided costs. The latter requirement broke up the connection between energy production and consumption that typified California law. The provisions of PURPA took the standpoint of the small producer, set the small power producer free from the traditional utility regulations, and offered decentralized investment opportunities. All these aspects corresponded well with the political views of the libertarian coalition.

Guided by the political tenor of PURPA's section 210, the emphasis in policy in California shifted from using renewable technologies solely as a means to achieve local energy self-reliance and free the public from utility dominance, to promoting independent power producers. This move towards a style of politics that smacked of laissez-faire liberalism was further dictated by the experience that the growth of solar, and public awareness of solar development was much lower than anticipated by the communitarian coalition (Drager 1979: 393). Besides, the explosion of corporate interest in solar energy across the country made clear that 'Corporate America' was able to use these technologies for their own benefits (Pursell 1993: 633). This was at odds with the argument of the communitarian coalition, at the system vindication level, that the production and use of appropriate technology would guide community development.

The understanding that a new technology base could encourage both a centralized and decentralized lifestyle replaced the views of the communitarian coalition. Visible lack of community effort made things look gloomy for them. In contrast, the libertarian coalition became encouraged by the wave of entrepreneurial activity unleashed by PURPA.

5.1.4 The Pro-Utility-Involvement Alliance

In the second half of the 1970s, the ideas of the anti-utility-involvement alliance dominated alternative energy policy in California. This was to the complete satisfaction of the corporate capitalist coalition, which did not hold with conservation and the decentralized use of cogeneration and renewable energy plants. Pragmatic actors within the public interest coalition made several attempts to convince the utilities of the benefits of these technologies.

The CEC's R&D proposal to develop large-scale wind turbines for utility use represented such an effort. At the situational validation level, the program reflected the public interest coalition's wish to diversify and clean up California's electricity supply system. This coalition was also convinced that wind energy systems represented a short-term cost-efficient energy supply option for the utilities. The California Legislature rejected the idea of locating the authority for developing large-scale wind technology in the state's central government, and supported only partial funding for developing wind sites.

The EDF tactics of attacking the grow-and-build strategy by means of the utilities' own second-order convictions serves as another prime example. For years, EDF argued before CPUC hearings that the small scale and modularity of renewables and cogeneration plants would give utilities greater flexibility and reduce their financial risks, and thus presented the path out of the financial and environmental difficulties associated with the utilities' current preference for large-scale nuclear, oil and coal plants. The utilities, however, tended to see their critics as "wild-eyed, bomb-throwing intervenors, out to take over management and beat the investors out of a fair return."[6]

Things changed at the end of the 1970s, when the combined effect of rising fuel costs, inflation and high interest rates had seriously weakened the financial position of utilities. The severe cash flow problems of the utilities, public complaints over electricity rates, and the recognition that time was too short for a surge of popular interest in solar systems and conservation to pay off, all helped to broaden the scope of California's renewable energy policy in the public interest. The electric industry's quandary made the utilities themselves conclude that the age of the large, central power plant may have passed, and that the 'renewable and alternative energy supply strategy' as proposed by EDF might not be a bad idea after all.

Such a prudent business strategy rejuvenated the compact between the public interest and corporate capitalist coalition that had vanished in the 'glory days'. Like the grow-and-build strategy, this new strategy represented a compromise on the situational validation level, which was able to relate to the second-order convictions of both coalitions. It differed from the post-war regime in that it aimed at incorporating the environmental issue. The message was that environmental protection was not necessarily the enemy of growth and profit after all (Roe 1984: back flap). In the early 1980s, the corporate capitalist coalition in California accepted the challenge of 'ecological modernization', as formulated by the pragmatic wing of the public interest coalition. Thus, the grow-and-build alliance had evolved into a *sustainable grow-and-build alliance* (Figure 5-1).

5.1.5 The Gradual Metamorphosis of the Two Alliances

In short, in the course of the decade the original 'one-track' policy, which was motivated by the communitarian idea that renewables should be reserved for small entrepreneurs and self-service, evolved into a 'two-track' policy, which furthered "the development of alternative sources of energy both by the utilities and by dispersed producers for purchase by the utilities" (Bryson 1980: 69). While the communitarian coalition was pining away, the use of renewables itself was no longer seen as a partisan issue, and the debate about who should be involved in their development had lost its political relevance.

The alliances that had formed around this issue, however, did not dissolve, but underwent a metamorphosis in the 1980s. As noted, the pro-utility-involvement alliance gradually changed into a sustainable grow-and-build alliance. Members of the public interest coalition within this alliance were especially engaged with the promotion of energy efficiency, and less with advancing renewable energy sources.[7] The anti-utility-involvement alliance developed into a so-called *regulated competition alliance* (Figure 5-1) that advocated the development of the independent power industry in combination with the use of renewable energy technologies.

5.1.6 The Deukmejian Bull in the China Shop

In 1983, Brown's successor proved unable to appreciate the learning processes that had taken place within and across coalitions, and severely challenged the alternative energy policy that had resulted from this. The new Governor labeled it as 'anti-development,' and 'anti-growth.' Deukmejian abolished the OAT, which had been the spiritual father of California's non-utility renewable energy policy during the mid-1970s. This act marked the end of the communitarian period.

In his effort to cut tax expenditures, Deukmejian also intended to terminate what he saw as another relic from the Brown era, namely the energy tax credit program. Within the state Legislature, pro-market Republicans argued that tax incentives had become crucial for the survival of the growing solar and wind industry, and, in this case, supporting business weighed heavier than the libertarian coalition's policy position on special interest breaks and balancing the budget (Mirviss 1983: 2-3). Facing broad support for the tax credit by both the public interest, and the communitarian coalition, as well as his own libertarian party members, the Republican Governor chose not to use his veto, and extended the tax credits. Deukmejian also tried to explain away the utilities' involvement in renewables, which he saw as forced upon them by a radical state government, and not as a modification of their business strategy. In spite of this, the alternative energy option came to dominate the overall energy picture at the end of Deukmejian's first year in office.

At that time, the standard concept approach of the CPUC had led to the ISO4 contract. This had been necessary because, in order to protect their monopoly position, the utilities had failed to negotiate power purchase agreements with independent power producers (IEP 1993: 9). When this lucrative contract came available in the autumn of 1983, planned nuclear plants were still not on line. The urgent need for new power forced the utilities to change their attitude, and they became fairly cooperative in signing up contracts with third-party developers. These proved very attractive to the independent power industry, and their unlimited availability led to a boom in the number of contracts being signed up. By 1984, the state had more than ten thousand megawatts of QF non-utility capacity on contract, and by 1985 the number exceeded fifteen thousand. Meanwhile, the utilities' nuclear power plants had finally come on line, and instead of facing capacity shortages, the state was now facing overcapacity.

Encouraged by both the libertarian (IEP) and corporate capitalist coalition, the CPUC suspended the availability of the ISO4 contract in April 1985. But by then the wind industry had ample contracts. The success of the independent power industry and the affluence of contracts had seriously weakened the fear that firms would move out of California if the credit would be repealed. In contrast, the cry for a balanced budget and reduction of the taxpayer's burden had become more powerful. Thus the libertarian coalition concluded that the energy tax credit had lost its urgency and state energy tax credits expired at the end of 1986.

5.1.7 The Regulated Competition Alliance

The largely unanticipated result of California's energetic implementation of PURPA was the creation of a non-utility industry. The utilities were not pleased about the fact that they had to share the electricity generation market with the independent power producers. At the level of technical verification, the corporate capitalist coalition argued that the independent power producers' existence was only due to the fact that they were being overpaid. At the situational validation level, this coalition defended the existing natural monopoly rationale for regulation. The corporate capitalist coalition (headed by PG&E) launched a series of legislative initiatives to scuttle the independent energy business.

This had little effect, since the libertarian and public interest coalition regarded independent power policy as a California success story. This *regulated competition alliance* claimed that this policy had eliminated California's oil dependence, increased the efficiency and diversity of the electric generation system, created a competitive non-utility industry that could build power plants at a savings to ratepayers, provided jobs and environmental benefits. The CPUC was encouraged to set up a more permanent bidding program, which became known as the BRPU process.

Developing it became a lengthy procedure, which incorporated a complex political compromise between and within the sustainable-grow-and-build alliance and the regulated competition alliance. The former promoted energy conservation. Environmentalists within the public interest coalition (e.g., NRDC, Sierra Club) preferred renewable energy projects over fossil, but found that programs to promote renewable resources were secondary to a commitment to energy efficiency (Sierra Club 1994: 10). Through the so-called 'state-wide collaborative process', the pragmatic side of the public interest coalition managed to get the utilities interested in conservation.[8] The result was that almost seventy-five percent of the anticipated need for electricity could be met by demand-side management. This shared prospect was incorporated within the BRPU process. When, finally, in 1990, the CEC determined that, over the decade, there would be a need for approximately 10,000 megawatts of new resources, it was decided that only a small portion of it – some 1,350 MW – was actually going to be out to bid. The rest was to be covered by demand-side management (DSM).

Members of the regulated competition alliance argued that the only reason that there was that much DSM was to avoid new contracts being offered to independent power producers and feared that this might impede development of the alternative energy industry (Mazmanian 1992: 211). Still, the regulated competition alliance seemed to be given a small piece of the BRPU cake. The independent power industry could earn its piece only if it proved to be more competitive than the utilities. For renewables, this seemed to be a lost battle. However, within the drawn-out development process, the public interest coalition (CEC, Democratic legislators led by Byron Sher, and CEERT[9]) was successful in incorporating two instruments to ensure resource diversity and reduce the air pollution of the electric system, namely environmental surcharges, and the renewable set-aside. Hence the BRPU can be characterized as a compromise at the situational validation level between the libertarian coalition and the public interest coalition.

The outcome of the BRPU at the end of 1993 was very positive for the independent energy business, but was challenged by the corporate capitalist coalition (in particular, SCE, and SDG&E). According to this coalition, the BRPU violated PURPA because it was a QF-only bidding, and because the bidding rules incorporated the above two environmental requirements. When the wish of the corporate capitalists to annul the bidding results fell on deaf ears, they went to the federal level for relief. Finally, FERC declared the bid illegal at the beginning of 1995.

5.1.8 The Deregulation Alliance

> Government is not a solution to the problem, government is the problem.
> *President Ronald Reagan in his First Inaugural Address, 1980*

But while the libertarian coalition was losing this battle it was making a lot of progress in the utility restructuring debate. This was triggered by the California Large Energy Consumers Association, which, during a major recession in the state, complained on high electricity prices compared to other states. Frustrated with the BRPU experiences, the issue of price quickly evolved within the CPUC to being the most visible 'symptom' of a fundamentally flawed regulated industry. It was concluded that, "the current regulatory approach conflicts with the Commission's policy of encouraging competition in the electric services industry." (CPUC 1993: 4) A policy of deregulation was the logical extension of the second order beliefs underlying the libertarian coalition's position on government intervention in the marketplace. A conservative political mood is one of the key factors which helps to explain why restructuring of the electricity sector emerged as an acceptable policy option in the 1990s.

Since the advent of regulation in the Progressive era at the beginning of the century, this possibility had not been seriously considered within the electricity sector. But in the early 1990s, the so-called *traditional regulatory compact* came under attack. Since 1967, Governor Ronald Reagan had made deregulation a popular political rallying cry. With his rise, conservatism emerged in the Golden State (Harvey 1985: 22). In the 1980s, President Reagan introduced this deeply anti-statist doctrine that emphasized the virtues of laissez-faire at national level. Reagan and the long-time libertarian intellectual, Milton Friedman, are the two current names most frequently linked with this ideology (Lipset 1996: 36). Together with other classical liberals, they helped to revive American conservatism, transform the Republican Party, and refurbish belief in the free market system (*ibid.* 202). This all paved the way for Reaganism, which reached a peak in November 1994 when the American electorate gave control of Congress to the Republican Party, whose campaign, 'Contract with America', promised to drastically cut taxation and the scope of government. The 'conservative revolution' has created a political context in the United States in which classical laissez-faire liberalism represents the dominant ideology, and determines the political heuristic.

At the situational validation level, deregulation represents a major political strategy of the libertarian coalition. Since the 1970s and 1980s, a number of other industries had undergone restructuring and deregulation, including telecommunications, airlines, and the natural gas industry. Being by far the greatest industry in America in terms of capital assets, even the Reagan Administration had regarded the electric utility industry as a 'special case'. The demonstration effect of deregulation in other infrastructure industries diminished the fear of such drastic operations.

Moreover, a powerful competitive independent power production (IPP) business had already grown, which, particularly in California, had become an integrated part of the state's overall economy. Consisting of small firms and large corporations, the IPP industry brought about half of the new U.S. electric generating capacity into service, and the technologies it had fostered – modular generation, energy efficiency, renewables, and storage – had made restructuring technologically feasible. It thus took decades of strengthening of the ideology and self-confidence of the libertarian coalition, together with the combined development of a new IPP industry and technology, before the option of electric energy restructuring became politically viable.

The *Blue Book proposal* (CPUC 1994a) that was concocted within the CPUC challenged the utilities' entrenched interests. The corporate capitalist coalition, however, felt presented with a *fait accompli*, and took to the idea of deregulation. It accepted that the issue was no longer whether or not the industry would be restructured, but how a transition to a fully competitive electricity wholesale market, and direct access for all classes of retail electricity customers would be introduced. As a result, the former archenemies within the energy debate over the last twenty years entered into discussions in May 1995. The forum was provided by the Republican Governor Wilson, and, after a couple of months, the two coalitions reached an agreement, known as the *Memorandum of Understanding* (MOU). A new and dominant *deregulation alliance*, consisting of the corporate capitalist and libertarian coalitions, had come into existence.

The deregulatory approach did not agree with the second order conviction's of the public interest coalition. This coalition was very concerned about how restructuring would effect renewables and energy efficiency. The public interest coalition had achieved two decades of notable environmental and social policy success by alternately entering into alliances with both the libertarian and corporate capitalist coalitions, which paradoxically are concerned with economic efficiency instead of the public interest. But now the two previously antagonistic coalitions had embraced each other, the public interest coalition suddenly became isolated. Since the beginning of the restructuring debate, the members of this coalition had actively sought collaboration among them, but had been unable to come out with a combined proposal. The arrival of a powerful alliance, that represented all the business interests in the state, forced them to come up quickly with a joint response to the *Memorandum*, to safeguard some political influence. After a few months, the public interest coalition members[10] submitted their mutual plan, the so-called *Framework for Restructuring in the Public Interest*.

With respect to renewables, the public interest coalition wanted the CPUC to "ensure that any industry restructuring restores and improves upon the level and diversity of renewables that existed in California in 1993 (before restructuring discussions began)". However, there was a "surprising lack of agreement about the means to achieve this policy goal".[11] The brand new issue of deregulation awakened the ancient disagreement within the public interest coalition between members who wanted to maintain public control of the electric industry, and impose explicit public policy requirements on all power suppliers, and others who had no faith in regulatory control, and wanted to deal with the public interest in a market oriented way.

Groups like AWEA and UCS exemplified the first wing. They had a social democratic tinge to them, and argued that all retail electric service suppliers should be required to possess a diverse portfolio of generation resources, including a minimum fraction of renewables to be determined at the state level. EDF and Sierra Club California were representatives of the other segment. They held that a prescriptive approach was neither economically efficient, nor politically viable, and instead favored a tax subsidy approach, which in fact was a pale shadow of their wish to internalize externalities.

In the past, its makeup of both pro-market and pro-regulation members had enabled the formation of alliances with both the libertarian and corporate capitalist coalition, and get the best from both worlds. In the new situation, however, this same diversity seriously weakened the public interest coalition. The *Renewables Portfolio Standard* (RPS) policy was lost, and only some funds were secured to maintain the existing base of renewables in California.[12]

5.1.9 Looking Back on the Political Debate

The visualization of the changes in alliances that has taken place within the last four decades in Figure 5-1 offers a look into the past as well as the future. It shows that, in the second part of this century, the public interest coalition represented the pivotal coalition, but it was later sidelined. In the post-war period, the corporate capitalist and public interest coalition agreed upon the universal benefits of the grow-and-build strategy. After the mid-1960s this rationale came under attack from the public interest coalition, but also from two 'newcomers': the communitarian and the libertarian coalition. The sectarians within the public interest coalition held that the democratic control of the electric utilities needed to be restored. The communitarian coalition argued that the control over the production of electricity should be decentralized. Both coalitions promoted the local use of renewable energy technologies. The libertarians disliked the monopoly structure of the industry, and wanted to restore competition. These three coalitions formed an influential anti-utility-involvement alliance during the 1970s.

In the early 1980s, the question of democratic political control, as raised by the public interest and communitarian coalition, dissipated, and the latter left the political scene. It was succeeded by questions of how to bolster competition and shift from government planning to market solutions. The anti-utility-involvement alliance took on a practical attitude, and slowly involved into a regulated competition alliance. The grow-and-build alliance also transmigrated. Encouraged by both environmental as well as business arguments of pragmatic members within the public interest coalition, this old alliance was gradually renewed into the sustainable grow-and-build alliance. The regulated competition alliance and sustainable grow-and-build alliance had different political agendas. The first was interested in developing a competitive independent power industry, and, as a part of this, a renewable energy industry, that would lead to a flexible, cheap and clean electricity supply system. The second was interested in the technological and ecological modernization of the existing system in order to maintain its power structure. While the sustainable grow-and-build strategy concentrated on conservation programs, the regulated competition alliance has been the political force behind the development of the renewable energy industry in California. The development of this alliance had been made possible by PURPA.

In the 1990s, the classical laissez-faire liberalism of the libertarian coalition came to dominate the energy debate. The growth of its influence ran parallel to the rise of American conservatism since the early 1980s. At the height of the conservative revolution in America, deregulation of the electric industry became a viable policy option in California. The outlook of such a fundamental regulatory change created a new sphere of influence that put under pressure all existing political coalition behavior, and caused the formation of a new strong deregulation alliance. A compact followed between the corporate capitalist and the libertarian coalition, from which the public interest coalition was shut out. This new political situation seems to mark a new era in which the 'market' is going to determine the dynamics of the electricity sector, and regulators are going to forego their ability to impose public policies. It appears to end a transition period of two decades, in which the public interest coalition had been successful in fostering the integration of energy and environmental policy by means of energy efficiency, conservation, and the use of alternative energy technologies.

5.2 Policy Change in the Field of Wind Energy in Denmark

> Grundtvig was opposed to domination in a way that stands outside of the European emancipation tradition, with its emphasis on liberal individualism. ... He wanted to substitute for violent revolution a peaceful transformation of all elements in society based on a mutual recognition that all had the right to exist. Yet his views on *vekselvirkning*[13] went further than the mere tolerance of diversity. He was really insisting upon a mutual recognition that each institution, each power center, and indeed each individual could both teach and learn in dialogue predicated on mutual respect. *Steven M. Borish, 1991: 169*

In the Danish case, only two relevant policy belief systems and ideological coalitions can be distinguished: the dominant established coalition, and a less politically powerful, but, in the long run, very influential ideological coalition, the cooperative coalition. Over the last two decades, a comprehensive learning process has taken place across these two coalitions. This has been organized around the story line of ecological modernization. The two distinct ideological coalitions in Denmark can be said to have formed a long-term learning alliance (see Figure 5-4).

After a short historical introduction, the beliefs of the established and cooperative coalition will be introduced. In the following subsections, it is reported how the cooperative coalition has constantly challenged the established coalition, its ideas creating both opportunities and threats. Finding an optimum implied a search for policy instruments and strategies that assured a proper balance between distinct core values, such as the need for economic growth versus the demand for centralized control of the electricity sector. The enduring interaction with the cooperative coalition also enticed the established coalition to accept the idea that, to sustain economic growth, economic growth itself needs to be (environmentally) sustainable.

5.2.1 A Short Historical Introduction

The energy crisis regenerated an old political controversy surrounding the centralization of the electricity system. In the 1920s and early 1930s, this debate culminated in the passing of the Power Current Law (*Stærkstrømslov*) of 1935 which fostered the technical and organizational concentration within the power generating system (Rasmussen 1982: 135-144). The government was not its prime guiding actor; rather the electricity sector was largely allowed to regulate itself, while the state acted as an economic safety net. The oil crisis had little direct impact on the Government's stance toward the centralization issue, but was a turning point with respect to the state's role. A few years later, the change in attitude toward state interference within the *Folketing* led to the almost unanimous adoption of the new Electricity Law which aimed to secure societal influence in the development of the electricity sector.

This revealed the reformist 'social engineering' mentality that had come to dominate Danish political culture during the 'happy sixties.' (cf. Jamison and Læssøe 1990: 68) The social democrats had long given up their original aim to socialize the means of production, while the non-socialist parties had come to accept state intervention in the economy. The liberals, conservatives, and radicals were 'capitalist' parties, but without clear ideologies. Time had erased much of the original ideological differences among the 'four old parties,' and although the Liberal and the Social Democratic Party had remained the focal points for contending alliances, "this antagonism was at least as much political competition as it was ideological hostility" (Miller 1996: 230).

The absence of clear-cut differences in opinion between these mainstream parties enabled a politics of reconciliation, since "there was always the prospect of compromise across party lines on most political issues" (*ibid.*). Using the familiar distinction between 'pragmatists' and 'sectarians', it is fair to say that within Danish 'socialism' and 'capitalism' the pragmatic style tends to prevail. This was strengthened by the political earthquake of 1973, when Danish politics became characterized by both a seriously fragmented *Folketing*, as well as minority Governments. Kenneth E. Miller, in his book, *Friends and Rivals,* on coalition politics in Denmark, explains that, because of the lack of a majority, successive Governments are forced "to seek compromises with other parties from issue to issue or, if possible, to work out longer-term commitments of support from policies." (*ibid.* 150)

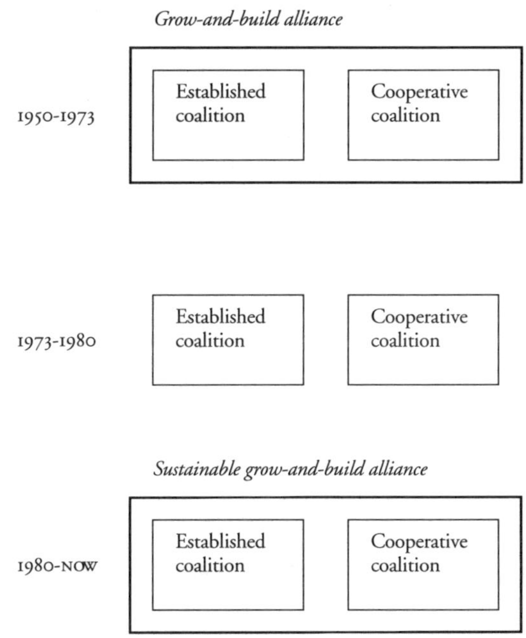

Figure 5-4
Coalition behavior within the Danish energy debate. The existence of an alliance between the established and cooperative coalition in the periods between 1950 and 1973, and 1980 till now is depicted by the thick-lined boxes.

5.2.2 The Established and Cooperative Coalition

> The term 'local democracy' is a recent arrival in Denmark, a slogan which in itself expresses a protest against the tendency for democracy in a centralized and bureaucratic social system to recede into the distance. But the phrase does not denote anything other than democracy; indeed, the concept has no direct antithesis and 'democratic centralism' (as referred to in the new constitution of the Soviet Union) is a contradiction in terms. *Niels I. Meyer, K. Helveg Petersen, and Villy Sørensen, 1978*[14]

These general remarks apply to the field of energy policy. Here, viewing the party system as consisting of a socialist and a non-socialist bloc is neither helpful nor enlightening. Key events, like the Power Current Law of 1935, the Electricity Law of 1976, the Easter and June agreements of 1986, and the Energy 2000 agreement of March 1990, are all legislative compromises that demonstrate a broad and long-established political consensus concerning the development of the electricity system that goes beyond the two-bloc competition. The long-lasting agreement between the 'four old parties' on energy policy issues seems to justify the view that the socialist and non-socialist parties could well be considered members of one ideological coalition. Since its ideology agrees with the mainstream political ideas that dominate Danish politics, this coalition will be denoted as the *established coalition* (Figure 5-5). The electric utilities, commonly represented by DEF, and the Ministry of Energy, are also members of this ideological coalition.

The established coalition's preferences about the social order can be lumped together under the tag 'The Danish model' (*Den danske model*),[15] which includes the notion of a social-liberal welfare state organized through the principles of representative democracy and centralized bureaucracy, and belief in technological progress. At the system vindication level, electricity is seen as a central source in such a highly technological society. Electricity is viewed as the motor of the related processes of technological modernization and economic growth, which are the pillars of prosperity and thus of the welfare state. At the first order level, the grow-and-build strategy was seen as the proper way to satisfy consumer demands and supply electricity at the lowest possible costs. Not only was the situational validation concept of large central power stations seen as the most cost-efficient way to generate electricity and secure its supply; also the rational social choice argument of long-term centralized planning was seen as an absolute precondition for a stable and cheap electricity system.

The oil crisis did not really challenge the belief system of the established coalition. At the technical verification level it was readily recognized that it was dangerous to rely so much on oil. This admission brought coal back into focus, created the need to temporarily curtail the increase in energy consumption, and revived Juul's idea of developing large-scale wind turbines that could contribute to the centralized electricity generation system as fuel savers. More important, it accelerated the already existing plans of the utilities to build nuclear power stations.

A far less powerful coalition, rooted in the environmental movement and the progressive folk high schools, saw "the plans of the Danish electricity companies in

Argumentation concerns	Established coalition	Cooperative coalition
Preferred social order	The Danish model: a social-liberal welfare state that is organized through the principles of indirect democracy and centralized bureaucracy.	A humane, ecologically sustainable society that is characterized by an active local (direct) democracy.
(Empirical and normative) Background theories	Belief in economic growth and technological progress.	Belief in a limit to growth., gradual social progress, and 'appropriate' technological progress.
	Electricity is the motor of economic growth, and energy consumption contributes to welfare.	Energy services contribute to welfare. The use of indigenous resources helps to create jobs and lower the balance of payment deficit.
Problem definition and the meanings of technological artifacts in the light of the problem	No problem. The grow-and-build strategy is the proper way to satisfy consumer demands, and to supply electricity at the lowest possible costs	Zero-growth strategy to avoid a global catastrophe.
	Through the construction of central power stations and the use of cheap oil the grow-and-build strategy can be maintained	Strong and lasting conservation efforts, combined with the slow but sure introduction of renewable technologies.
Perceived policy solutions	More of the same	Grassroots activities, energy information campaign, anti-nuclear demonstrations, alternative energy plans (AE-plan 1976, AE-plan 1983, Energy 2000)

Figure 5-5
Policy belief systems of the established coalition and the cooperative coalition engaged in the Danish energy debate in the 1970s.

1974 to build centralized, life-threatening, mammoth nuclear plants as a provocation that could not go unanswered".[16] The *cooperative coalition* – consisting of OOA, NOAH, OVE and the ecological practitioners, the engaged scientists who drew up the alternative energy plans, progressive journalists like Torgny Møller, DV, SF, RV, and the folk high schools – became frustrated about the fact that the established coalition considered this a normal bureaucratic decision based on rational economic and technical criteria (Figure 5-5). The nuclear power issue quickly evolved within this coalition to one of the most visible symptoms of a fundamentally flawed expertocratic and centralized democratic system.

The second-order convictions in the policy belief system of this coalition were based on the principle of direct democracy. Referring to Grundtvigian concepts like 'the living word' (*det levende ord*), and 'people's enlightenment' (*folkeoplysning*), this approach emphasized the need for a truly open democratic debate. This implied that the general public should be given access to the information upon which political energy planning decisions might be based. It was anticipated that with professionals sticking to the role of giving advice, energy planning would become a truly public (*folkelig*) activity. The cooperative coalition wanted the *Folketing* to postpone deciding whether to build nuclear power plants, demanding more public information and a moratorium till a public debate was held. These demands met broad popular support and were swiftly accepted by the established coalition.

Just like the established coalition, the cooperative coalition "maintained an attitude of extreme caution, not wanting to upset anybody with dogmatic, ideological rhetoric" (Jamison and Læssøe 1990: 97). This was illustrated by the AE-plan it published in reaction to the governmental energy plan in 1976. In this alternative plan it was argued that, also from the point of view of the established coalition conservation, the use of renewable energy sources made perfect sense, since it would offer considerable advantages in terms of more jobs, less use of foreign currency, and greater reliability of supply. The cooperative coalition used the precondition of economic growth as a mathematical input parameter within its alternative scenario, despite the fact that it challenged the dominant belief that economic growth automatically leads to higher welfare. Its argument on the technical verification level was that no evidence could be found that confirmed the opinion of the established coalition that growth led to fairer distribution of wealth.

On the system vindication level, it was held that there existed absolute physical limits to growth. The cooperative coalition concluded that growth was not sustainable, and thus would not lead to a fancied 'humane, ecologically sustainable society'.[17] The parallel reasoning with respect to energy was that it was not energy consumption, but energy services that directly contributed to human welfare. Based on its conviction that fossil fuels were limited in stock, and that the environment constituted real constraints, the cooperative coalition favored a low-energy society that would rely on renewable energy sources.

5.2.3 Ecological Activity Followed by Industrial Activity

The practical activities developed by the cooperative coalition (in particular OVE) during the 1970s to demonstrate the technical feasibility of its proposals presents an ultimate illustration of its pragmatic attitude. In line with the adage, 'action speaks louder than words,' members of this coalition experimented with zero-energy houses, biogas and solar energy systems, and windmills. These can be regarded as a practical form of argumentation on the technical verification level inspired by higher order beliefs. There is little doubt that the public enthusiasm aroused by these activities helped in triggering the established coalition's interest in the decentralized use of renewable energy technologies. The fact that many craftsmen and small industries followed the actions of the ecological practitioners played a decisive role. In fact, during the broad political negotiations in August 1977, the established coalition formulated industrial involvement as a necessary precondition for the introduction of an investment subsidy on renewable energy projects.

While, on the situational validation level, the cooperative coalition saw decentralized energy systems as a means to shift the authority to decide energy matters to the local level, the established coalition saw these technologies as a means to induce novel industrial activities which would create new jobs. Although initially, the energy subsidy was an industrial policy instrument, it was thus rapidly integrated into Denmark's energy policy. Since 1981, the Government announced that it aimed at 60,000 small wind turbines at the end of the century. The adoption of this target was only one element within a comprehensive learning process, in which the established coalition embraced first-order beliefs of the cooperative coalition.

The sectarian part of the established coalition (DEF, Elsam, Elkraft) persevered with support of the grow-and-build strategy, and its argument that a sufficient supply of electricity was a precondition to solve the economic crisis. Its dominant pragmatic wing, however, definitively replaced the old strategy by a so-called *modified grow-and-build strategy*, which included the need for permanent energy savings (Figure 5-6). Such a revision became possible after the coalition abandoned its belief in the linkage between growth in energy consumption and economic growth.[18]

On the technical verification level, this assumption had been confirmed during the 1960s, but falsified during the 1970s. Inspired by their faultfinders, the established coalition replaced its original hypothesis with the more subtle idea that there existed a link between growth of energy services and economic growth. Thus the modification of the grow-and-build strategy can be regarded as a deliberate attempt to "reduce the coherence between increasing energy consumption and national economic growth", so as to sustain the second order belief in economic growth. This policy-oriented learning process was accelerated by the second oil crisis, and stimulated by the elaborate views of the cooperative coalition.[19]

Apart from the Electricity Law, the establishment of a Ministry of Energy in 1979 showed that the government was consolidating its position within the field of energy. Before, Danish administrative practice was characterized by corporatist arrangements that, in practice, were specified by the electric industry itself. After the Government had built up its own expertise, it had become a worthy interlocutor in relation to the representatives of the industry. Though the *Folketing* had become authorized to unilaterally impose regulations through public laws, the law making process kept its cooperative trait. But the mere possibility served as an important means for the Danish Parliament to influence the utilities' behavior within the corporatist negotiation process. Cherishing the idea of self-dependence and self-regulation, the utilities were inclined to compromise themselves with societal objectives in order to avoid law interference. The setting of the rules for wind turbines serves as an interesting case to take a closer look at the relationship between 'the' *Folketing* and the relevant interest organizations (i.e., DEF, DV, and FDV).

Argumentation concerns	Established coalition		
	1973-1980	1980-1985	1985-1990
Preferred social order (Empirical and normative) Background theories	The Danish model Belief in economic growth and technological progress. Electricity is the motor of economic growth, and energy consumption contributes to welfare.	The Danish model Belief in economic growth and technological progress. Economic growth and energy consumption do not need to be linked.	The Danish model Belief in sustainable economic growth through technological innovation. The growth of energy services increases welfare.
Problem definition	The temporary modified grow-and-build strategy.	Modified grow-and-build strategy.	Sustainable grow-and-build strategy.
The meanings of technological artifacts in the light of the problem	Nuclear power will phase out Denmark's vulnerability to foreign oil supplies, and recapture the old grow-and-build strategy. Wind turbines are potential fuel savers.	Wind turbine technology is a potential industrial export product and can contribute to economical growth and the creation of jobs.	Wind turbines are export products and represent the environmentally most acceptable way of generating electricity.
Perceived policy solutions	Fuel switch to coal, energy savings campaigns, nuclear power (HM-plan 1976)	Lasting conservation efforts combined with the introduction of renewable technologies (Renewable investment subsidy, tax exemption, first 100-MW agreement)	Integrated Resource Planning (Energy 2000, second 100-MW agreement)

Figure 5-6

The gradual upward dissolution of the protective belt of the policy belief system of the established coalition in the 1970s and 1980s.

5.2.4 The 'Wind Energy' Cooperative Movement

> Only a newspaper can put the same thought at the same time before a thousand readers. ... So hardly any democratic association can carry on without a newspaper.
> *Alexis de Tocqueville, 1835*

When, in 1976, DEF published its first provisional guidelines for small grid-connected wind turbines, the wind organizations, DV and FDV, were nonexistent. At the time, the growing number of requests for connecting private wind turbines to the public grid incited some DEF members to articulate their wish for unambiguous rules. More important, the cooperative coalition used the press to coerce DEF to take such action. Despite DEF's official policy, most members did not live up to it. Among other things, discontent with this situation led to the establishment of DV.

Within the *Folketing*, the cooperative coalition (i.e., RV, SF) had a ready ear for DV's complaints. Claiming that the outcome of possible private negotiations between DEF and DV was likely to be unfair because these two parties differed so strongly in power, the cooperative coalition proposed to lay down legal rules. The established coalition, however, thought that legal initiative unnecessary. Trusting to the benevolence of the electric utilities, it ordered DEF to revise the old regulations as soon as possible. Since DEF did not want to lose confidence, it quickly came up with the provisional 1979-rules.

The new rules encompassed cooperatively owned wind turbines, which was a demand from DV. Cooperatives had had a long tradition in Denmark.[20] Solidary collaboration in cooperatives reminded large parts of the rural population of the Golden Age of Danish agriculture in the years preceding the First World War. At that time, the cooperative form had given farmers the economic benefits of large-scale production without effecting their independence. The fact that the Danish people were familiar with the advantages of this way of organizing helps to explain why the idea of windmill cooperatives struck root.

To some extent, the large number of local initiatives to set up a *vindmøllelaug* can be regarded as a collective and nostalgic journey back to the 'good old days' of self-organization. Moreover, the birth of a 'wind energy' cooperative movement meant that the rural population could recognize and appreciate qualities in the argumentation of the cooperative coalition (cf. Jamison and Læssøe 1990: 107). Many people were able to identify with the ideas of the cooperative coalition, which in many ways reminded them of the well-known ideological basis of the agricultural cooperative movement. This included the principle of direct democracy, and was expressed in words like autonomy, independence, self-awareness and personal responsibility (cf. Hansen and Larsen 1983). The cooperative tradition of rural Denmark provided a historical repertoire of forms of collaboration that, having proved their worth in the past, were available to citizens for addressing new problems of collective action, for example, the energy quandary.

The fact that the alternative energy movement could draw on an earlier tradition, provides an important key to the success of collective ownership of wind turbines in Denmark. It should be stressed that, just like the earlier voluntary associations, the contemporary cooperatives signified less an idealistic altruism than a pragmatic readiness to cooperate in order to be economically better off. Regardless, in the 1980s, the decentralized use of wind energy encountered widespread active public support. Its popularity laid the foundation of a moderate home market for medium-sized wind turbines. This contrasted to the concurrent situation in California, where the grassroots movement (communitarian coalition) was rapidly losing political credibility.

5.2.5 Industrial Growth at the Expense of Centralized Control

From an industrial-policy point of view, the success of the investment subsidy gave the established coalition an incentive to continue this demand-stimulation instrument. The cooperative coalition also praised the subsidy program, which promoted both the use of renewable energy sources, as well as local decision making. Although the private development of wind energy thus worked both ways, it created a tension between two core beliefs of the established coalition. The coalition wanted to sustain a sufficient home-market for the wind turbine industry, but it also believed in the central control of the electricity system. The utilities were extremely sensitive with respect to the latter second order belief, and were opposed to every minor policy, which might harm their absolute control over the power generating system. Anticipating that the public interest in wind energy would be short-lived, the utilities regarded the grassroots activity as a temporary 'experiment'.[21] They controlled the experiment by imposing rigid siting and consumption criteria.

Interestingly, their strategy to limit the private use of wind energy was in line with the second order belief of the cooperative coalition that such a development should be a local affair. The cooperative coalition argued on the technical verification level that the 3-km rule was too strict and advocated that owners should be allowed to set up their wind turbine within the community they lived in. In private negotiations with the wind organizations, DEF did not budge an inch either on the siting issue, or on the demand for less stringent consumption standards. Once again, the cooperative coalition within the *Folketing* came to the wind organization's aid, proposing to settle the rules by law.

At the start of 1984, the threat of legal interference had considerably more impact on the utilities than four years before since, in the meantime, the pragmatic wing of the established coalition had become happily surprised with the steady development of the home market. Moreover, as Denmark was on the verge of becoming one of the world's leading producers of wind turbines, the 'windmill' developed into one of the show-pieces of Danish industry. Because the pragmatists regarded an adequate home market as an essential precondition for competing on the export market, they were willing to tolerate the private use of wind energy at the expense of utility control and ownership in order to stimulate industrial growth.

Although the bill did not lead to state regulation, it provoked a political discussion that made clear that the established coalition also wanted to sustain private development of wind energy. The existence of broad political support for their demands strengthened the negotiation position of the wind organizations. Soon, a 10-year agreement was reached that embodied many demands of the cooperative coalition. It was agreed that wind turbine cooperatives could recruit members from within the supply area of the electric utility company that served them. Moreover, the electric utilities agreed to buy *all* surplus power at a favorable rate from cooperatively owned wind turbines.

5.2.6 The Unification of Industrial Growth and Centralized Control

In the meantime, the cooperative coalition had also managed to exempt wind power from electricity tax. The tax exemption, investment subsidy, loosening of siting criteria, and the fact that the connection between private production and consumption of electricity was abandoned in the 1984-agreement, made it profitable for municipalities to set up wind turbines, but also created lucrative opportunities for businessmen to develop wind farms. A boom in turbine sales on the domestic market resulted. The utilities lambasted the involvement of public organizations and entrepreneurs in the field of wind energy and argued that the 1984-agreement was not intended to set up investment projects. This was accepted by the established coalition because a large-scale private development of decentralized wind energy severely challenged the centralized character of the electricity system. The reaction of the cooperative coalition was two-fold. It welcomed the municipalities' interest in wind energy, but unmistakably disapproved of non-local private investors getting involved, fearing speculative forces would lead to chaotic development and turn West Jutland into a 'Wild West.'[22]

Such widespread antipathy allowed the Government to impose strict siting- and consumption-criteria on the state subsidies for wind energy, so as to stop the capitalist development of wind farms. The investment subsidy thus became a custom-made policy instrument for the promotion of wind turbines with *local attachment*. In a way, this was a mark of honor for the development path, which had been envisioned by the cooperative coalition, and which had led to the establishment of a successful industry. Nevertheless, for both practical (first order) and ideological (second order) reasons, the established coalition expected that local cooperatives would play a limited role in the coming development of wind power and the wind turbine industry. The first *100-MW agreement* at the end of 1985, was an attempt to secure a stable future home market. This agreement represented a compromise between the sectarian and pragmatic part of the established coalition, in which an industrial-growth scenario would go hand in hand with maintaining centralized control of the planning and production system.

5.2.7 The Blend of Economics and Ecology

> It is encouraging that such a plan (*Energy 2000*) can achieve broad political support. But it is discouraging and frightening that most politicians still refuse to even talk about the last factor for reaching sustainability, i.e. bringing the economy to a halt at an optimum level instead of striving for a maximum level. *Jørgen Stig Nørgård and Bente L. Christensen 1993: 330*

The 100-MW agreement was another signal that the Government was strengthening its hold on the utilities. It showed that, in the mid-1980s, the authorities no longer limited their role to discussing proposals made by the industry, but had begun enforcing specific capacity choices on the electric utilities. Besides wind turbines, the industry agreed to build a specified number of small local cogeneration plants, and to use natural gas instead of coal. These corporative arrangements were usually justified for either industrial (e.g., renewable energy technologies) or national economic reasons (e.g., the use of natural gas). In contrast, the environmental regulation of the electricity sector was set down in public laws.

During the 1970s, the cooperative coalition had set the pace of the ecological debate. Initially, the established coalition followed their track in slow-motion. Focusing on employment and economics, it regarded environmental degradation a localized and technically solvable negative side effect of growth. This changed in the early 1980s, when the Danish public began to rank environmental protection as the number one political issue. In reaction to opinion polls, the established coalition tried to give the environmental issue a more central position within its policy belief system. A first major result was the SNOX-law in 1987, developed in close cooperation with the utilities, which established emission quotas for SO_2 and NO_x. The coalition's attempt to blend ecology and economy coincided with the activities of the Brundtland commission. The established coalition accepted the United Nations' commission's concern for global warming, and eagerly adopted the concept of sustainable growth as a guiding principle on the systems vindication level to counteract the environmental quandary (Figure 5-6).

As we have seen, the idea of sustainability was already introduced in the 1970s by the cooperative coalition. At the rational choice level, this coalition preferred a humane, ecologically sustainable society. On the systems vindication level it was argued that economic growth was inextricably bound up with environmental degradation. The cooperative coalition therefore held that limits in economic and material development in the industrialized countries were needed to make the transition to real sustainable society. But, although some members of the cooperative coalition saw sustainable growth as "just a fashionable concept,"[23] they welcomed their apparent success in moving the belief system of their rival coalition in their own direction. They were eager to cooperate in designing the Government's plan of action for sustainable development in the energy sector.

Despite disagreement on core values, the established coalition more or less incorporated all of the cooperative coalition's first order elements within the plan, *Energy 2000*. At the technical verification level, the overall objective of this plan was to provide a basis for a 20 percent reduction of the CO_2 emissions of the Danish energy sector up till the year 2005. To achieve this target, the modified grow-and-build strategy at the situational validation level was replaced by *integrated resource planning* (IRP). In general, IRP serves as a method of assessing whether energy savings and load management are economically more attractive measures than supply-side expansion through construction of new power capacity (IRP 1993: 10). In the Danish situation, IRP should be seen as a planning tool for determining, at a given demand for energy services, the least cost route of reaching environmental targets for the control of greenhouse gases (*ibid.* 2).

The sectarians within the established coalition were very upset with the Green Plan. They complained that it was too ambitious and the IRP planning tool was very loosely described. On the second order level, they feared that aggressive pursuit of environmental targets would weaken the long term planning capacity of the electricity sector, which could harm the whole Danish high technological society. The schism within the established coalition had already become apparent during the plan's final preparation stage, when the Government had stopped its cooperation with the utilities. The existence of a broad political consensus with respect to *Energy 2000* gave the Government the nerve to break with the corporatist tradition, and to force its views upon the electric utilities. The fracture was only temporary. Shortly after publication of the Green Plan, the utilities acknowledged the consensus within the field of energy politics, and expressed their willingness to cooperate with the authorities to implement the energy plan.

The increased utilization of wind energy was part of the action plan. In line with the claim that "measured against the greenhouse effect wind energy [was] the environmentally most acceptable way of generating electricity" (EM 1990: 101), the Government compelled power companies to enter into a second 100-MW agreement. This aimed to support and strengthen the position of the Danish wind turbine industry abroad. The Government also introduced an export guarantee scheme, set up a new approval procedure for wind turbines, and introduced a law that set out rules for connection to the electricity grid. This was at the request of the interest organizations, DV, FDV, and DEF, which had not been able to reach agreement on the succession of the 1984-rules on their own. In the Wind Turbine Law, the two principles that had guided the development of private wind turbines – the principle of local attachment and the connection between private consumption and production – were preserved.

5.2.8 Toward a Synthesis between Centralization and Decentralization?

> The fundamental task is to achieve smallness *within* large organization. Once a large organization has come into being, it normally goes through alternating phases of *centralizing* and *decentralizing*, like swings of a pendulum. Whenever one encounters such *opposites*, each of them with persuasive arguments in its favor, it is worth looking into the depth of the problem for something more than compromise, more than a half-and-half solution. Maybe what we really need is not *either-or* but *the-one-and-the-other-at-the-same-time*. *E. F. Schumacher, 1973*[24]

While centralized planning has always been the hallmark of the development of utility-owned wind turbines, analysis shows that, through the years, also within the private development path, centralized decision making has come to play an ever-increasing role. This contradicts the original aim of strengthening local decision-making. Several reasons can be attributed.

The increased scale of development led to clashes of interests that could not easily be settled by self-governance within networks of interest organizations. Defining the decentralized development as an 'experiment' initially helped to disguise the existence of a real conflict between wind turbine owners and utilities. But this evasive strategy lost its attractiveness for the utilities when local involvement in wind energy proved to be a concrete and permanent success.

Second, the demand for an active local democracy was soon eclipsed by industrial policy aspects, which became intermingled with energy policy considerations and environmental facets. The growing embedding of wind energy in Danish society created the need for a central actor who had the competence and authority to achieve the right balance between myriad related aspects.

The established coalition's strong preference for centralized control offers an important third reason. Directly, because the established coalition stimulated the development of private wind turbines for its own industrial policy reasons. Indirectly, because the established coalition favored and incited the development of utility-owned wind turbines, which, from a political perspective, represented a totally different development path. Within that 'path,' the utilities' reluctance regarding renewable energy resources also imposed a central, active role for the national Government so as to overrule the planning authority of the utilities.

It is interesting to see, however, that the Danish physical planning process counteracts the strong trend towards centralization. Only at the end of the 1970s, the siting of cooperatively owned wind turbines posed a problem to local and regional governments. They asked the Physical Planning Agency to provide more precise rules to deal with this new phenomenon, and, after a few rounds of consultation, the planning issue was quickly solved. In the second half of the 1980s, finding suitable sites for utility-owned wind farms as part of the 100-MW agreements proved to be a more wicked problem since many proposed utility projects evoked strong local protests.

The sectarian wing of the established coalition (in particular the Social Democratic party) blamed the complex Danish three tier – State, Counties and Municipalities – administrative system for hampering centrally planned development. At the end of 1992, the sectarians, therefore, proposed to work out a national planning directive to secure the development of utility wind turbines. Obviously the cooperative coalition strongly opposed the idea of centralized physical planning. But the pragmatic wing of the established coalition also argued, at the second order level, that the siting of wind turbines should be a local competence. Broad support for decentralized physical planning confirmed the longstanding tradition of extensive self-regulation at the local level, and excluded a dominant position of the central Government with respect to physical planning issues.

5.2.9 Looking Back on the Political Debate

> Being small is no excuse, especially if you are rich and well educated. *Niels I. Meyer, K. Helveg Petersen, and Villy Sørensen, 1978*

The above indicates that the two types of development that, during the 1970s, were presented as two ideologically opposite routes eventually lost their distinct character. In the course of time, the viability of the development path that was envisioned to stimulate an active local democracy has got more and more dependent on the support from the central Government. In the opposite way, the development path inspired by the belief in centralized planning has become increasingly dependent on decisions made on the local level.

Besides the loss of ideological purity, the two paths have become mutually dependent in practical terms. Since access to suitable locations is limited, more co-ordination is required to ensure the viability of each development path. This raises the intriguing question of whether there are still two ideologically separate frames of meaning left to shape the future development of wind energy in Denmark. In other words, has policy-oriented learning within and across the two coalitions led to a final synthesis between the previous polar rational choice concepts of centralization and decentralization?

To avoid leaping to conclusions, it seems prudent to first recollect the joint search of the two coalitions for both the right balance between core values, as well as for the proper policy strategies and instruments to achieve those core beliefs. With respect to energy policy, a steady learning process within the established coalition can be discerned that leads to the will to ensure sustainable development (Figure 5-6). In 1973, the learning process was triggered by the oil crisis. To safeguard society and industry from a future lack of energy supply, several alternative supply options were considered to counteract Denmark's dependence on imported oil. After the established coalition had made its nuclear intentions known, the cooperative coalition started to articulate an alternative belief system. When it became clear that nuclear power was also unacceptable to the majority of the Danish public, the established coalition gradually incorporated elements from the belief system of the cooperative coalition within its own. The established coalition first replaced the idea that energy consumption is a measure of welfare with the assumption that there is a direct relationship between the growth of energy services and economic growth. At the situational validation level, the grow-and-build strategy was replaced by a modified grow-and-build strategy. During the 1980s, increasing concern about global warming and other emissions from electricity generation motivated the established coalition to adopt the goal of a sustainable energy system. A sustainable grow-and-build strategy, in which integrated resource planning played a central role, thus replaced the old modified grow-and-build strategy. One might therefore speak of a *sustainable grow-and-build alliance* (Figure 5-4) between the established coalition and the pragmatic side of the cooperative coalition.

The plan, *Energy 2000*, and its follow up, aim to reduce the environmental impact of the use of energy but not at the expense of economic growth. This keeps members of the cooperative coalition worried about the 'predicament of mankind.' Instead of using the concept of sustainable growth as an opportunistic idea to tie growth and sustainability together, the cooperative coalition prefers to look at development aspects from the utopian view of a New World order based on equity and global solidarity. It considers equity on a global scale, "not only as an integrated part of sustainable development, but also as a necessary condition for achieving political acceptance for such a development" (Benestad *et al.* 1993: 1225). In its '*economic saturation*' scenario, global social and material equity is reached through economic growth in the developing countries, and economic and material saturation in the industrialized countries. In contrast to the '*economic growth*' scenario of the established coalition, in which overall man made capacity to produce goods determines the size of consumption, in the saturation scenario the size of production is shaped by people's demand. In the latter case, people's demand for material consumption is gradually saturated, "reaching an optimal level, where preferences are shifted toward non-economic welfare like leisure time, job satisfaction, and a better environment" (Nørgård, Nielsen, and Viegand 1994: 406). The coalition holds that, by combining a conscious no growth policy with the necessary technology development, Denmark could, within a couple of decades, reach a stable consumption of electricity around half the present level. This would make it possible to phase out nuclear power and coal power plants, and would pave the road to a real sustainable energy system (*ibid.* 399).

Through the years, wind energy technology has acquired several meanings (Figure 5-6). During the 1970s, wind turbines were thought to be potential fuel savers for the electric utilities. Frightened by the nuclear vision of the established coalition, the cooperative coalition saw wind turbines as an alternative technology that would keep the future open to human intervention. Moreover, the availability of decentralized energy systems would enable the decentralization of the political decision making process. The practical attitude of the cooperative coalition led to all kinds of experiments with conservation and renewable energy concepts that inspired entrepreneurs to get involved in this field. Confronted with an economic recession, the established coalition looked with favor on this novel industrial activity that might create new jobs. This led to an acceptance of the decentralized, private development path of wind energy. As long as private development was still moderate, the economic growth argument was seen as more important than the absolute preservation of centralized control over the electricity system by the utilities.

Public support for the private development path and the success of the Danish wind turbine industry on the California market created a very favorable political context for wind energy, which briefly made it possible for private investors to set up wind farms. The established coalition perceived this 'capitalist' path as a real danger to the central control of the electricity system, and, with support from the cooperative coalition, the speculative development was rapidly killed. The established coalition forced the utilities to get involved in setting up wind projects in order to both promote the wind industry and maintain centralization. Besides being helpful to the environment, the increased utilization of cleaner energy sources also offered excellent opportunities to the industry. The integration of energy, environmental and industry policy was completed within *Energy 2000*. The plan confirmed Denmark's ambition to play a key role in guiding the world toward ecological modernization. Hence it can be concluded that a comprehensive learning process between the established and the cooperative coalition has taken place, which only leaves the question whether economic growth and the resolution of the ecological quandary can be reconciled a matter in dispute. However, the controversy between centralization and decentralization has been carefully avoided by both coalitions. Thus, the answer to the provocative question that was raised at the beginning of this section, about whether policy-oriented learning across the two coalitions has led to a final synthesis between the polar rational choice concepts of centralization and decentralization, is a clear-cut 'no'.

5.3 Comparative Conclusions

It is time to return to the central research question of this Chapter and see what we have learned.

> What were the politically significant ideological coalitions and their associated policy beliefs within the political debate related to wind energy innovation in California and Denmark, and how did the interaction between the various ideological coalitions change their policy beliefs over the last two decades?

Four, by now familiar, coalitions were identified within the energy debate in California, and two in Denmark. In addition, the change in policy beliefs that took place over the past two decades was described in detail. Here, we will focus on the way policy change is realized in interaction between the various ideological coalitions.

5.3.1 Competitive and Cooperative Forms of Policy Learning

> The conflict in nature between different kinds of organisms has been popularly expressed in phrases like the struggle for existence and the survival of the fittest. Yet few people realized that mutual cooperation between organisms – symbiosis – is just as important, and that the fittest may be the one that helps another to survive.
> *William Trager, 1970: vii*

First, each ideological coalition was found to be engaged in long-lasting cooperative relationships with one or two other coalitions. This implies that the policy making process should not only be seen as the product of competition between coalitions. To comprehend policy change over long periods of time, that is, a decade or more, a more sophisticated image of coalitional behavior is needed. Often, long-standing partnerships between coalitions exist that direct the policy process. It can be concluded that: *Policy changes over periods of a decade or more is best understood as the product of both competition as well as cooperation between several ideological coalitions.*

Competition and cooperation need not to be mutually exclusive. By definition, ideological coalitions differ from each other with regard to their preferred social order. This ideological conflict does not exclude cooperation. Coalitions may enter into an enduring relationship with other coalitions to look for agreement on first order beliefs that fit the basic beliefs of both. These relatively stable alliances, so-called learning alliances, are based on mutual confidence and compound interest, and lay the foundation for comprehensive, long-term policy-oriented learning across various coalitions. As indicated and confirmed within the analysis, such learning is usually confined to first order elements of the belief system. This is in accordance with the view that second order beliefs are quite resistant to change, whereas beliefs at the first order levels are given up more easy.

Our study demonstrates that coalition learning should not solely be conceived as a strategy to 'outlearn' oppositional coalitions in order to seek domination or to better one's bargaining position. In contrast to these *competitive* forms of learning, interactive *cooperative* forms of policy learning within alliances were often found that aimed at incorporating different points of view within a public policy program. It follows that learning can serve multiple purposes. *Coalition learning can be aimed at domination, compromise, and integration.*

5.3.2 Denmark: Ecological Modernization

The analyses of the energy debates in California and Denmark show remarkable similarities as well as striking differences (cf. Figures 5-1 and 5-4). Both debates more or less opened up at the same time. In the 1960s, a broad political consensus existed in both states concerning the appropriateness of the grow-and-build strategy. Prior to the oil crises and the advent of the environmental movement, both Danish and California energy policy aimed at minimizing energy prices and maximizing the availability of energy supply. The grow-and-build doctrine was accomplished at the expense of major environmental pollution and human health costs, and dependency on foreign oil. The hidden cost of becoming dependent on foreign oil was made very clear through the oil crises. In combination with the strong antinuclear movement in both Denmark and California, these events crystallized a consensus about the need to develop domestic energy resources (in particular, oil, natural gas, biomass, and wind energy) and the imperative to stimulate energy conservation.

In Denmark, a long-term, comprehensive learning process started around the story line of ecological modernization. During this process the established and cooperative coalition downplayed their ideological motives, and reached first-order level consensus at the end of the 1980s. At that time, the established coalition came to promote a sustainable grow-and-build strategy that had been advocated for a long time by the cooperative coalition. The alliance between the Danish established and cooperative coalition exemplifies a long-term interactive and integrative learning process. In the end, the ideas of the cooperative coalition on energy policy, and its wish to minimize the environmental impact of the electricity sector, were no longer perceived as being at odds with economic growth, but seen as a prudent policy to maintain economic growth (that in its turn sustains the social-liberal welfare state). *The two Danish coalitions can be said to have formed a learning alliance that was organized around the story line of ecological modernization.*

5.3.3 California: Entrepreneurial Modernization

In California, a similar type of learning process with a similar outcome took place. Here, the interaction between the public interest and corporate capitalist coalition eventually led to a sustainable grow-and-build alliance. This tried to change the grow-and-build strategy into a sustainable grow-and-build strategy, while leaving the regulated monopoly structure of the electric industry untouched. *The corporate capitalist and public interest coalition in California can be said to have formed a learning alliance that, like the Danish alliance, was organized around the story line of ecological modernization.*

However, beside this comprehensive learning process, a parallel interactive process took place in California that was more or less absent in Denmark.[25] *That learning process was organized around the story line of entrepreneurial modernization.* While the public interest coalition pushed the notion of ecological modernization, the libertarian coalition represented the main driving force within the learning alliance that was organized around the concept of entrepreneurial modernization. In its long effort to introduce competition into the electricity sector, the libertarian coalition looked for and found support from different ideological sides.

In the 1970s, the public interest and communitarian coalition joined the learning alliance, which then had the guise of the anti-utility-involvement alliance.[26] That alliance tried to improve the economics of the electric sector and decrease its environmental impact by introducing competition in electric generation. After the communitarian coalition dropped out of the picture in the early 1980s, the learning alliance gradually changed into a regulated competition alliance. At the beginning of the 1990s, the two parallel learning processes had come far in integrating economic, environmental and social aspects within the field of energy policy. Still, at that time, the libertarian rhetoric that assumed the superiority of private sector solutions to public solutions became dominant within the energy debate.

A key event presents the agreement that was reached in 1995 between the independent power industry and the investor-owned utilities on how to restructure the electric industry. The compromise was set down in the so-called Memorandum of Understanding (MOU), which paved the way for establishing a competitive electric industry in California. By signing the MOU, the corporate capitalists abandoned the learning alliance with the public interest coalition and reconciled themselves to the story line of entrepreneurial modernization. As a result, the public interest coalition was sidelined, and the story line of entrepreneurial modernization became dominant. As a result of this 'radical' shift, customer and environmental interests got a clearly weaker position within the California energy debate. This looked like turning back the clock, as it negated, and to a large extent undid, the first-order learning processes that had taken place across ideological coalitions within both the sustainable grow-and-build alliance and the regulated competition alliance.

6 Policy Tradition and Learning

> Political culture is transmitted from generation to generation, but it is not transmitted unchanged, nor is it transmitted without question or by chance. ... To stay as we were requires vast energy. There is no reason why political culture, conceived as ways of life that are continually being negotiated, tested, and probed by individuals, cannot make sense of political change, long considered the Achilles heel of cultural theories.
> *Michael Thompson, Richard Ellis, and Aaron Wildavsky, 1990: 218-219*

In the former Chapter, we described the process through which ideological coalitions in California and Denmark, in interaction, tried to grasp the constantly changing and ambiguous energy field in order to further their policy objectives. By identifying coalitions and their changing beliefs, and by detecting long-term learning alliances and the story lines that bound them, the development of the energy debate since the early 1970s was analyzed in some detail. One might argue that, from the perspective of each specific country, everything has now been said and done regarding the energy policy field. From a comparative policy perspective, however, some important and interesting questions are still unanswered.

It was found that the behavior of California coalitions was far more diverse and complex than that of the Danish ones. In California, therefore, policy change became less predictable and harder to analyze. Theodore White was correct in claiming that "California politics squirm with a complexity and intrigue that defy reasonable analysis."[1] Obviously, the higher degree of complexity in California politics is related to the number of relevant ideological coalitions which offers a wide selection of potential partners. In California, two learning alliances came to exist at the same time. One was organized around the story line of ecological modernization, the other around entrepreneurial modernization. In Denmark, only one learning alliance was found. In this Chapter, we will look for an historical explanation of this, and examine why certain coalitions enter into long-term alliances with certain others, while treating others with contempt.

In Chapter 4, it was acknowledged that countries constitute of a mix of political cultures.[2] It was also argued that the deep core values of the various ideological coalitions are connected to distinct historically defined political traditions. In the next two sections, a brief description of the political history of America and Denmark is given both to support the notion of a national mix of political cultures as well as to place the various ideological coalitions within a specific political tradition.

It was also noticed that the particular national mix of traditions limits and molds coalitional behavior within a particular policy subsystem by determining the universe of legitimate policy opportunities and constraints. One would expect, therefore, that the national political culture would both include and exclude certain learning alliances. In section 6.3, we will analyze the way in which the mix of cultures has evolved in both countries over the period under study. In the final section, we will examine how the national mix of political traditions enabled and constrained policy-oriented learning processes within the energy policy subsystem of California and Denmark.

6.1 American Political Traditions

> An American attends to his private concerns as if he were alone in the world, and the next minute he gives himself to the common welfare as if he had forgotten them.
> *Alexis de Toqueville, 1840*

> American political thought ... is a veritable maze of polar contradictions, winding in and out of each other hopelessly: pragmatism and absolutism, historicism and rationalism, optimism and pessimism, materialism and idealism, individualism and conformism. But after all, the human mind works by polar contradictions; and when we have evolved an interpretation of it which leads cleanly in a single direction, we may be sure we have missed a lot. *Louis Hartz, 1955*

> In the American republic conservatism and reform, capitalism and democracy, private interest and public purpose, join to define the political tradition.
> *Arthur M. Schlesinger, Jr., 1986*

In his book, *Americans and the California Dream*, Kevin Starr (1973: 46) describes the prophetic patterns that emerged from travel literature before California won statehood in 1850. "Certainly an ideal California – a California of the mind – underwent composite definition: the elusive possibility of a new American alternative; the belief, the suggestion (or perhaps only hope), that here on the Pacific shore Americans might search out for themselves new values and ways of living. In this sense – as a concept and as an imaginative goal – California showed the beginnings of becoming the cutting edge of the American dream." In a similar way, America was not only 'discovered,' but also 'invented.' (Lemaire 1986) At the end of the Middle Ages, the image of a country in the West that could serve as a solution for the problems that swept Europe turned up everywhere. Territorial expansion of Europe was seen as a prerequisite for solving the crisis of feudalism.[3] Speculations on a 'New World' enticed an era of voyages of discovery. In his attempt to find Asia, Christopher Columbus in 1492 ran aground on a sandbar that proved to be only a small part of the great landmasses that stretched from the Arctic to the Antarctic. Perversely, both continents that were later labeled *America* blocked the way to Asia's riches.

6.1.1 The Yeoman Farmer

> The small-land holders are the most precious part of the state. *Thomas Jefferson (1743-1826)*

The American wilderness held out a variety of attractions to prospective settlers. For some, the new land offered the hope of material riches – gold, fish, furs. For others, it represented a religious haven in an uncorrupted society.[4] As European presence became more firmly established in North America, new colonists arrived, whose interests lay in the more mundane pursuits of farming and trade. The English, Dutch, Swedes, Germans, and French who populated the new continent did not stay the people they were in Europe. They became American. "The American is a new man, who acts according to new principles." (Schulte Nordholt 1963: 9) Ownership of fifty or a hundred acres of land dramatically changed settlers' view of themselves. In this nonfeudal world, the class-bound Europeans swiftly shed their humble demeanor, and started to probe what it meant to be an American. In defining their identity, the natural-right's philosophy of John Locke (1632-1704) formed a significant source of inspiration (cf. Hartz 1955).

The English philosopher's basic social norm was the concept of free individuals in a state of nature. In the original state of nature all men were equal, but this ideal situation had been distorted by feudalism. For Locke, America symbolized this original abundant state of nature. This explains his words "In the beginning all the world was America." The idea of social liberty had a revolutionary tenor in feudal Europe. In contrast, in the United States the reality of atomistic social freedom became the primary assumption of political thought (*ibid.* 62). The peculiar meaning that American life in the 18th century gave to the Locke's words might be designated the *agrarian myth* (Hofstadter 1955: ch. 1) Its hero was the yeoman farmer, its central conception the notion that he was the ideal man and the ideal citizen. Its argument was that the land is the common stock of society to which every man has a fundamental right to labor the earth. In Europe, this idea was loaded with dynamite. In the United States with its apparently endless frontier, however, it became a standard argument.

Thomas Jefferson (1743-1826) made the agrarian myth the basis of a strategy of continental development.[5] He anticipated "that the great empty inland regions would guarantee the preponderance of the yeoman – and therefore the dominance of Jeffersonianism and the health of the state – for an unlimited future". (*ibid.* 29) In this way, "an empire of small farms"[6] would be established. The conception of the independent farmer as an ideal free man within a veritable state of nature had wider implications for the legitimacy of political power.[7] The only reason for the farmer to give up his freedom, and submit himself to government rule was to protect his own property. This implied two things. First, the government rules in agreement with the people and for the common welfare of the people. Locke called this the social contract. Second, because the American democrats feared an absolutist national government that would severely undermine the noble role of the farmer in society, they concluded that the power of the state must be limited.

The agrarian myth had virtually universal appeal among the intellectual classes in the

18th century, and logically played an important role in the revolt against the authority of England. On the fourth of July of 1776, America threw off the yoke of the centralized monarchy with the Declaration of Independence. Thomas Jefferson wrote it, clearly inspired by the political philosophy of John Locke (cf. Achterhuis 1988). Also during the first party battles under the Constitution, the Jeffersonians appealed again and again to the moral primacy of the yeoman farmer in their attacks on the Federalists. As a result, the Constitution apportioned political power between the national government and the states in such a manner as "to minimize the ability of both levels of government to engage in policy adventures that would intrude upon private property rights and to make it difficult for popular majorities to mobilize on the basis of shared political interests." (Robertson and Judd 1989: 16) The American Constitution established a divided form of government, the presidency, two houses of Congress, and a federal high court, and reflected a deliberate decision by the country's founders to create a weak and internally conflicted political system by means of checks and balances. The second and present one, ratified in 1789, divided the government into many divisions, each one selected differently for varying terms of office.

It can be concluded, that during the colonial period, American identity was closely related to the hero within the agrarian myth: the self-sufficient farmer, who left to his children a strong penchant for craftsmanlike improvisation and a tradition of a household industry. According to Louis Hartz (1955: 58), the almost universal appeal of the American version of the Lockean theory forms the basis of America's moral absolutism: "the sober faith that its norms are self-evident." Because American moral absolutism arose out of contact with an opposing way of life, "it was so sure of itself that it hardly needed to become articulated, so secure that it could actually support a pragmatism which seemed on the surface to belie it." (*ibid.* 58-59)

6.1.2 The Small Entrepreneur

From the beginning of the 19th century until the Civil War (1861-1865) commercial farming gradually got the upper hand. The cash crop converted the yeoman into a small entrepreneur. During the colonial period, inadequate transportation and a very limited domestic market had held the commercial potentialities of agriculture in check. But in the first half of the 19th century the strong growth of the home market, and an emerging network of canals and railroads transformed the character of American agriculture. It came to consist predominantly of isolated farmsteads in the midst of great acreage. The frequent movements and the absence of village life lowered the chances of association and cooperation. Organic community values were weakened, and the individualistic and anti-statist ones were enormously strengthened. Commercial farming fostered a new social outlook that made the agrarian myth a fiction, an anachronistic idea.

The Americans adapted their identity to this change in business style. As they embraced liberalism, the self-made, acquisitive entrepreneur became their new hero. After the Civil War, the laissez-faire doctrine became hegemonic. This again intensified the mood of America's moral absolutism. A successful attempt was made to link this classical form of liberalism (normally denoted in this book as *libertarianism*) with the term *Americanism* (*ibid.* ch. 8), suggesting a completion of the American identity, and the end of ideology. Being an American had become an ideological commitment by itself (Lipset 1996: 31). The ethos of survival of the fittest also became a characteristic part of the California temperament.

6.1.3 The Plutocracy of the Robber Barons

At the time the United States captured the territory of California from Mexico, the newborn state was a sparsely settled land. This changed dramatically after 1848, when gold was first discovered in the Sierra Nevada. California became charged with human hope. Drawn by the frontier opportunities and driven by the cult of individual success, thousands and thousands poured into California. In short, California was born out of an enormous materialism. The *Gold Rush* made the state "a land of adventuring strangers, a land characterized by an essential selfishness and an underlying instability, a fixation upon the quick acquisition of wealth, an impatience with the more subtle premises of human happiness." (Starr 1973: 65-66) The Gold Rush intensified these American liberal traits and consolidated them as part of the regional experience.

After completion of the transcontinental railroad in 1869, waves of settlers also moved West to establish a small farm, big enough to ensure a good life. In fact, the dream of owning land came true for few Californians. Instead of the 'empire of small farms' envisioned by Jefferson, a plantation-like pattern developed.[8] The railroads also brought and perfected the corporate form of business organization, and pulled the Golden State into the international market.

The industrial revolution that took place in America during the last three decades of the 19th century was both cause and effect of the arrival of corporate capitalism. The captain of industry fulfilled the old dream of personal ascendancy and became the hero of this form of capitalism. California's railroad magnates, Leland Stanford, Collis P. Huntington, Mark Hopkins, and Charles Crocker, are all examples of this new type of hero. Each of these Big Four – as the founders of the Southern Pacific Railroad were known – became major economic powers. By means of bribery and extortion, they established a railroad monopoly, became the largest landholder and employer in the state, and gained political control over the Democratic and Republican parties. Their greed and corrupt politics became the primary target of Kearneyism, California's version of Populism, and Progressivism (cf. Bell and Price 1980: ch. 4).

By the end of the 1870s, the economic recession brought the Promised Land to the brink of social upheaval. Led by Denis Kearney, the Workingman's Party railed against the problems of the day. Kearneyism was a kind of racist socialism[9] that came to stand for "establishing state regulation of railroads, utilities, and banks; a fair tax system; an eight-hour workday; compulsory education, and the direct election of United States Senators." (*ibid.* 54)

The Populist Party that emerged toward the end of the 19th century more or less advocated the same kind of reforms. Besides socialists, this minor third party consisted of farmers, who saw the best acreage in the state turned over to the railroads. Heavily indebted farmers, who had become strongly dependent upon the export market, lashed out against banks that foreclosed farms and the railroads that charged exorbitant freight rates. The fact that the agrarian myth was overrun by commercial realities was one of the original sources of Populist grievances. While neither third-party movement gained sufficient strength to implement its programs fully, each did help pave the way for later successful reform. Third parties have often played an important role in American politics. They supply the dynamic element in the political life of the United States. Hofstadter (1955: 97) maintains "When third party's demands become popular enough, they are appropriated by one or both of the major parties and the third party disappears. Third parties are like bees: once they have stung, they die."

In California, a small group of liberal Republicans continued along the lines of Populism and Kearneyism.[10] The Progressive movement was not a local phenomenon, but a nation-wide revolt against monopolies and special privileges in both economic and political spheres, and against limits upon the avenues of personal advancement. At the turn of the century, the reformers, who were affectionately familiar with the world of individual enterprise, saw an extraordinary acceleration in the trustification of American industry. Major local mergers simultaneously took place in the fields of the telephone, gas, and electric power and light. The California Progressives criticized this modern, bureaucratic style of capitalism. The well-fixed Progressive leaders were opposed to "the impersonal, concentrated, and supposedly privileged property represented by the behemoth corporation. Looking backward to an older America [they] sought to recapture and reaffirm the older individualistic values in all strata of political, economic, and social life." (Mowry 1951: 89)

Hofstadter (1955: 8-9) maintains that the struggles of the Progressive era in many ways were influenced by a conflict between two thoroughly different systems of political ethics. One was founded upon the indigenous Yankee-Protestant political traditions, which were described above. "The other system, founded upon the European backgrounds of the immigrants, upon their unfamiliarity with independent political action, their familiarity with hierarchy and authority, and upon the urgent needs that so often grew out of their migration, took for granted that the political life of the individual would arise out of family needs, interpreted political and civic relations chiefly in terms of personal obligations, and placed strong personal loyalties above allegiance to abstract codes of law or morals." (*ibid.* 9) It was chiefly upon this system of values that the political life of the immigrant, the boss (a dealer in public privileges and support), and the urban machine was based. The machine of San Francisco's most notorious political boss, Abe Ruef, captured city hall in 1901.

Among others, the corporate executives of PG&E used Ruef's assistance to expand their networks of power. PG&E had been founded by John Martin and Eugene J. de Sabla, Jr., who had a common interest in the gold mines of the Yuba River region to the north of Sacramento (see Hughes 1983: ch. 10). In search of an economical power source for their mines, they turned to hydroelectric power. With the success of their first six-megawatt Nevada City plant, they became confident there was a market for more power in nearby Sacramento, and even in San Francisco. The nascent coastal industry relied on very expensive coal from Australia as its source of energy. The San Francisco Bay area, with its immense need for new facilities offered a magnificent market for the electricity business. Privileged by Ruef, in 1901, Martin and de Sabla decided to construct the world's longest transmission line – a distance of 140 miles – in order to reach the rapidly developing Oakland region.

After completion, PG&E entered a new phase of development. Through mergers and the acquisition of distribution systems, the Power Company changed into a company presiding over an integrated regional system. This transition was completed before the First World War. The history of PG&E is in many ways representative of the history of other utility companies operating the major California transmission lines and connected generation and distribution systems. Like PG&E, these utilities resulted from mergers of power companies and urban utilities.

Abe Ruef was eventually convicted of bribery (Bell and Price: 55-58). As a prosecutor, Hiram Johnson had helped secure Ruef's conviction. Building on his Ruef trial publicity, the Progressive leader Johnson won the general election in 1910. Besides, Progressive candidates won control over both the State Assembly and the State Senate. This gave them the power "to enact the most comprehensive and far-reaching political reforms in California history." (*ibid.* 57)

In short, after the Civil War, the dominance of the laissez-faire ideology made way for the advent of modern capitalism, and, in its wake, the industrialization of America. The trend toward management, toward bureaucracy, and toward bigness led to superior systems for the production of goods and services on a massive scale, but clashed with the old notions of individual economic success. Hofstadter (1955: 11) explains that during the Progressive era, "All of society was felt to be threatened, not by economic breakdown, but by moral and social degeneration." The rapid and sometimes stormy transition from the conditions of an agrarian society to those of modern urban life made the American society find itself in an identity crisis. Americanism had lost its universality. To make up the loss, the Progressives started a reform movement "against monopolies and special privileges in both the economic and political spheres, against social distinctions and the restriction of credit, against limits upon the avenues of personal advancement." (*ibid.* 10) All over America, Progressives sought to secure passage of legislation aimed at weakening political parties, reducing the power of special interest, rooting out corruption, and expanding the citizen's role in the political process.

6.1.4 The Consumer

The Progressive reforms in California were successful and sharply reduced bossism and corruption in the state. California became known for its honest government. Among the many reforms, a new railroad Commission – the later CPUC – was created by State Constitution to protect the consumer against unreasonable rates and see that the railroads and private utilities provided adequate facilities and services. A real effort to restore, maintain and regulate competition rather than regulate electricity monopoly seems to have been more congenial to the ideology of most of the Progressives. There were at least four reasons for this.

First of all, the use of hydroelectric power in combination with long-distance power transmission had given the electricity system a very centralized character in a technical sense. Second, and related to the first, it was recognized that in some circumstances a business grew big through superior efficiency instead of by circumventing competition. The acceptance that big business sometimes would be technologically more progressive than small business indicated that corporate capitalism itself had become an ideology. Embracing technology as the source of progress, corporate capitalists promoted that through the proper application of corporate power a new kind of society could be built.[11] Another reason why corporations within the electricity sector were not dissolved was related to their political power. In fact, the industry itself sought and achieved regulation as a means of advancing its own economic self-interest. A fourth reason is related to the fact that in the Progressive era, the urban consumer first stepped forward as a serious and self-conscious factor in American social politics. "Consumer consciousness became ... a focus for the common interest of all classes that had to concern themselves over family budgets, it cut across occupational or class lines, and did a great deal to dissolve the 19th century habit of viewing political issues solely from the standpoint of the producer." (Hofstadter 1955: 173) Although afraid that the trusts would squeeze him, and undermine democracy, the common man was not really interested in the organization of capitalism. The public was worried about its own disorganization, and more and more sought in governmental action a counterpoise to the power of private business.

A silent revolution in public opinion with respect to government responsibility took place during the Great Depression. After the crash on Black Thursday (October 29, 1929), the statist sentiment experienced both an upswing as well as a qualitative change. "The powerful labor movement, coupled with the interests of the unemployed, gave the later New Deal a social-democratic tinge that had never before been present in American reform movements." (*ibid.* 308) It was demanded that the government should not only prevent exploitation by virtue of superior power, but that it should take responsibility for unemployment insurance, social security, wages and hours, and housing. Franklin D. Roosevelt's New Deal changed the national government from a passive protector of property rights and capitalism to the senior partner, in cooperation with business and powerful interest groups, in maintaining the nation's economic health.

Two years after Roosevelt's landslide presidential victory, the long-time socialist Upton Sinclair[12] ran for governor. Sinclair had switched to the Democratic Party and ran for its gubernatorial nomination on the platform *End Poverty in California* (EPIC). He won the Democratic primary, but lost the general election. Four years later in 1938, an EPIC State senator from Los Angeles became California's first Democratic governor in the 20th century. Culbert L. Olson tried to set up California's own New Deal. His attempt, however, was obstructed by an economy bloc within the assembly formed by Republicans and conservative Democrats. (Harvey 1985: 19) They found in the New Deal an outrageous departure from what they saw as 'the' American way of life. Nevertheless, after the Second World War, the conservative politics of private interest gradually gave way to the liberal politics of the collective good that was advocated by the reformative wing in the Democratic Party.

6.1.5 The American Post-war Settlement

To definitively embed the New Deal in the American political order, President Harry Truman (1945-1953) initiated the Fair Deal. During the same period, the nonpartisan Earl Warren was governor in California. He supported policies he called 'social progress.' He advanced higher welfare and unemployment benefits, and compulsory health insurance (*ibid.* 20). In the quiet Eisenhower years (1953-1961), also on the national level a kind of consensus came to exist. This could be derived from the way Eisenhower described himself as 'dynamic conservative' and 'moderate progressive.' "In this way, the New Deal in the hands of historians like Arthur M. Schlesinger, Jr., and Richard Hofstadter became the standard, the golden mean between collectivism and unrestrained capitalism." (Van Rossem 1991: 177)

A new rush of commitment followed in the 1960s, when John F. Kennedy (1961-1963) launched the New Frontier, and Lyndon B. Johnson (1963-1969) initiated the Great Society, with the aim to complete and extend the Fair Deal. In California, the Democratic governor Edmund (Pat) G. Brown (1959-1967) expanded the role of the state government. He initiated a state water project, highway development, increased funding of education, and more stringent consumer protection laws. The election of Johnson (and Brown) brought the post-war consensus to its climax. In that period the liberals were the real rulers, and one could speak of a *liberal establishment* (*ibid.* 223-225).

6.1.6 The Ideological Roots of the Californian Coalitions

The above exercise supports a plural conception of the American political culture. It was shown that classical liberalism eclipsed Jeffersonianism as the 19th century wore on. In the 20th century, libertarianism itself was surpassed by consumerism and corporate capitalism. But it never disappeared completely. It was precisely the mix of these political cultures or ideological traditions that comprised American political culture. The above description makes it a straightforward task to clarify the historically developed political undercurrents, which the ideological coalitions that played an active role within the California energy debate related to (see Figure 6-1).

Jeffersonianism and its ideal of self-sufficiency lives on to inspire the communitarian or grassroots coalition.[13] Jeffersonian theory relied on the civic virtue of the enlightened liberal yeoman and made land the indispensable base of liberal democracy. In a similar way, the grassroots coalition put its trust in people's own responsibility, and made decentralized production the essence of a quiet movement toward community self-reliance and grassroots democracy.[14]

The rising entrepreneur of classical economics was and still is the hero of the libertarian coalition. Their belief system is rooted in 19th century liberalism with its ideal of opportunity and the notion that success is a reward for energy, efficiency, perseverance, ambition, and insight. The independent small businessman represents its hero. Anti-statism represents an important dimension of classical laissez-faire liberalism or libertarianism. As with Jeffersonianism, this ideological tradition strongly believes that less government is better.

It goes without saying that the corporate capitalist coalition and corporate capitalism form a unity. The successful corporate businessman or urban professional is its hero. Within this tradition, a lot more government is favoured. Especially when it favors corporations.

But the coalition most in favor of government interference is the public interest coalition. This coalition finds its ideological roots within the consumer consciousness that arose, in the first decades of this century, out of fear of political and economic power of great corporations. The support for consumerism received an upswing during the depression, and peaked in the first half of the 1960s.

	American anti-statism		**American statism**	
Period	Colonial period	19th century	Around the turn of the 20th century	Around the turn of the 20th century
Social group	Self-sufficient farmer	Commercial farmer, small business man	Corporate professional	Consumer, man-in-the-street
Political tradition	Jeffersonianism	Jacksonianism / Classical liberalism	Corporate capitalism	Consumerism
Ideological coalition	Communitarian coalition	Libertarian coalition	Corporate capitalist coalition	Public interest coalition

Figure 6-1
The ideological sources of the four ideological coalitions engaged in the energy debate in California.

6.2 Danish Political Traditions

> The Danish way of life harbors a sense of respect for the individual. The Danish democratic social system is based on liberty, equality, care and responsibility. The Danes are regarded as informal, unpretentious, ironic and anti-authoritarian. They are individualists, but united by a multiplicity of fellowships. They are regarded as both conformists and as 'anarchists.' *Per Himmelstrup, 1992: 5*

In contrast to America, Danish society in the 18th century was still feudally organized, with the large estate as the economic and administrative unit. For centuries, a form of serfdom had existed in Denmark that not only bound peasant farmers to the feudal estate of their birth, but also forced them to work the particular farms and holdings to which they had been assigned. The peasant class was subjected to a number of duties, and if a tenant farmer was not able to fulfil the required donations of rent, produce and labor, the landed farmer could loose his tenancy and become a landless farmer.

Around 1770, the European Enlightenment had reached Denmark, and made the age conducive to reform. Jean-Jacques Rousseau's romantic conception of nature heightened the appreciation of the Danish elite for the peasant life around them. More important, Adam Smith's *The Wealth of Nations* influenced many with its view of the economic and social importance of free enterprise (Borish 1991: 145). In this period, the *Great Agrarian Reforms* set free the landed tenant farmers (*bønder*), and fundamentally reorganized agricultural land use.[15]

Throughout Europe, the end of the 18th century was colored by the French Revolution and the ensuing Napoleonic Wars (1796-1815). Initially, the Danes profited from the war, and experienced an economic boost that enabled the rapid introduction of land reforms. By 1807, already more than 60 percent of Danish *bønder* lived on self-owned farms (*ibid.* 151). Nevertheless, Denmark also became caught up in the deadly enmity between France and England. In August 1807, an English fleet bombarded Copenhagen for three days until the Danes surrendered, and handed over their navy to the English. Denmark entered the continental alliance that denied the English access to European harbors. The consequence was fatal to Denmark. A harsh recession led to national bankruptcy in 1813. Even worse, a year later, when Napoleon was finally defeated, the Danes were too. The fact that Denmark had to give up Norway to the Swedish king was a final staggering blow to Danish pride. All this not only halted the reforms. It even reversed them for several decades. Some observers described the Danes as "a lazy, sluggish and disconnected people," and attributed this to the shocking lack of political freedom (*ibid.* 156).

6.2.1 The Polity of the Estates

Although the 19th century started out dismal and lean, by the 1830s, Denmark awakened to a cultural revolution in the arts, philosophy, natural sciences,[16] and literature. Danish cultural life was enriched with two persons who still are world famous today: the philosopher Søren Kierkegaard (1813-1855) and the fairy-tale writer Hans Christian Andersen (1805-1875). But the personality who probably exerted the most marked influence on Danish society during that period is little known internationally. He was the clergyman, historian, educator, writer and poet Nikolaj Frederik Severin Grundtvig (1783-1872).

In the 1830s, the country also took its first steps toward democracy. In 1831, the King introduced Provincial assemblies of various social groups or 'estates' of the kingdom. Representatives of the nobility, clergy, bourgeoisie, and peasantry were to assemble every other year to discuss public affairs and to present their advice to the King. Grundtvig feared that this would increase the power of the estate owners at the independent farmers' expense. To ensure the equality of the peasant population with other groups, he suggested the idea of a new kind of folk high school. Three study trips to England inspired Grundtvig, and convinced him of the enormous importance of dialogue.

The advisory assemblies gave rise to the formation of political parties. The independent farmers and the urban upper middle class joined forces to form a United Liberal Party (*Forenede Venstre*). This party wanted more than the polity of estates, and organized a mass demonstration on March 21, 1848, to Christiansborg Palace to demand a constitutional government. King Christian VIII acceded peacefully to their demands, and abolished absolutist rule. Grundtvig held that the introduction of a Constitution (*Grundlov*), in 1849, made the need for a folk high school doubly urgent. He thought that it was "vitally important that every Dane should know the meaning of the fellowship of the Danish people, i.e., the values uniting all Danes in some fundamental way behind class differences and special interests." (Slumstrup 1984: 4)

The key words of Grundtvig's political program were *freedom* and *folkelighed*. Inspired by British liberalism, he advocated almost unlimited freedom in every intellectual and spiritual field, as well as in economic matters. He stressed that the new democracy should be in fullest agreement with the people themselves, and the culture and outlook that characterized them.[17] Grundtvig was afraid lay people would have scanty power unless they were given access to such a degree of education on social affairs that the people's voice might speak freely and strongly. However, things took another turn.

The independent farmers had supported the bourgeoisie in their struggle for a constitutional monarchy, but this alliance disintegrated almost as quickly as it had been formed (Borish 1991: 179). The urban upper-middle class joined the large estate owners in the National Liberal party. In 1864, under the rule of this party, Denmark was drawn into war with Bismarck's Prussia. As a result, Slesvig-Holsten was lost to Prussia. This further erosion of its domain opened the question of the very survival of Denmark as a nation (Bendure and Friary 1996: 21). But, "Along with the inferiority complex this engendered, the will to 'win inwardly, that which outwardly is lost' was born." (Himmelstrup 1992: 5)

6.2.2 The Struggle for Parliamentarism

The loss of Slesvig-Holsten was of great symbolic importance for the infant folk high school movement. The period 1864-1876 saw an explosion of new schools over the entire Danish countryside. The folk high school leaders taught their pupils independence, devotion to the common good, and stimulated the self-confidence of the rural population. As such, "the new type of school became the effective social instrument of a single, relatively homogeneous group of the rural elite, a class of self-owning farmers whose existence was made possible by the Land Reforms of the late 18th century." (Borish 1991: 202)

After the fiasco of 1864, the right-wing alliance (*Højre*) between bourgeoisie and nobility attempted to drastically limit the Constitution. The Constitution of 1866 provided for a bicameral *Rigsdag*, with the *Folketing*, elected by 'independent' males over the age of thirty, and the upper house, the *Landsting*, chosen mainly by citizens in the upper-income groups but with some members appointed by the King. The Constitution authorized the King to appoint and dismiss his ministers without reference to party strengths in the *Rigsdag* (Miller 1996: 1). As a consequence, *Højre* came to control the *Landsting*, and the Government, until the end of the century. Although political reforms came to a standstill, the conservatives guided a number of economic advancements. The railroad was extended throughout the country, and Denmark's major industries – shipbuilding, brewing and sugar refining – came to maturity (Bendure and Friary 1996: 21).

The liberal *Venstre* was frustrated by this conservative monopoly. In their quest for reform and governmental power, the independent farmers soon found allies. Since, the industrial revolution had come to Denmark, many landless farmers moved to the towns to offer their labor to the expanding industry. In this way an urban working class came into existence, which formed the basis of the Socialist Democratic Party that was founded in Copenhagen in 1871. The socialists joined the liberals in their struggle for Parliamentarism. From 1872 on, the reformers had a majority in the *Folketing*. When the elections of 1901 almost eliminated the *Højre* party from the Lower House, the King gave up his political backing of the landed interest and appointed the first Liberal Government. This so-called system shift brought about the democratic change the opposition parties had been fighting for, namely "the principle of cabinet responsibility to the majority in the *Folketing*." (*ibid.* 2)

6.2.3 The Agricultural Cooperative Movement

To a large extent, the political success of the farmers resulted from a philosophy and practice of cooperation that had a long tradition. In Denmark, land was limited and costly, and small farmers lived in villages that provided a community and a cooperative environment. In the feudal era, it was customary that the landed farmers of a village met in a council to discuss common affairs. The fellowship that surrounded the old peasant village faded away after the Land Reforms, to be gradually replaced in the course of the 19th century by new forms of fellowship in communal work, and later, in the Danish agricultural cooperative movement (Borish 1991: 144-145).

The latter had started in the 1870s, after railroad and steamship transport had enabled American and Russian farmers to export enormous masses of grain to Western Europe. Grain prices went down drastically. Decades of organizational experience combined with cultural awareness provided the means to avoid a catastrophe. The farmers turned toward animal husbandry, and chose to improve their products with the aid of new techniques. They also decided to collaborate in cooperative societies. All kinds of small and primitive cooperative societies were established, such as dairies, retail societies, and bacon factories, which gave the farmers all the benefits of large-scale production without affecting their independence.

Also, in the emerging field of electricity, all kinds of local cooperatives were set up to establish rural power stations. The folk high school teacher Poul la Cour, nicknamed the 'Danish Edison,' played a special role. With a team of scientists, he started, in 1891, to build a test windmill at the Askov folk high school, "to establish the technological basis for electrification of the rural areas, securing the small societies there a share in the blessings of the new electrical energy" (Rasmussen and Øster 1990: 8). In 1903, they started the *Danish Wind Power Society* with the participation of dignitaries within the electricity field. The team assisted in establishing rural power stations, and when La Cour died in 1908, they had been involved in 95 new plants, 32 of which used wind power. The cooperative philosophy combined with practical constancy furthered a widespread decentralization within the Danish electricity production system.

Within the cooperatives, all members had an equal say, regardless of turnover: 'votes were for heads, not head of cattle'. This rule was distinctly democratic and partially directed against shareholder capitalism, and the ongoing political struggle between capitalists and socialists in the towns. It was thought that the right combination of the virtues of community and commerce would lead to progress toward an equitable environment for all instead of a struggle between the have and the have-nots. Thus the developments in Denmark sustained the ancient form of independent small-scale local cooperation. The success of the cooperative movement proved that the cooperative way was suited to deal with the problems of modern time. An unexpected result was that this '*folkelig*' strategy, and the philosophy it was based on, got an even wider appeal within the Danish society.

6.2.4 Intermezzo: Grundtvigianism and Jeffersonianism

From a comparative perspective, Borish (1991) gives an interesting assessment of the Danish rural society in the 1910s. The American anthropologist holds that, "By all means it was a society that had moved significantly in the direction of that earlier version of the American dream that went under the name of 'Jeffersonian democracy'" (*ibid.* 72).

Borish's evaluation suggests a strong resemblance between the ideology of Grundtvig and Jefferson: the spiritual fathers of Denmark and America. Both shared the ideal of a life lived close to nature and the soil, the esteem for the primary contacts of country and village life, and the cherished image of the independent farmer. So, while the agrarian myth became a fiction in America because it was overrun by commercial realities, it became reality in Denmark.

6.2.5 The Danish Post-war Settlement

After the system shift, the liberals and social democrats quickly abandoned their former alliance. In reaction to the new situation, the *Radikale Venstre* broke away from the folk high school democrats within *Venstre* in 1905. The party consisted of the Copenhagen cultural radical intelligentsia (pacifists and liberals who thought the older party was no longer committed to reform), and of small farmers who believed the Liberals increasingly spoke for the interest of the larger agricultural landholders. Like the Social Democrats, the Radicals proposed social insurance, unemployment compensation, improved old age pensions, public take-over of monopolies, reduction in military expenditures, constitutional and political reforms. The *Radikale Venstre* saw it as its historical mission to be the non-socialist party that could cooperate with the worker's party.

With the formation of the *Radikale Venstre,* the 'Four Great Constants' in Danish politics had all appeared on the political scene. Till now, these 'four old parties' have determined the character of politics. Since the Social Democratic party first received responsibilities of government in 1924, this party has, more than any other, left its mark on Danish society, particularly through building up the welfare society during the inter-war years and after the Second World War. During the 1960s, the principles of the welfare state gradually gained the support of other parties.

In his essay on the creation of the Danish welfare state, Henning Fonsmark (1990: 92-94) argues that the adoption of Grundtvigianism by most parties provides the main clue to questions like: what made *Venstre* and the Social Democrats abandon their original aggressive laissez-faire policy, and Marxist doctrine, respectively? In other words: How could a broad consensus about the Danish model come about? As Borish (1991: 211) puts it "Instead of continuing to act as the vehicle of a single group, Grundtvig's views were gradually absorbed into the general fabric of Danish society."

Starting in 1910, the ideas of Grundtvig came to function as an ongoing critical corrective to the socialist philosophy that formed the Social Democratic party's point of departure. This made the party clearly reformist, rather than revolutionary. In its turn, the approbation of the Grundtvigian mindset by the Social Democratic party made it easier for the staunch liberal high school democrats within *Venstre* to accept and identify themselves with the growth of state power. Still, *Venstre's* support for the social democratic welfare project only gradually grew during the first half of the 1960s (Fonsmark 1990: 206-215).

The impetus for this change came from a number of folk high school principals. They aimed to prevent a division in Danish politics into two hostile ideological blocs, into a rigid socialism and a reactionary conservatism. To avoid that, "the Danish people would be divided into two camps," (*ibid.* 209) they argued that the party should be willing to work with the Radicals and the Social Democrats, not just with the Conservatives (Miller 1996: 76-77). The political role of the Radikale Venstre was also guided by this theory of 'the cooperating democracy' (*det samarbejdended folkestyre*). Since the welfare state was given a Grundtvigian trait, and the Grundtvigian idea of 'democracy as dialogue' (*samtale*) prevented bloc politics, the Danish model could find trust among voters and parties with divergent political views.

On the one hand, it can be concluded that the basic assumption behind the Danish welfare state is related to the ideology of the cooperative movement, namely that solidarity can be combined with total individual freedom, and self-interest with good will (Fonsmark 1991: 93). On the other hand, Grundtvig's insights of the 1830s – people's enlightenment (*folkeoplysning*), equal and balanced dialogue (*vekselvirkning*), the ordinary people over the learned ones (*folket overfore de dannede*) – have slowly reached out their *folkelig* tentacles to most segments of Danish society. His ideas have saturated the political culture in Denmark with cooperation and compromise.

6.2.6 The Ideological Roots of the Danish Coalitions

In the American situation, different ideological traditions stemmed from different periods in time, and to a large extent maintained their original meaning over time. The same counts for cultural radicalism that, in combination with Grundtvigianism, has lived on to inspire the cooperative coalition (see Figure 6-2).

Something different seems to have happened with conservatism, liberalism, and socialism. In the 20th century, these political cultures have strongly mingled and transcended their original 19th century intentions. The result is the Danish model.[18]

Achieving widespread acceptance only in the second half of the 1960s, the Danish model is a recent phenomenon. It created a powerful political bloc that came to dominate the political climate. The established coalition's rational social choice arguments embody the Danish model.

	Danish Grundtvigian welfare statism			
Period	Beginning of the 20th century	Second half of the 19th century	Feudal era	Second half of the 19th century
Social group	Small farmer and urban intellectual	Commercial farmer	Large land owner, aristocrat	Landless farmer, worker
Political tradition	Cultural radicalism / Grundtvigianism	Grundtvigianism / Cooperative capitalism	Conservatism (and Grundtvigianism)	Socialism (and Grundtvigianism)
Ideological coalition	Cooperative coalition	Established coalition	Established coalition	Established coalition

Figure 6-2
The ideological sources of the two ideological coalitions engaged in the energy debate in Denmark.

6.3 Change in National Political Culture over the Last Three Decades

Although fundamental cultural values are expected to be 'relatively stable,' it is important to analyze how the national political cultures in California and Denmark have changed during the period under study, which covers the last two to three decades. This will give a good insight into the way in which the public support and political acceptability of various sets of beliefs have changed across time. Hence this effort enables us to address the question, central to this Chapter, of how the mix of political traditions enabled and constrained policy-oriented learning processes within the energy policy subsystem. In other words, it makes it possible to study the reflective dynamic, that is, to examine how changes in political culture influence the dynamics of first-order beliefs. In the following, the different political cultures or traditions and their adherents will be referred to with the same names.

6.3.1 Public Trust and the Establishment

Several scholars talk about the *post-war settlement* as the 'consensus' that grew after the Second World War both in most West European countries as well as in the United States around ideas such as social democracy, the welfare state, corporatism and Keynesianism (cf. Taggart 1995).[19] It was already indicated that, in Denmark, the established coalition embodied such a broad consensus. It may well be argued that the American post-war settlement resulted from an alliance between the corporate capitalist coalition, and the public interest coalition that both favored the proper deployment of state power in order to raise living standards and stabilize capitalism (cf. Harvey 1990: ch. 8). The alliance represented a *liberal establishment* that defined the political climate in the United States during the 1960s, and determined the universe of legitimate policy opportunities and constraints. Though the market conservatives or libertarians strongly disapproved of the idea that the state should take on Keynesian roles, their political power was kept weak by the post-war economic boom.

The social, political, and economic crises of the post-war settlement served as the facilitators of alternative political ideas that ranged from the far left to the far right. During the 1970s, a loss of establishment credibility was evident throughout the modern Western society. The credibility gap between the establishment and the public was multi-faced. At the one extreme, it reflected "a growing disenchantment with the consumption-oriented Horatio Alger society, and the even more disgusting by-products of [the] historical preoccupation with growth *per se*."[20] At the other extreme, the credibility gap indicated a significant rise of cynicism about government, and the longing for a laissez-faire type of consumption society. As a result, both Denmark and the United States experienced a renaissance of 'populism.'[21] Along with the above two extremes, and inspired by an idealized past, populism could be said to have a so-called soft and hard side (cf. Hofstadter 1955: 47).

The communitarian coalition in California represented the soft side of American populism. As we have seen, it was inspired by the agrarian myth. The cooperative coalition fitted within the soft side of Danish populism that looked back to the sort of cooperative society that had existed at the beginning of this century. The libertarian coalition represented the hard side of populism in the United States. In Denmark, that side of populism was most prominently represented by Mogen Glistrup's Progress Party (Fonsmark 1991; Taggart 1995).

6.3.2 The Hard Side of Populism

> American politics is not in the least free of ideology. Perhaps strong ideological views play a more important role in American politics than in the politics of most West European countries. *Maarten van Rossem, 1991: 177*

Glistrup's call for the abolition of income taxes, as few and as simple laws as possible, and restrictions on bureaucracy and reduction in government programs (Miller 1996: 128), would have sounded very familiar to libertarian Americans. However, while right-wing populism in California got a warm reception, as it was seen as a true part of the American Creed,[22] it got an extremely cool reception in Denmark. In fact, the 'radical populist right' in Denmark was not regarded as *folkelig*. For example, its political spokesman, the Progress Party, was seen as "an improper Danish deviance."

On first sight, it seems rather hypocritical to say such a thing of the party that had taken over the aggressive laissez-faire rhetoric from the typically Danish *Venstre*.[23]

Venstre's skepticism about state power was related to the preference for a popular democracy, which was braced by their historical struggle against the state during the 19th century. The idea of social responsibility, which was expressed in Grundtvig's words 'Then have we progressed far in wealth, when few have too much and fewer too little,'[24] eased *Venstre*'s acceptance of government action. The party's views on *vekselvirkning* and the need for open political dialogue seem to form an even more important explanation. Thus, what really was un-Danish about the Progress party was not only their political stance, but also their way of doing politics that was at odds with the Grundtvigian policy style of balance and moderation.[25]

The fact that, after the elections of 1973, Glistrup's Progress party became the second largest party in Denmark reflects that taxes were certainly on the mind of the Danes. But although they had become the most heavily taxed people in the world, the vast majority plainly showed their solidarity with the welfare system, and saw the imposition of high and progressive taxation as a necessity to uphold it. Most Danes abhorred Glistrup's call for a burning of tax records, since they perceived the State as a joint arrangement which they had chosen to build up themselves (Gundelach 1993: 9). This "solidarity is connected with a strong emphasis on individual freedom," and the idea that "the individual can choose for herself to give part of her freedom to the collective" exactly in order to strengthen individual freedom (*ibid.*).

As such, the right-wing populist demagogy – for example, Glistrup compared tax evaders to Resistance fighters during Nazi occupation – was perceived as a threat to the Danish consensus on the welfare society (Miller 1996: 128, 231). None of the other parties wanted to see Glistrup's group in office, and treated it as a political pariah (*ibid.* 145). The fact that the establishment faced a common enemy had "the ironical consequence that the political protection of the Danish model remained untouched" (Fonsmark 1991: 341). To repair the welfare system, a tight fiscal policy was adopted in the early 1980s. Themes like privatization and deregulation also reached the political agenda. They provoked strong negative reactions and vanished long before an attempt to implement a privatization and deregulation policy was really tried (March and Olsen 1989: 102).

Toward the end of the 1970s, when, in Denmark, popular support for the hard side of populism was already falling, "an antigovernment backlash was mobilized in the United States that was more intense than in countries where public spending was much higher" (Robertson and Judd 1989: 127). Public trust in the Danish Government was maintained by the belief that it was possible to improve the service, economy and efficiency of the public sector. It was generally held that by 'modernizing' the public sector, "It shall become easier to be a Dane"[26]. In contrast, "the American public's trust in the national Government and its policy performance plummeted. Three out of four people surveyed in 1980 believed that the Government could rarely ever be trusted 'to do what is right'". (*ibid.*)

In the Golden State, Proposition 13 reflected a public mood and political environment that demanded a slow-down of governmental machinery (Harvey 1985: 218). During the next two decades, the demands for a drastic cut back on taxation and the scope of government found widespread support within the GOP – "the most ardently anti-statist major political party in the world" (Lipset 1996: 27) – and among the so-called Reagan Democrats (the conservative side of the Democratic Party). While the confidence in the *conservative establishment* developed and expanded, the liberal establishment experienced a spiraling diminution of confidence. In the terms of Talcott Parsons, its power deflated. As such, the anti-system oriented right-wing populism that arose at the end of the 1960s, has gradually come to dominate the political culture of the United States. A conservative establishment has firmly taken over the position of the liberal establishment that defined the political climate during the 1960s. As a result, the political agenda in America is currently dominated by libertarian beliefs.[27]

6.3.3 The Soft Side of Populism

> Face to face with the two anarchic extremes, democracy all too often forgets – and this is its weakness – that it, too, has a radicalism to present: the radicalism of the center.
> *Hermann Broch, 1959*[28]

As the hard side of American populism became more dominant, its soft side eclipsed and totally lost political influence in the early 1980s. In Denmark the situation was different. As we saw, the Progress Party that represented the Danish hard side of populism in the *Folketing* was doomed to be a political spectator.[29] For several reasons, the social-liberal establishment did tolerate the soft side of Danish populism.

First, the complete ignorance of both sides of populism would be simply too much, because it would strengthen the populist argument that the people were not the real rulers of Denmark. Since there was a clear lack of trust in politicians among the Danish people, the establishment could not simply disregard this essential point of critique.[30] The ruling class was forced to deal carefully with grassroots activities to regain credibility and maintain its political power.

Second, the beliefs of the cooperative coalition could serve as an important source of inspiration for the established coalition to address the social and economic crisis and assure the viability of the system.

Third, the pragmatic side of the cooperative coalition saw it as its 'mission' to renew the system from within. It tried to envision an alternative to the traditional Western type of industrial society that might appeal to the democratic center parties. It wanted to represent a radicalism from the center.[31] More bluntly, Grundtvigian cultural radicalism represents a type of populism within the Danish establishment itself.[32] As a critical corrective, it has helped the social-liberal establishment to restore its public trust by producing and reinforcing a new legitimacy.

6.3.4 The Dialectic of Control in America and Denmark

While the American and Danish political cultures during the post-war period resembled each other, they seem to have gradually diverged in the last three decades. In America, the liberal establishment has experienced an ongoing loss of public credibility. In Denmark, the social-liberal establishment has managed in the course of time to regain the political support of the public. This was done with the help of the cooperative coalition that can be regarded as some sort of accepted critical corrective to the established coalition.

While soft populism modernized political culture in Denmark, political culture in the United States was renewed by populism's hard side. In the period under study, the public support and acceptability of *Americanism* has gradually increased. Anti-statist laissez-faire libertarianism has gradually taken over the dominant position of statist consumerism. The dialectic of control in America has led to a fundamental change in character of its national political culture.[33] The dominant and, for this analysis, most relevant distinguishing feature between the liberal and the conservative establishment in the United States concerns trust in government. The liberal establishment favors governmental action as an important instrument to control and facilitate capitalism. The conservative establishment, however, is distrustful of government.

While, in America, a political climate has developed that embodies the Jeffersonian belief that less government is better, the Danish government has managed to sustain its public trust. The hard side of populism was rapidly banned by the Danish social-liberal establishment. By portraying the Progress party as a dangerous common enemy, its rise was instrumental in closing the ranks of the establishment. This strategy was not effective with respect to the soft side of Danish populism, which comes from within the establishment. It is a kind of Trojan horse, but a benevolent one. The social-liberal establishment used its ideas in a selective manner to renew Danish society. Although the soft side of Danish populism has played an essential role, it is still marginal.[34] Since the Danish consensus about the welfare state and economic growth is still widely intact, it can be concluded that the political culture in Denmark has not undergone any fundamental change since the early 1970s.

While the Danish fundamental sociocultural values have been relatively stable during the last three decades, the American political culture has changed drastically. The liberal establishment lost its hegemony through the 1960s and in the decade that followed, a struggle between the liberal and the conservative establishment took place, which was won by the latter at the beginning of the 1980s. In other words, while the political culture in Denmark gradually evolved during the period under study, it took on a revolutionary nature in America.[35] Using Kuhn's words, the political culture in the United States seems to have experienced a paradigm shift.

6.4 Comparative Conclusions

> 'What passes for social reality stands in immediate relation to the distribution of power; not only on the most mundane levels of everyday interaction, but also on the level of global cultures and ideologies, whose influence indeed may be felt in every corner of everyday social life itself. *Anthony Giddens, 1984*

As part of the broad crisis of confidence, both the California and Danish power industry also experienced a credibility crisis. Since all the preparatory work has now been done, it seems time to see how the national mix of ideological traditions has enabled and constrained political learning processes within the energy policy subsystem in California and Denmark over the last three decades. Put somewhat differently: In what manner does the national political culture demarcate the universe of ideologically appropriate coalitions and alliances, and legitimate policy-oriented learning processes?

6.4.1 California: Two Establishment Alliances

To reveal the policy dynamics in the very long term, it is essential to grasp that the alliance of the public interest coalition and the corporate capitalists dates back to the Progressive era at the beginning of the 20th century. This learning alliance has stood the test of time remarkably well. The oil crisis, however, brought an end to complacency. In response to the crisis, a new alliance was born that forced the old alliance to modernize itself. Here, it is argued that the two separate California learning alliances are reflecting, on the policy subsystem level, the battle between the liberal and conservative establishments for cultural hegemony on the level of the entire political system. This puts a new complexion on the concept of learning alliance, as one might then speak of two distinct establishment alliances (Figure 6-3).

On the one hand, a liberal establishment alliance can be distinguished that consisted of the corporate capitalist and the public interest coalition. In the 1970s, this alliance attempted to preserve the *status quo* by gradually renewing it. The 'success' of such an effort would help to restore the political hegemony of a liberal establishment. On the other hand, a competing conservative establishment alliance tried to undermine the existing system. Its success would thwart the liberal establishment. At that time, there was little talk about deregulating the electric utility industry. Much of the blame was put directly on the industry itself. This clearly put the liberal establishment alliance on the defensive. It looked for more stringent democratic control and better regulation. Within other policy subsystems (e.g. the airline industry) libertarian items, like deregulation and restoring competition, were surely floating around. But a proposal to abandon the traditional regulatory compact within the electricity industry was totally out of the question within the political context of that time.

Nevertheless, a proposal like PURPA strongly embodied these types of beliefs. It marked a new era; the beginning of what would end in a paradigm shift with respect to the regulation of the electric industry. As said before, the introduction of PURPA was a very special event. First, the fact that this law initially was not noticed by its later opponents made it look as if it was snuck in through the back door by members of the communitarian or libertarian coalition. For the corporate capitalists the law came as a real surprise. Second, the national law gave the libertarians in California a power foundation on which to build their alliance with the communitarian and public interest coalition.

As a result, two alliances came to exist within the California energy debate enabling two parallel learning processes (see Figure 5-1). Each consisted of two or more ideological coalitions, and although they transformed, their make-up was more or less stable until the early 1990s. It is beyond the scope of this study to explore the process by which the beliefs of the various ideological coalitions within the energy subsystem are diffused into the wider national political culture. Nevertheless, two tentative remarks can be made.

The first concerns the impact of experiences from other policy areas with deregulation. Initially, the economic size of the energy sector and the enormous political interests involved kept libertarian policy makers from deregulating the electric utility industry. Through the years, experiences in other subsystems (or industries), however, have encouraged policy makers to do the same trick within the power industry. Second, and related to the former, the mental inertia of the energy subsystem as caused by its huge impact on the entire socio-economic system made it about the last in the row to be deregulated. As a result, it came to represent one of the last bulwarks of the liberal establishment in the United States.

Here, the reverse process of the national political culture influencing the interaction of ideological coalitions within the energy subsystem is at stake. In the 1990s, the ideology of the Progressives finally won the day. Instead of regulating electricity monopoly, a real effort is now being made to restore competition. The dominance of the libertarian anti-system critique of centralized planning within the California political culture eventually made the liberal establishment alliance a dated one.

The hegemony of the libertarian coalition forced the public interest and corporate capitalist coalition to adapt their first order beliefs in accordance to the new political climate. The corporate capitalists made the best of a bad situation. During the whole

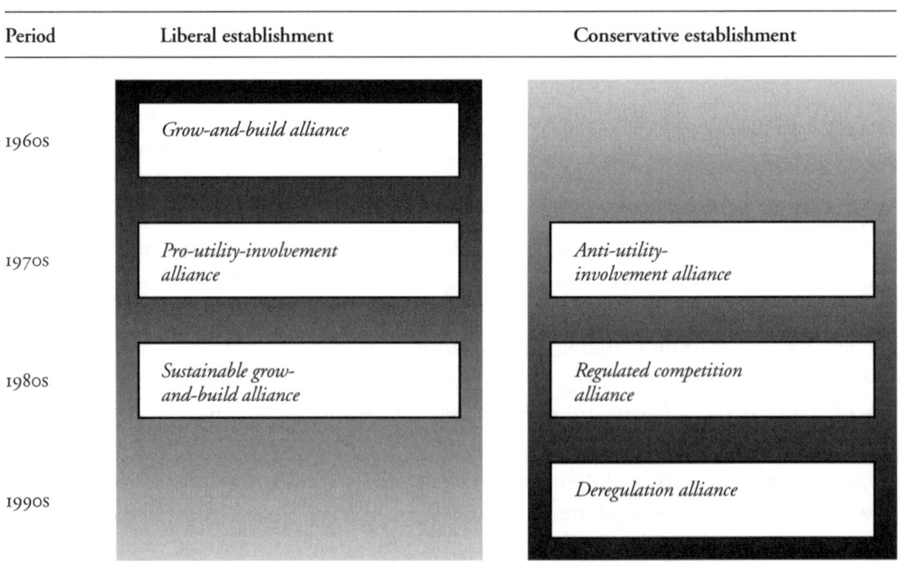

Figure 6-3
The liberal and conservative establishment alliances within the energy debate in California. (The relative strength of the two establishments at a certain point in time is depicted by shades of gray, with black representing the total dominancy of one establishment.)

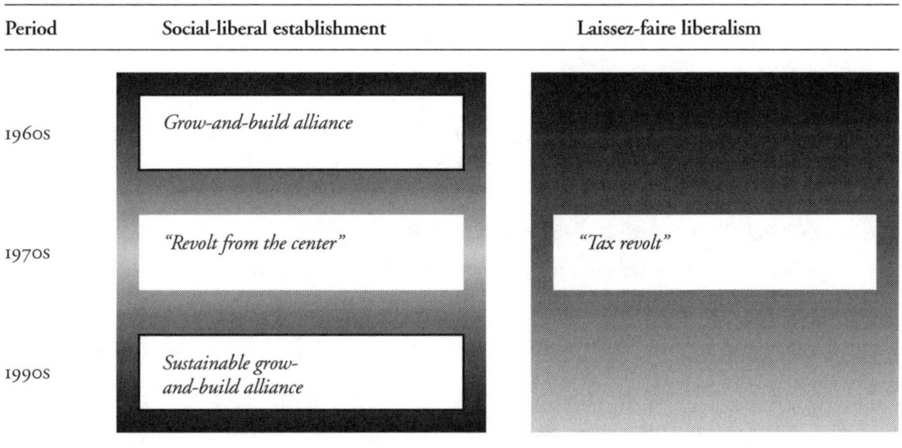

Figure 6-4
The social-liberal establishment alliance within the energy debate in Denmark. (The relative strength of the two establishments at a certain point in time is depicted by shades of gray, with total black representing the total dominancy of one establishment.)

process, this coalition and the libertarian coalition had been very big adversaries, but now they chose to enter the new conservative establishment alliance. On the one hand, this new alliance enabled policy-oriented learning across its constituent coalitions. On the other hand, it brutally ended the learning process within the liberal establishment alliance. Moreover, as its second order arguments were suddenly perceived as 'outdated,' the public interest coalition now, to a large extent, got excluded from the learning process within the conservative alliance. A learning process that the public interest coalition itself had started and had influenced for over two decades! From the perspective of this coalition such a policy-oriented 'learning' process even seems to be self-destructive.

6.4.2 Denmark: One Establishment Alliance

The liberal establishment in America had a European social-democratic tinge to it. In fact, in the 1960s, the Danish establishment in many ways resembled the establishment in the United States. As seen, in Denmark the establishment has stayed in power through the years in spite of the crisis of confidence in the first part of the 1970s. Maintaining the same position is a very active process. When, in the United States, the liberal establishment came to face an increasingly popular conservative tide, it was forced to tolerate this because both a large part of the general public as well as a growing conservative elite found the anti-statist arguments legitimate.

The Progress party in Denmark was effectively shut out of any direct political influence. In contrast, the established coalition actively used the cooperative coalition's ideas to maintain and strengthen its political hegemony. It seems reasonable to argue that the learning process that has taken place in Denmark across the established and cooperative coalition is to a large extent comparable to the learning process across the corporate capitalist and public interest coalition in California. As was described above, the starting points of both processes (grow-and-build) were about the same, and their results at the beginning of the 1990s (sustainable grow-and-build) were very similar (cf. Figure 5-1 and 5-4). It seems fair to argue, therefore, that the established and cooperative coalition formed a so-called social-liberal establishment alliance that is quite similar to the liberal establishment alliance in California (Figure 6-4).

6.4.3 Comparative Conclusions

Our study shows that national political cultures include some types of learning alliances and exclude others within the energy policy subsystem. *Throughout the period under study, the cultural mix in Denmark accommodated only one learning alliance.* During the period under study, the political culture in Denmark offered room for only one establishment alliance. In Denmark, a libertarian coalition existed, but the seeds of its rebellion were rigorously nipped in the bud by political dissenters. So, although the social-liberal establishment was severely challenged in the 1970s, it effectively beat off the right-wing attack. The establishment, thus, actively prevented the libertarian story line of entrepreneurial modernization from gaining a firm foothold in Danish politics. Within this establishment the cooperative coalition played both a lubricating and creative role. Hence the cultural mix remained unchanged. Within this stable political climate, cooperative coalition behavior within the learning or establishment alliance organized around the story line of ecological modernization prevailed. In contrast to the Californian situation, the Danish government has increased its command and control over the electric utilities through the years.

The liberal establishment alliance in America resembles the social-liberal one in Denmark. The similar establishment alliances cover similar types of policy-oriented learning processes. However, the shift in the political culture in America that took place during the 1970s came to offer room for a second establishment alliance. In the 1960s, the liberal establishment (representing a mix of core values of the public interest and corporate capitalist coalitions) dominated the political scene. Its hegemony was challenged during the 1970s, and was lost in the early 1980s. A conservative establishment contending that 'government presented the problem not the solution' took over.

The overall shift in the cultural mix created a special situation. While the energy policy-making subsystem was still dominated by the alliance between the public interest coalition and the corporate capitalists, it now came under the influence of an adversarial political culture. This situation backed up the alliance organized around the story line of entrepreneurial modernization, which overshadowed the whole political system. The change in political climate caused that the ecological modernization alliance had to run against the libertarian wind. *The change in the cultural mix in combination with the inertia of the energy policy subsystem explains the existence of two competing learning alliances at the same time in California.* This in turn implies an adversarial style of politics that characterizes the American political system.

In the 1990s, the conservative establishment achieved political hegemony. The persistence of the conservative political climate made the corporate capitalist finally decide to leave its strategy of integration with the public interest coalition. This ended the liberal establishment alliance that had dominated the energy policy subsystem since the turn of the 20th century. As a result, the learning-process within the liberal establishment alliance organized around the story line of ecological modernization became irrelevant and meaningless. The corporate capitalist coalition abandoned its strategy of domination toward the libertarian coalition and opted for compromise. This, to a large extent, sidelined the public interest coalition and opened up the political possibility of deregulating the electric industry. The dominance of the story line of entrepreneurial modernization was a fact.

Part Three

The Innovation Game

> Although there is a widespread belief that goals should and can be set independently of the plans to attain them, there is overwhelming evidence that the more immediate objectives are – possibly more often than not – the result of opportunities that newly discovered or perceived alternatives offer rather than a source of such alternatives. For instance, the objective of landing a man on the moon did not arise until technology made it a feasible attainment. *E. S. Quade, 1975: 87*

In Part Two, the theoretical framework as developed in Chapter 4, presented the apparatus that pulled the policy string out of the seamless web of society, industry and technology.

First, it was analyzed how policy-oriented learning within and across ideological coalitions affected policy change over time. It was found that, besides competing each other, coalitions also showed cooperative behavior. Cooperation was both guided by, and led to agreement over, first-order beliefs. Cooperative structures that promoted learning across coalitions were named alliances.

Second, it was realized that change and tradition are two sides of the same coin. Each new proposal stems from, and is tried out against a background of, fixed structures. In particular, it was investigated how the belief of each ideological coalition was related to a distinctive policy tradition. The mix of these traditions define the national political culture. The dynamics of this cultural mix within the last three decades was scrutinized, and its relationship with the political interaction and learning processes within the energy policy subsystem was discussed. It was found that learning alliances presented so-called establishment alliances. National political cultures both included and excluded certain types of learning alliances, c.q. learning processes.

The attentive reader has probably noticed the tension that exists between the metaphor of a seamless web and the image of pulling the policy string. Picturing the seamless web as a plate full of sticky spaghetti, the policy string should then be visualized as a single strand of spaghetti. Pulling one strand of spaghetti, however, will bring in motion all other strands. An Italian cream sauce will solve this problem. Its lubricative properties reduce the interaction between the strands.

In Part Two, we pretended that our theoretical framework worked like a cream sauce. It detached policy change from change in industry and technology. Effectively, it treated actions within the industrial and technical world as *external events*. Although this goes against the idea of a seamless web of society, it offers a strategy for drawing out the policy string and analyzing it separately. In this way, and given our interest in the public policy view, the focus on the energy policy subsystem represented an ideal theoretical starting point for an attempt to unravel the seamless web from a policy perspective.

When, however, technological and industrial development themselves become objects of public policy, and crucial for its success, the divide between the political system and the techno-economic system gets blurred. This is exactly the case in the energy policy field, where, for example, the political wish to deregulate the utility industry presupposes the availability of both decentralized electricity production systems, as well as an independent power industry. Then the question of the relationship between political and economic power becomes an extremely relevant one. In Part Three, we will examine in what way the policy process (as analyzed in Part Two) has constrained and enabled technological innovation and industrial development within the field of wind energy.

In order to analyze wind energy innovation in California and Denmark, a theoretical framework is set up in Chapter 7. The innovation process will be conceptualized in terms of interacting *innovation networks* each populated by different types of innovation actors (policy makers, technologists, firm managers, and consumers) who are all guided by different (professional) frames of meaning.

In Chapter 8, the relationship between national political and techno-economic traditions is studied historically. It is found that these different types of traditions rely on each other and, as such, have commonly developed into connected wholes, which are named *national innovation traditions*. This implies that different ideological coalitions have different ideal types of innovation networks in mind.

In Chapter 9, the interaction of policy makers with other innovation actors in wind energy innovation networks is studied. In addition, the interaction between various wind energy innovation networks will be examined.

7 Theoretical Framework of Innovation, Tradition and Learning

> [T]he policy process can be seen as learning by different actors through a series of games. To be successful, actors not only need to know about responses of other actors in the debate, but also must understand these responses well enough to anticipate them. *John Grin and Henk van de Graaf, 1996: 92*

In the introductory Chapter it was suggested that policy makers play two interrelated roles. First, there is the legitimizing role they exercise in modern democratic societies. In Part Two, the democratic struggle for the legitimization of wind energy innovation was examined in detail. Various ideological coalitions were identified, each with a distinct vision as to what the right course of action was, philosophy of government, and ideas on what kind of nation and society should be aimed at. The energy debate not only focused on such second order policy beliefs, but also centered on the effectiveness and efficiency of government actions, and their fairness and appropriateness.

Besides acting as a political watchdog, interpreter or advocate, policy makers may also play a constructive role in the development of wind energy technology and industry. This raises the central question: How should the government act so as to realize innovation? In other words: How to implement wind energy innovation? In the *innovation game*, innovation programs get their definitive form. These public policies do not by themselves lead to aspired policy outcomes. Preferred policy results, like the building and use of wind turbines, are produced by actors outside government who in one way or the other respond to the policy (cf. Van de Graaf and Hoppe 1989, Grin and Van de Graaf 1996). At best, policy programs may aspire to 'orchestrate' the actions and interactions of the other actors within a policy area. In order to contribute in a sensible way to the development and use of wind turbine technology, and the creation of a viable wind power industry, policy 'conductors' have to have some intelligible ideas about the relationship between the political process and the process of techno-economic innovation.

In section 7.1, we will discuss this relationship by elaborating on the famous concept of techno-economic network. This exercise will lead to the concept of *innovation network*. We will describe the structure of innovation networks, the actors involved and their guiding (professional) beliefs. It is argued that successful innovation requires a learning process among innovation actors. This so-called *joint innovation learning* can be regarded as a major driving force behind the dynamics of innovation networks.

In section 7.2, several additional concepts are introduced to properly study the dynamics and statics of innovation processes over a decade or more. In particular, it is speculated that national political traditions and techno-economic traditions rely on each other, and as such jointly develop into connected wholes, which will be referred to as *national innovation traditions*.

In the final section, the plan of the remainder of Part Three will be set out.

7.1 Innovation Networks and Learning

7.1.1 The Structure of Innovation Networks

Innovation has long been described as a linear process moving from one step to another once the initial decision is made (Larédo and Mustar 1996: 146). The initiative could come from engineers – technology push – or from demand – market pull. But recent economic and sociological findings reveal that such a linear model provides a too simplistic perspective on innovation. It became realized that there seldom exist clear-cut borders between different phases and/or actors in the innovation process. For example, Von Hippel (1988) confirmed that in certain areas (e.g., computer aided design systems), it is users who form the true developers of successful new products.

To include and study the iterative dimension of the innovation process, Callon *et al.* (1992) have introduced the concept of *techno-economic network* (TEN). No longer is innovation perceived as only taking place within isolated public or private research laboratories. Instead, the whole techno-economic network determines the locus of innovation. A TEN is defined as "a coordinated set of heterogeneous actors – public laboratories, technical research centers, industrial companies, financial organizations, users, and public authorities – which participate collectively in the development and diffusion of innovations, and which via many interactions organize the relationships between scientific and technical research and the marketplace." (*ibid.* 220)

Techno-economic networks are organized around five major poles, three of which are the supporting pillars: a scientific, a technical, and a market pole. Poles can be distinguished both by the actors constituting them as well as by the nature of their production. Within the *scientific pole* (S) certified scientific knowledge is produced by scientists and researchers, who work within universities and public or private research centers. The main actors within the *technical pole* (T) are engineers and technicians working in technical laboratories in companies, cooperative research centers, or pilot plants, where they conceive of, develop or transform artifacts destined to serve specific purposes.

In the terminology of Callon *et al.*, the *market pole* corresponds solely to the universe of users. In our discussion 'market' will be used in the classical sense as a place where supply meets demand. Thus, it seems elegant to rename the 'demand-side-oriented' market pole that Callon *et al.* use. That pole will be called the consumption pole. In addition, the supply-side of the market will be named the business pole.[1] The business and consumption poles jointly cover the market. Within the *business pole* (B), general managers either try to anticipate new consumer demands or translate demands expressed by users into products. Moreover, they organize the production, distribution and marketing of these novel products. The *consumption pole* (C) corresponds to the universe of the consumer, who ultimately buys, uses, and thus economically values the artifact.

The notion of techno-economic network embodies several important insights about techno-economic innovation. It involves the idea that innovation can come into being at any point along the network. For example, in some cases science acts as the leader and gives rise to new technology, and in other situations technological developments may take science in tow (cf. Nelson and Rosenberg 1993: 6). In their search for profit, companies often play a leading role in the innovation process, but in some cases the innovation pattern is user-dominated (Von Hippel 1976). More important, the very concept of 'network' implies that the success of technological innovation relies on the interactions between *heterogeneous* actors.

In particular, three types of actors are present within a techno-economic network as defined by Callon *et al.*. Technologists are engaged in the generation of new technologies. General managers of firms are involved in the producing, marketing, and selling of new artifacts. Finally consumers are buying and using the artifacts. Although by definition (see above) 'public authorities' are assumed to play a role within techno-economic networks, they seem not to be an integral part of them. Policy makers do not populate any particular pole of the techno-economic network as conceived by Callon *et al.*; they appear to play their role beside, outside or above techno-economic networks.

Our case descriptions and policy analysis, however, show that in the field of wind energy there exist a strong interdependence between the actors within the techno-economic network and the actors within the political system. Policy makers played a crucial, integrated role in this field, by legitimizing and actively supporting wind energy innovation in various ways. Whereas policy makers politically legitimize techno-economic development, the reverse is also true.

Anticipated techno-economic success of a politically desired innovation serves to legitimize policy intervention from a feasibility point of view. Moreover, actual technological and industrial success strengthens and solidifies original political arguments on preferred paths of innovation. For example, the technological and export success of the Danish wind turbine industry gave strength to the alternative energy plans of the cooperative coalition. In contrast, poor development of wind and solar energy at the local level severely undermined the power position of the communitarian coalition in the California energy debate.

Our research thus shows that the force of political arguments is both influenced by the perceived future feasibility of a certain technological option as well as the accumulated practical experience with that option. To emphasize and visualize the continuous interweaving of political aspects with technical, managerial, and consumer interest aspects in the innovation game, we like to add a fifth *political pole*. By introducing such a pole, the term innovation looses its narrow definition as techno-economic challenge. Instead, innovation is conceived of as a human solution to human problems with scientific, technical, economic, user-related, and political aspects. We are in favor of a broad definition of innovation that both involves technical and industrial change, the transformation of consumer behavior, and political change, as well as embodies the idea that political innovation and techno-economic development are mutually dependent and are constantly shaping each other.

Because of the need for introducing a political pole and a broad conceptualization of innovation, we, instead of employing the concept of techno-economic network, prefer to speak of an *innovation network*. As indicated, innovation networks will be described as organized around five poles (see Figure 7-1): a scientific (S), a technical (T), a business (B), a consumption (C), and a political pole (P). Technologists play a dominant role within the science and technical poles, firm managers are the main players within the business pole, users reside in the consumption pole, and policy makers occupy the political pole.

Pole	Science	Technics	Business	Consumption	Politics
Actor	Scientist, researcher	Engineer, technician	Manager	Consumer	Policy maker
Role	Production of scientific knowledge	Design and development of an artifact	Production and marketing of a product	Consumption and economic valuing of the artifact	Stimulation and regulation of the innovation process

Figure 7-1
The innovation network.

7.1.2 The Frames of Meaning of Innovation Actors

The notion of heterogeneity implies that an innovation network consists of different types of actors, often driven by different types of (professional) beliefs. Each of the four types of innovation actors has an interest in the development of a certain artifact. The interests, however, differ. For example, a policy maker may look upon wind turbines as a means to produce clean electricity to stop the depletion of the ozone layer. The manager might see the wind turbine as a way to diversify his product range and/or as a product to make profit from. An innovation network presents a modern version of the Tower of Babel, in which different actors speak different languages.[2]

Not only do the languages differ: even perceptions on what the 'tower' actually is differ. The policy maker will tend to see the innovation game as a *political game*, and the innovation network as a kind of *policy network*. Engineers may perceive it as a technological development game, and perceive the innovation network as a technology network. In their turn, managers and consumers may be specifically concerned with the process of production and consumption, and see the network as an institutional setting to sustain market development and fulfil customer needs, respectively.

By using the terms innovation network and game, we do not want to ignore the existence or disregard the importance of the different ways of looking at the network and its activities. On the contrary, the term innovation game is chosen to stress the fact that it consists of several concurrent, non-hierarchical games, which have unique rules, styles of play, and mental demands on the players.

In Part Two, we, so to speak, concentrated on the political poles of the innovation networks in the field of wind energy, which, in correspondence with the idea that policy makers are members of an ideological coalition, are populated by ideological coalitions. What is at stake in the political game is to what extent a particular ideological coalition is able to further its policy objectives. Ultimately, the primary issue is whether wind energy innovation is able to bring closer the preferred social order. However, since the political game is an integral part of the innovation game, we need to look for the politics of innovation and its democratic legitimization not only within the political subsystem, but also within the confines of the technological communities, the management of firms, and consumer groups. This is what we intend to do in this part of the book.

To examine the innovation network as a whole, we need to know something about the languages of the various innovation actors involved; how do various innovation actors perceive problems, objectives, goals and means?

In Chapter 4, we used Fischer's four-level evaluation scheme to structure the belief systems of policy makers. It was found that the second-order policy beliefs ultimately define what is at stake in the political game. Grin and Van de Graaf (1996) argue that similar things can be said for other innovation actors, and use the same four-layer model to reconstruct their frames of meaning (Figure 7-2).

The authors argue that technologists and firm managers can be seen as disciplinary professionals. Accordingly, it is possible to describe the frames of meaning that guide their actions. Whereas policy makers are guided by their policy-oriented belief system, technology actors are led by their *technology-oriented belief system*, which is usually denoted as *technological paradigm*. The actions of firm managers are guided by their *management-oriented belief system*, or *management paradigm*. Both types of paradigms are reconstructed in a way similar to Fischer's four-level policy evaluation scheme.

As we saw in Chapter 4, the highest level of a policy-oriented belief system is related to the policy maker's fundamental preferences about the social order. Within the frame of meaning of technologists the highest level is related to the preferred meaning of an artifact. While policy programs are intended to change social reality in the direction of the preferred social order, technological development paths are meant to lead to the envisaged artifact.

The choice of these paths are also evaluated against lower level second-order convictions that include overarching theories in the form of scientific and technological theories and bodies of practical knowledge along with broad guidelines for sensible technology development. These theories are shared by the technological community involved and provide the language and repertoire for reasoning and communicating about the technological innovation process.

General managers of firms make technology choices and other decisions based on their preferred position within the market. To give an example, according to Nordtank (1995: 17), its "overall objective is to maintain and further enhance its position as one of the world's leading manufacturers and suppliers of wind turbines and turnkey wind farms". Such market preferences define the highest level of a management-oriented belief system. Managers are concerned with pursuing a business strategy that is expected to contribute to the achievement of their firm's preferred market position. Management strategy deals with circumstances that are external and internal to the firm. General managers interpret these strategies from their appreciation systems and, sometimes tacit, overarching theories that deal with the market, internal organization, innovation strategies, *et cetera*.

It is thus possible to define frames of meaning that guide the actions of the dominant actors within the science, technical, business and political poles. To define the frame of meaning which guides the behavior of the consumer seems to be a less obvious task.

	Private consumer	Technologist	Manager
Ultimate preferences	Desired level of well-being and comfort for own household	Preferred meaning of the artifacts	Firm's preferred competitive position
Empirical and normative background theories	Beliefs about hygiene, health, upbringing, et cetera.	Technological and scientific insights; aesthetic assessment frameworks	Business economics, marketing, micro- and macro-economic insights
Problem definitions	What is needed to make sure there is food, shelter, clothes, et cetera?	How can I design or improve the artifact?	What challenges my company here and now?
Evaluation of solutions	Assessments of effects, side-effects and costs of means available for meeting own needs	Assessment of the effectiveness of technological development paths	Causal evaluations of the effectiveness of different business activities and strategies

Figure 7-2
Structure of the frames of meaning of technologists, corporate managers and consumers. The information in this table is based on Grin and Van de Graaf 1994, and Grin et al. (1997).

Still, some remarks can be made. Sometimes, users themselves are guided by one of the above professional beliefs. For example, the decision of the electric utility industry in America to buy and set up wind turbines is a business decision that is guided by their specific management paradigm.

If this is not the case, one might argue that the second-order beliefs are existent but unknown. Nevertheless, so-called *consumer preferences* exist. These preferences can simply be treated as first-order beliefs.[3] On the highest first-order level, users are said to have a certain problem or demand. In the light of this problem the user gives meaning to the artifact. The artifact that is self-built or offered by the manufacturer presents (part of) the instrumental solution to this problem.

7.1.3 Breakthrough Actions and Continuity Operations

Building on the work of Callon *et al.* (1992), we may give a rough characterization of innovation networks. These may be incomplete or chained, and convergent or dispersed. If, for example, the technical pole is underdeveloped whereas the scientific and market poles are strong the network is said to be *incomplete* with respect to the technical pole. If, on the other hand, all the actors and poles are in position and strongly structured, the network is called *chained.*

A second characteristic is the network's degree of integration. In a *dispersed* network, relationships exist between actors (stemming from a single pole or distinct poles), but they are of low density and characterized by dissension and lack of understanding. In a dispersed network it will be difficult for a given actor to mobilize the rest of the network. In contrast, in a strongly *convergent* techno-economic network, the activities of each of the poles can easily be tied to each other, and there is also a high degree of integration within each pole. "Such a network would be particularly efficient, since it would dispose at once of the force of the collective and the synthetic capacity of the individual. Any particular actor would be able to speak in the name of all, to mobilize at a point all the skills and alliances in the network." (*ibid.* 224)

Two types of government actions can now be envisioned. If the actual network is incomplete, the logical policy aim is to create or strengthen the missing link(s). In this case, one might speak of "a *breakthrough action*, to indicate the creation of a discontinuity with respect to the current situation. ... [T]he aim is to bring out new combinations – a radically new piece of technology, a clientele that did not exist before or a company, etc." (*ibid.* 229)

If the actual network turns out to be dispersed, then the public policy should aim to increase the innovation network's degree of integration. The interventions that aspire to foster the development of pre-existing activities and their interactions are called *continuity operations*.

7.1.4 Innovation Learning

> A totally convergent network would be a sort of Tower of Babel, in which each would speak in their own language and everyone else would understand them, and each would have skills that all others would know how best to use. *Michael Callon et al., 1992: 224*

How can policy makers assess what kind of continuity operations are necessary to heighten the innovation network's degree of convergence? Or, with respect to wind energy, how can policy makers coordinate the actions of scientists, engineers, wind turbine manufactures, and users in such a way that some correspondence comes to exist between the technology that is developed, produced, and used?

Grin and Van de Graaf (1996) address this question in some detail in an article titled *Technology Assessment as Learning.*[4] The authors argue that differences in interest and perspective do not necessarily prevent a fruitful interplay among the various types of actors. They maintain that "*whereas different meanings will be found more often than not*, they need not be incongruent; that is, *they need not exclude joint action.*" (*ibid.* 89) Grin and Van de Graaf claim that the different innovation actors all share an interest in one notion: *the meaning of a technological artifact.* For policy makers, managers, and consumers, the artifact is the instrumental object of their activities. It represents a part of the solution to the policy problems of policy makers, serves as a solution to a challenge facing firms, and to consumer needs. For technologists, however, the meaning attributed to the artifact serves as a normative guide for their actions.

Thus, an innovation network is dispersed when the different meanings the artifact has for different actors are *incongruent*; that is, they exclude joint action. For example, technologists make artifacts that firms do not want to produce, or products are made that users do not want to buy or use. In contrast, in convergent innovation networks the different meanings are *congruent*. Thus, establishing congruent meanings among actors with different types of frames of meaning increases the innovation network's degree of convergence.

Improving the innovation process through public policy, not only requires learning among ideological coalitions, but also between policy actors and other innovation actors. A form of mutual learning is needed that "aims to discover how the belief systems or paradigms [or preferences] of other actors generate their perceptions of problems, objectives, ends, and means." (*ibid.* 92)[5] We will name this *joint innovation learning*. Depending on the degree of completion and convergence of the innovation network, successful joint innovation learning can have different aims and results. This will be discussed in the next subsection.

7.1.5 Four Types of Innovation Learning

The interaction between public policy makers and other innovation actors can take on many complex forms. Policy makers aim to align the activity of the other actors in the network with their own political aspirations. Hence they try to manage the convergence between political and other relevant aspects of innovation within the network. In other words, they seek *internal political legitimization* of the innovation network.

At the same time, the policy maker has to ensure that the politically desired innovation becomes techno-economically feasible. The mutual search for 'a proper match between the desirable and the feasible' (Grin *et al.* 1997) was named joint innovation learning. Depending on the match between political and techno-economic support successful joint innovation learning will have different aims and results.

To structure our discussion we will distinguish between networks that do or do not receive political support, and networks that do or do not receive techno-economic support. This leads to a simple two two matrix that illustrates four distinct states of innovation networks (Figure 7-3), which can demonstrate the different aims and roles innovation learning can have. In each state four 'changes' can occur. First, the character of the network remains unchanged. Next, the nature of the network changes and ends up in one of the other three cells. Successful innovation learning implies an innovation dynamics that ends up either in the upper-left cell of the matrix (that is, the networks receives both political and techno-economic support), or in the cell below at the right side of the matrix (that is, the network receives neither political nor techno-economic support). In both cases, the end result is a match between political and techno-economic support.

The first cluster of situations is characterized by the absence of both political and techno-economic support. In this case, the network would logically be nonexistent. In one case, successful innovation learning would preserve the *status quo*, and no action would be taken. However a dramatic event may lead to a change in attitude among both policy actors as well as other innovation actors towards the suitability of an innovation network. Successful innovation learning then would lead to the *creation* of the suddenly politically and techno-economically desired network. In these two cases, the actions of policy makers and techno-economic actors come to agree with each other. In the next two cases, however, techno-economic development is at odds with political change. First, policy actors who first rejected an imaginary network may come to embrace it. If, at the same time, this venture experiences little private support, the imaginary network is bound to stay imaginary. Although policy-oriented learning is at stake, no innovation learning takes place. Second, the fact that policy makers have little sympathy for a certain network does not exclude private actors from setting up that network. This is characterized by techno-economic learning, but not by joint innovation learning.

In the second cluster the techno-economic network is politically disliked, but desired from a techno-economic perspective. Again four dynamics can be traced. First, policy actors may choose to sit on the sidelines and grind their teeth. While no political action is taken, firms may be working to maintain the network. Although it grieves policy makers, the *status quo* is preserved. A second, rather hypothetical, situation arises when policy-oriented learning leads to a positive change in attitude towards the existing techno-economic network, while at the same time techno-economic actors withdraw themselves from the network. In both cases, techno-economic developments are at odds with political change. Two different dynamics can be envisioned in which innovation learning leads to a match between techno-economic and political demands. On the one hand, policy makers may try to actively '*terminate*' or '*destroy*' the politically disliked techno-economic network. In this case, they have to 'learn' how to thwart the activity of actors within the politically disliked innovation network. Private actors will most likely try to offset the policy countermeasures. Besides this strategy of confrontation, the various actors may try to match the existing techno-economic network with the political demands. The search for integrated solutions requires joint innovation learning. On the other hand, joint innovation learning might lead policy makers to politically accept and support the activity of other innovation actors.

The third group of innovation situations is characterized by the fact that the network is politically desired but encounters little techno-economic support. Innovation learning is needed to secure support from innovation actors outside the political system. Successful learning then may lead to the creation of an innovation network. The result of learning might also be that policy makers abandon their initial ideas to set up a certain network, and give up their effort.

Within the fourth cluster, politically and techno-economically desired innovation networks are situated. The problem is to '*maintain*' these networks. Some politicians will argue that the network has taken on such a robust form that political support is no longer needed and the market can sustain the network itself. In this dynamic world, however, everything changes, and new dangers will turn up to challenge the network's viability. To counteract threats, public policy needs to constantly adapt to new circumstances. This requires a continuous joint innovation learning process. If the existing innovation network is not regularly repaired, it will disintegrate. Standing still means decline, which may be caused by a loss of political and/or techno-economic support. Successful innovation learning would mean that these processes go hand in glove and hence speed up the destruction of the network.

		Political support	
		High	*Low*
Techno-economic support	*High*	Sufficient political and techno-economic support *Cluster 4*	Sufficient techno-economic support, but little political support *Cluster 2*
	Low	Sufficient political, but little techno-economic support *Cluster 3*	Insufficient political and techno-economic support *Cluster 1*

Figure 7-3
Four distinct states of innovation networks.

The above shows that joint innovation learning has to do with matching the political and techno-economic legitimization of innovation activities. The desired result is an innovation that is both politically legitimized as well as techno-economically feasible. If such a politically and techno-economically desired network already exists, successful learning will cause the *maintenance* of the network. In case the desired network is non-existent or incomplete, successful innovation learning leads to the *creation* of a techno-economically well-functioning and politically desired innovation network.

Successful innovation learning can also lead to the conclusion that a certain type of innovation is legitimate neither in a political, nor in techno-economic sense. If the disliked network does not yet exist, successful learning would preserve this preferred status quo. If the disliked network does exist, successful innovation learning would result in the smooth and rapid *destruction* of the existing network.

Finally, successful learning may cause all innovation actors to *abandon* their desire to establish a network because they come to realize that such an effort would be unrealistic either from a techno-economic or political point of view.

7.2 The Dynamics and Statics of Innovation

> Fundamental to our approach is the introduction of the concept of desired network, since it underlines the voluntarist nature of any programming and the irreducible uncertainty surrounding it. *Callon et al. 1992: 227*

If it is clear what kind of innovation network is politically desired, the implementation of wind energy innovation can focus on the establishment and convergence of that network. If this is not clear, it is hard to define purposeful government interventions. As the innovation learning process becomes blind, the concept of learning loses its significance. Policy makers, therefore, have to address the question of what kind of innovation network is politically desirable?[6]

In this section, we will examine this question theoretically. In the first subsection, we explore the complexity of the innovation process in the event that rival networks exist. Next, we inquire whether there are indeed different innovation networks or approaches a policy maker can choose from. And, if this is the case, whether these distinct options have different political implications. In other words: Does every ideological coalition have its own favorite approach toward innovation?

7.2.1 The Seemingly Seamless Innovation Implementation Web

> Considered as a whole, a program can be conceived of as a system in which each element is dependent on the other. ... A breakdown at one stage must be repaired, therefore, before it is possible to move on to the next. The stages are related, however, from back to front as well as from front to back. ... Program implementation thus becomes a seamless web. *Jeffrey L. Pressman and Aaron Wildavsky, 1979: xxi*

Imagine a situation in which there are several innovation networks or approaches to chose from. Moreover, assume that each ideological coalition prefers a different type of innovation network. This implies that the various political poles are populated with different ideological coalitions. We saw that one task of the actors within a political pole is to seek internal political legitimization of the innovation network. Moreover, they have to find political support for the innovation network activities within the political system at large. Thus, *the ideological coalition's task is to seek both internal and external political legitimization of the innovation network it prefers to create or maintain.*

This resolves the perceived duality between the legitimizing and constructive role of policy makers. Both roles are legitimizing as well as constructive. In finding external political support, ideological coalitions are confronted with other coalitions that promote other innovation networks. In order to bring techno-economic development within an innovation network in line with their political beliefs, ideological coalitions are confronted with the frames of meanings of other types of actors. To be successful, ideological coalitions have to be active on two fronts: within the political as well as the innovation (implementation) game.

Within the political game it is decided which (set of) network(s) is to receive support from government. From the policy maker's point of view, the political struggle then continues with new means and new (non-political) actors outside the policy subsystem. In other words, implementation represents the struggle over the realization of political ideas (Pressman and Wildavsky 1979: 194). Outside the political subsystem, the skills and powers of technologists, firm managers, and consumers will co-determine which innovation network is going to be established and whether it will be thriving or not. Not policy change by itself but techno-economic change into the right direction becomes the measure of political success. To achieve this, ideological coalitions have to buttress and elaborate their preferred innovation networks.

An ideological coalition achieves ultimate success when its favored network takes on a robust form and no longer needs direct political support. At that time, the network has become a sustainable policy outcome that supports the political culture of the associated ideological coalition. In other words, the belief system of the ideological coalition has become institutionalized within the innovation network (March and Olsen 1989). These types of remarks bring back the metaphor of a 'seamless web' of politics, industry, technology and society. From a policy view one could speak of a *seamless innovation implementation web*.[7] An attempt to unravel the seamless innovation web from a policy perspective should take into account the intended and the unintended consequences of policy interventions that result from the fact that policy makers depend on the willful cooperation of various actors within the innovation network, who act according to their own legitimate frames of meaning.

Several factors complicate the implementation of innovation. First, the political debate around innovation is often surrounded by political dissensus. Second, the fact that a network is 'politically desired' does not immediately imply that the network will receive sufficient private support to be successful. Thirdly, the irreducible uncertainty that surrounds the establishment of an innovation network is not only caused by political dissensus, but also by technological and economic uncertainties. Fourth, the fact that several innovation networks may simultaneously exist also introduces complexity. For example, policy measures that intend to stimulate the development of one particular politically desired innovation network may unintentionally stimulate another network that is politically disliked. Moreover, while innovation networks may exclude each other in a political sense, they may be less distinct from a technological or managerial perspective. Cross-fertilization may occur between the scientific, technical, business and consumption poles of separate politically desired or disliked innovation networks. In other words, interaction between networks in the making or existing networks will complicate the description and analysis of the dynamics of innovation and the role of innovation learning herein.

The notion of innovation network tries to do justice to uncertainties, surprises, the role of existing institutional techno-economic structures, and intended and unintended consequences of actions of a variety of actors. The concept enables us to unravel the 'seamless web' in its constituting, interdependent parts (actors, perceptions, and actions) which in concert form a seemingly seamless web of society and technology.

7.2.2 National Innovation Traditions

We would like to make two assumptions that will be studied and documented in Chapter 9. First, we speculate that, when policy makers want to implement innovation, they can choose from a kind of standard repertoire of innovation approaches, which relates to a set of or mix of national, historically developed techno-economic cultures. From this perspective, techno-economic innovation exists in continuity with the past, which in turn supplies the means of its initiation (cf. Giddens 1979). By choosing a certain innovation approach, policy makers become involved in the reproduction of the techno-economic culture concerned. Thus, the choice for a certain innovation approach is in fact a choice for a certain techno-economic culture.

The second assumption is that several ideological coalitions will have their favorite approach toward techno-economic innovation approach. Here, we theoretically explore these issues. In particular, we will look at the relationship between techno-economic and political traditions.

During the last decade, the research on national techno-economic traditions has received widespread attention. Much of this research centers the concept of *national system of innovation* (NSI), which has been generally adopted to describe and compare the innovative competencies of nations.[8] Basically, 'innovation' stands for techno-economic innovation, or "what is required of firms if they are to stay competitive in industries where technological advance is important" (Nelson 1993: 509). For example, progress of conventional technologies was vital for the well-being of the electric utility industry in the United States in the post-war period. Hirsh (1989) describes the role of *technological stasis* in its decline in the late 1960s and 1970s. In the same way, technological advance is crucial both for the wind industry as a whole as well as for individual firms. Continuous proof of technological progress leading to declining costs of wind power is crucial to maintain political support. For firms staying competitive also requires continuing innovation.

The basic notion of a 'national innovation system' is a complex and somewhat problematic one (Nelson 1993: 506). In defining the national system of innovation, various scholars have focused on different things. It is shaped by factors such as size and the distribution of natural resources that effect comparative advantage at a basic level, and reflects differences in economic and political circumstances. In particular, beliefs about the kind of role governments should play in shaping industrial development seems to be a powerful variable. However, a well-articulated and verified analytical framework linking socio-institutional arrangements to technological and economical performance is absent (Nelson and Rosenberg 1993: 4).

Recent historical descriptions of the American and Danish innovation systems can help to orient a comparative analysis of the two countries. Mowery and Rosenberg (1993) consider the description of the U.S. national innovation system as a 'gargantuan' task. They focus on three elements. First, the role of industry, universities, and the federal government is studied as performers and as funders of R&D. Second, the part played by small start-up firms within the U.S. economy is scrutinized. They argue that such firms appear to have been more significant within the American economy than in other modern economies, with the possible exception of Taiwan and Denmark. Third, two public policies – the antitrust statutes and the important role of military R&D procurement – in particular contributed to contrasts between the structure of the U.S. national innovation system and those of other nations. For example, the increasingly stringent interpretation of the Sherman Antitrust Act (1890) in the late 19th century was in part responsible for the 1895-1904 U.S. merger wave. Finding that the legality of price-fixing and market-sharing agreements was under attack, companies resorted to horizontal mergers to control prices and markets. In addition, "a widening array of firms increased corporate reliance on industrial research and innovation to forestall or offset the effects of antitrust prosecution. Industrial research supported corporate diversification and the use of patents to attain or retain market power without running afoul of antitrust law." (*ibid.* 32)

Bengt-Åke Lundvall (1988; Edquist and Lundvall 1993) introduces the Danish system of innovation by giving a brief historical sketch of the processes of industrialization. Next, he describes the anatomy of the system of innovation in terms of the socioinstitutional set-up, the structure of production, and the R&D system. Among other things, the formation of the Danish agro-industrial complex is described. Facing a serious crisis during the last decades of the 19th century, the Danish farmers started to export high-quality standardized products, like butter, bacon, and eggs, from animal husbandry. This development induced many technological innovations in milk transport, dairy machinery, in feeding and breeding of animals, in veterinary services, and quality control instruments.

Lundvall (1988: 32) stresses that socioinstitutional change represented an important ingredient in the development of the agro-industrial complex. Small and medium-sized farmers had joined in a cooperative movement organizing the procurement of inputs as well as the processing and marketing of agricultural products in farmer-owned firms. These were based on both economic necessity as well as political visions, in particular Grundtvig's democratic ideas. The movement evolved into an important political force that led the government "to support research organizations, a consultant network geared for diffusion of techniques, and a Danish patent law that was especially negative towards property rights, so that inventions could easily be diffused." (*ibid.*)

The above illustrates that national systems of innovation often are the unintended consequence of combined political, managerial, and technological action. The Grundtvigian agricultural cooperative movement promoted collective forms of entrepreneurship, and technology development through knowledge sharing. The American public's antitrust sentiments, as crystallized within the Sherman Antitrust Act, enticed the creation of even bigger corporations, the creation of private R&D laboratories and the prioritization of knowledge protection. On its turn, the national system of innovation constrains and enables technical and managerial activities.

Often the influence of certain specific public policies on the dynamics of the national systems of innovation is taken into account. Seldom, however, the full range in which political cultures interact with national innovation systems is touched upon. In Chapter 8, we will try to do exactly that: to relate political with techno-economic traditions. It is assumed that these different types of traditions rely on each other and as such jointly develop into connected wholes, so-called *national innovation traditions*. Similar to national political traditions, national innovation traditions are considered as potentially plural, and subject to gradual change.

7.3 The Plan of Part Three

In Chapter 8, the mutual historical development of techno-economic and political traditions is described within the United States and Denmark. On the basis of this history of innovation in both countries, the techno-economic preferences of each ideological coalition will be studied. In this way, we will hope to answer the next question:

> What kind of wind energy innovation networks do the various ideological coalitions in California and Denmark ideally promote?

After having distinguished the various politically desired wind energy innovation networks in the two countries, the dynamics and interaction of these networks is scrutinized in Chapter 9. In summary, the central question in Chapter 9 is:

> How did the various politically desired wind energy innovation networks in California and Denmark develop over the last two decades, and what role has joint innovation learning (or the lack thereof) played within those developments?

8 National Innovation Traditions

From the brief discussion in the previous Chapter a long list of aspects flows which together defines national innovation traditions.

First, a methodological aspect that relates to a scientific community's view on the preferred style of innovating. Innovation approaches may vary their emphasis on either theory or practice as a prime source of innovation. A distinction is made between the theory-based *design-by-extrapolation* approach to create new products and the practice-based *design-by-experience* approach.

Second, dominant managerial ideas about proper organizational or production structure, innovation strategy, and their orientation toward markets, were indicated as relevant parameters.

Third, the influence of political ways of thinking upon national innovation styles was pointed out. These provide guidelines to deal with questions like: Who should guide the process of technological renewal, and to what purpose? Should science be autonomous? If not, should it be guided by elected politicians, by people at the local level, by small high-tech start-up firms, or by large corporations? The relationship between the technological, economic and political power is at stake here.

In describing the American and Danish innovation traditions in the next two sections, the above list of technological, political, and managerial features will be treated in a more or less random manner. Based on the brief historical description of the various national innovation traditions, the final section will address the question of what type of innovation network the various ideological coalitions in California and Denmark would ideally prefer to promote.

8.1 American Innovation Traditions

At the beginning of the 19th century, a technical consciousness appeared in the United States that differed from the one that had already evolved in Europe (Ellul 1964: ch.1). Whereas the Europeans tended to mechanize simple operations, such as spinning, the Americans began by mechanizing complex operations, which eventually produced the assembly line.[1] Invention became a part of the normal course of life. As Giedion (1948) maintains, "Everyone invented. Every entrepreneur dreamed of more rapid and economical means of fabrication." (Quoted in Ellul 1964: 52) Technological innovation relied largely on mechanical skills, rather than on formal scientific research. Productivity was enhanced in part through the development of the *American system of manufacturers* for the production of light machines and other mechanical devices (Mowery and Rosenberg 1993: 31).

Power industry's most famous pioneer, Thomas Edison (1847-1931) was such an inventor-entrepreneur, who focused on inventing new technology to profit from it (Hughes 1989: 18-19). He formed several firms to organize his invention and the introduction of the lighting system: an R&D company, others for manufacturing components, and another to preside over the control of the system. Edison managed the whole process of technological change, from problem identification to innovation and technology transfer.[2] Edison established the technological basis for the electrification of urban areas. In 1882, he initiated the world's first electrical power generating and distribution system in the heart of New York's downtown Pearl Street financial district.[3]

In the 20th century, in-house industrial research came to supplant the inventor-entrepreneur. As the American manufacturers became inspired by the theories and experiments of Frederick Taylor, and by the practices and ideology of Henry Ford (among others), mass-production systems came to typify much of U.S. manufacturing.

8.1.1 The Corporate Innovation Tradition

Henry Ford (1863-1947), who regarded his friend Edison as his role model, saw himself as a productive industrialist and relied heavily upon American faith in technological genius and industrial entrepreneurship (*ibid.* 293-294). He was fascinated by the idea of organizing production on a mammoth scale. The organization itself became seen as a machine and workers as parts of that machine. Frederick Winslow Taylor's (1856-1915) principles of scientific management provided the cornerstone for work design within such a mechanized organization. Taylor advocated five simple principles, which can be summarized as follows (Morgan 1986: 30):

1. *Shift all responsibility for the organization of work from the worker to the manager*; managers should do all the thinking relating to the planning and design of work, leaving the workers with the task of implementation.
2. *Use scientific methods* to determine the most efficient way of doing work; design the worker's task accordingly, specifying the *precise* way in which the work is to be done.
3. *Select* the best person to perform the job thus designed.
4. *Train* the worker to do the work efficiently.
5. *Monitor* worker performance to ensure that appropriate work procedures are followed and that appropriate results are achieved.

The effect of Taylor's system of management was enormous. For industrialists it offered the prospect of increasing productivity manifold. For the workplace it presented a danger, since it implied the replacement of skilled craftspeople by unskilled workers. As managers became seen as engine drivers, workers were reduced to automatons. Henry Ford passed lightly over the fact that for most people assembly-line work is simply boring or alienating. Instead, he was driven by democratic sentiment to mass-produced automobiles so that everyone might benefit from this luxury article (Ellul 1964: 211). According to Ford, standardization could be used to create social equality. Standardization was thought to reduce prices. Consequently consumption would increase, welfare would be more widely distributed, and living standards would be equalized.

Ford explicitly recognized that "mass production meant mass consumption, a new system of the reproduction of labor power, a new politics of labor control and management, a new aesthetics and psychology, in short, a new kind of rationalized, modernist, and populist democratic society." (Harvey 1990: 125-126) According to David Harvey (1990: 125-126) in his brilliant book *The Condition of Postmodernity*, this separates Fordism from Taylorism. Fordism symbolically started in 1914, when Henry Ford introduced his five-dollar, eight-hour day as recompense for workers manning the automated car-assembly line, he had established the year before at Dearborn, Michigan, to produce the Model T. By doubling wages to his famous 'five-dollar a day' Ford was able to stabilize the work situation and to persuade workers to accept the new production technology.

At the same time, it was meant to provide workers with sufficient income and leisure time to consume the mass-produced products the corporations were about to turn out in ever-vaster quantities. Ford believed so strongly in corporate power that he increased wages with the onset of the Great Depression in the conviction that this would boost effective demand, revive the market, and restore business confidence. However, the crisis tendencies of capitalism proved too powerful for even the mighty Ford. "It took Theodor Roosevelt and the New Deal to try and save capitalism by doing through state intervention what Ford had tried to do alone." (*ibid.* 126-127)

Along with the great corporations' demand for specialized skills, the number of technicians and salaried professionals multiplied. After the Civil War, the modern university had emerged in the United States.[4] Besides growing in numbers, the professorate acquired a lot of influence and prestige in and out of the classroom (Hofstadter 1955: 154). In line with a trend toward specialization and bureaucratization, the modern university promoted a division of labor between the university and the professions (Schön 1983: ch. 2).

The latter were to give their practical problems to the university. Then scientists at the university or in separate R&D laboratories were to create the fundamental theory which professionals and technicians within corporations would apply to practice. "Thus were planted the seeds of the Positivist curriculum, typical of professional schools in American universities, and the roots of the now-familiar split between research and practice." (*ibid.* 37)

Industrial research developed in parallel with academic research, and powerful linkages between the two existed. As noted before, federal antitrust policy contributed to industrial research investment by large firms. Mowery and Rosenberg (1993: 34) observe that this policy "paradoxically may have aided the survival of these firms and the growth of a relatively stable, oligopolistic market structure in some U.S. manufacturing industries." Among the manufacturers of turbine generators for the fledging electric power business, two stood out: the General Electric Company and the Westinghouse Electric Company (Hirsh 1989: 37).

The growing number of engineering personnel accelerated the use of scientific knowledge and problem-solving techniques within the industry. Before 1940, however, the training of these engineers was often elementary in character. A look at the engineering community in the American electric utility showed that the scientific approach to technological innovation was still in its infancy. The historian Richard F. Hirsh (1989: ch. 3) writes that the industry's rapid expansion during this period created a fertile environment for developing and introducing new technology.

The utility managers wanted the technology to be reliable, efficient, and cheap. As a result of the demanding market, manufacturers developed their new machines using a so-called *design-by-experience* technique (*ibid.* 38). This meant that, based on the experience of their customers, manufacturers incrementally improved on their designs. Such an approach was necessary because sophisticated theory simply did not exist. Because engineers could not understand everything about their hardware and the environment in which it operated, they typically compensated by building in extra margins for error. "The practice led to machines that the literature described as having 'generous' designs." (*ibid.* 39) Things changed after the Second World War.

8.1.2 The Civilian Innovation Tradition

> The United States, once criticized as the creator of a crassly materialistic order of things, is today throwing its might into the task of keeping alive the spirit of Man and helping beat back from the frontiers of the free world everywhere the threats of force and of a new Dark Age which rise from the Communist nations. In defeating this barbarian violence moral values will count most, but they must be supported by an ample materials base. *President Truman's Materials Policy Commission, 1952: 1*

> Paradoxically, the Manhattan Project's success in creating weapons of unprecedented destructive power contributed to rosy post-war perceptions of the constructive possibilities of large-scale science for the advance of societal welfare. *David C. Movery and Nathan Rosenberg, 1993*

The economist John Maynard Keynes (1883-1946) became the most famous advocate for using scientific managerial strategies and state powers to balance demand and supply in the economy, and thus stabilize capitalism. In 1940, Keynes reviewed the New Deal experience. Discontented, he concluded "It seems politically impossible for a capitalistic democracy to organize expenditure on the scale necessary to make the grand experiment that would prove my case – except in war conditions."[5] His words proved to be remarkably prophetic.

There had been under peacetime conditions an immense weeping and wailing over the budgets of the Roosevelt Administration. During the war, however, it became hard for either capitalists or workers to refuse large-scale state planning as well as thorough rationalizations of the labor process, which improved efficiency at a time of all-out war effort (Harvey 1990: 127). As a result, both the role of the government in determining the course of the economy as well as the use of Fordist production technology became widely accepted in America. Fordism became firmly connected with Keynesianism, and the capitalist planning welfare state was born (*ibid.* 131).

The Second World War also created the dominance of the view of professional knowledge as the application of scientific theory and technique to practical problems.

In the war technologists drew upon scientific research as never before.[6] The atomic bomb project, known as the Manhattan Project, yielded several major legacies. Its successful completion created a military-industrial complex. It also became the very symbol of the successful use of science-based technology for national ends. Its director, Vannevar Bush, became the main architect of America's post-war governmental approach to technology and science policy. Bush (1945) saw basic science as an *Endless Frontier*[7] and the university as the organization providing the necessary autonomy for scientists to explore and conquer those untamed realms (Slaughter 1993: 278). Basic research was thought to bear fruits for all, and was seen as a non-partisan issue. As such, it captured a bipartisan consensus to expand science funding. Scientific knowledge would be used to create welfare and solve societal problems.

In his 1953 *Atoms for Peace* address to the United Nations, President Eisenhower set in motion a greatly expanded civilian program for the development of nuclear power plants.[8] In the mid-1950s, the Soviet launching of Sputnik evoked another effective national technology project: the Apollo Project to place an American on the moon within the next decade. On July 21, 1969, Neil Armstrong became the first human being to set foot on the moon. The space program once again strengthened the faith politicians had put in the American science community. Moreover, it reinforced the technological confidence of American engineers and scientists themselves. The act of 'putting a man on the moon' became the new and ultimate symbol of success of the government-directed technology development based on fundamental science in combination with the normal engineering model.

The post-war rise of the scientific approach at the expense of craft was also reflected in the change of technology strategy within the engineering community of the electric industry. The craft-based practice was abandoned after the Second World War, in favor of a science-based *design-by-extrapolation* approach (*ibid.* ch. 4). Instead of waiting for experience – learning-by-doing and learning-by-using (Rosenberg 1982) – to accrue while observing how earlier machines operated in the field, manufacturers made abrupt jumps in design to the next stage. A Westinghouse engineer noted that the company extrapolated "on the basis of calculations, model tests, and the belief that if a small step is ok, the next one would be ok." (*ibid.* 63) Since designers had new and more resistant materials, and computer-based design tools had come at hand, both manufacturers and utilities felt confident that forsaking the design-by-experience approach would still lead to successful technologies (*ibid.* 64).

8.1.3 The Crisis of Keynesianism and Fordism

> More generally, the period from 1965 to 1973 was one in which the inability of Fordism and Keynesianism to contain the inherent contradictions of capitalism became more and more apparent. On the surface, these difficulties could best be captured by one word: rigidity. *David Harvey, 1990: 141-142*

Thus, during the postwar period, corporate managers began to take the supremacy of Fordism and technological progress for granted. Politicians also embraced Fordist mass production technology and came to see scientific knowledge as a way to create wealth, achieve national goals, improve human life and solve social problems. Though science represented a force for progress, its methodology was thought to be incomprehensible for political and managerial forces. Serene R&D islands were trusted to come up with unpredictable but beneficial results. Both the survival of the corporations as well as state and public well being were regarded to be dependent on science, the source of progress and technological wonder. But at the same time scientific management and technical professionalism declared their triumph, the limitations of the rigid *top-down* approach to management and technology innovation had become visible.

The detractors can be divided in two groups. The first wants to enlist government to achieve what it feels to be a fairer and more just society. The second is very suspicious of government. Within the first group, some focused on science and technology, and others criticized the role of large corporations. With respect to the latter, Ralph Nader stands out as the most important example. In 1965, he wrote the book *Unsafe at Any Speed*, on automobiles. The result was a great national demand for controls on car design, and the emergence of all kinds of consumer advocacy groups all around the country. The war in Vietnam prompted people within the scientific establishment itself to protest the misuse of scientific and technical knowledge, especially as it related to defense research at universities (UCS 1993). In 1969, a large group of students, scientists and engineers at the Massachusetts Institute of Technology (MIT) founded the Union of Concerned Scientists (UCS) that soon became one of the nation's foremost nonprofit organizations dedicated to advancing liable public policies in areas where technology plays a critical role.

The Appropriate Technology (AT) movement did not regard government involvement as a step in the right direction. It did criticize the autonomous role of scientists in the development of technology. Its so-called *soft* side emphasized the role of lay people in the formulation of technological needs and the insertion of values into the scientific process.[9] Its *hard* side stressed the healing effect of the market, the need for applied and product-oriented research, and the use of more flexible production technologies. Both sides will be briefly discussed.

8.1.4 The Communal Innovation Tradition

By the end of the 1950s, private groups started to address the failings of the postwar aid efforts in Third World countries. In 1949, President Harry S. Truman had called for the United States to provide the technical assistance necessary to overcome the poverty of the underdeveloped world. The aid consisted of two parts: "the technical, scientific, and managerial knowledge necessary to economic development," and "production goods – machinery and equipment – and financial assistance in the creation of productive enterprises" (Pursell 1993: 631). The so-called 'pre-industrial' countries were thought to lag behind. If only they would follow the same pattern of industrialization as the donor nations, development would automatically happen. (Reitsma and Kleinpenning 1991: 123) Thus, large factories, mechanized agriculture, the rapid exploitation of natural resources and good infrastructure were seen as critical. Often, such efforts ignored or misunderstood local environments, both natural and cultural. As the awareness grew that the use of professional knowledge and super-technology could not really alleviate world poverty, it also was noticed that the engineering model was unable to solve the problems of modern industrial society itself.

The soft side of the Appropriate Technology movement challenged the above hegemonic technocratic notion of technology, which was believed to be destructive of social and natural health. It was argued that industrial technology recognized no self-limiting principle – in terms, for instance, of size, speed, or violence (Schumacher 1973: ch. 5). A more reserved practicing of science was needed in which "thought, feeling and activity would form a unified whole" (Riedijk 1986: 82). The soft side favored a certain type of technology. For example, the Farrallons Institute at Point Reyes, California, defined appropriate technologies as those "that are: 1) cheap enough to be accessible to nearly everyone, 2) simple enough to be easily maintained and repaired, 3) suitable for small-scale application, 4) compatible with man's needs for creativity, and 5) self-educative in environmental awareness." (Pursell 1993: 632)

For developing 'technology with a human face,' a craftsmanlike epistemology of practice, to cover the divide between theory and natural and cultural context, was promoted. The Appropriate Technology advocates have often been falsely accused of being anti-technology. They neither were this, nor did they long for the primitive technology of bygone ages. Yet, they did wish for the old times of craftsmanship to return, when 'expertise' implied knowledge and craft, creativity and art. They regretted the fact that the artistic craft was rapidly disappearing in favor of the positivist epistemology of practice, and argued that the latter was not able to deal with phenomena like "complexity, uncertainty, instability, uniqueness, and value-conflict". (Schön 1983: 39)

Appropriate Technology proponents also claimed that the system of production by the masses was preferable to the capitalist mode of production. In line with Schumacher (1973: 163), they argued that, "The technology of mass production is inherently violent, ecologically damaging, self-defeating in terms of non-renewable resources, and stultifying for the human person". In the 1970s, private initiatives were soon followed by efforts of government at all levels to help develop appropriate technologies. The Office of Appropriate Technology was founded in California, and on the federal level the National Center for Appropriate Technology was initiated by President Carter. Within these governmental initiatives in particular the hard side of the Appropriate Technology movement could be discerned.

8.1.5 The Entrepreneurial Innovation Tradition

> People ... want to contribute to the quality of life in their communities. They want to create a network of cooperating producers that can help their communities become more self-reliant, to recycle consumer dollars locally, and to shift to a greater use of locally available renewable resources. *Ronald Reagan, 1979*[10]

Some sixty years after Taylor had published *The principles of scientific management* on how to order large-scale organizations, Ernst Schumacher (1973: ch. 2) also formulated five principles which he believed would restore the balance between order and creative freedom within large-scale organizations. He proposed that large corporations should consist of many semi-autonomous units. So-called *quasi-firms* were to be given a large degree of freedom, "to give the greatest possible chance to creativity and *entrepreneurship*." (*ibid.* 261) Clearly, the management culture that big was beautiful had come under attack.

In the postwar period American managers had become obsessed with huge size and mass production (Reich 1983). Economies of scale were evident in the automobile, shipbuilding, steel, petrochemical and rubber, and (consumer) electrical industries; the propulsive engines of economic growth after the war. Huge firms were thought to be technologically more progressive than small ones. They were able to set up large, specially equipped R&D laboratories generally unavailable to small firms. Moreover, they could afford the expertise and organization to make use of advances in technology that increased the optimum plant size.

But in the late 1960s, problems became apparent "with the rigidity of long-term and large-scale fixed capital investments in mass-production systems that precluded much flexibility of design and presumed stable growth in invariant consumer markets" (Harvey 1990: 142). Some people began to notice that large-scale enterprises found it difficult to cope with the new individualized demand for variety. They pointed out that, besides economies of scale, there were also certain economies of small-scale production technologies.

New technological developments boosted this view. In the field of energy, food, information, building, clothing, and manufacturing, advanced small-scale technologies were thought to have a strong decentralizing effect (Boes 1989: 69). California and Massachusetts represented states in which initiatives from new small and medium-sized technology-based firms, stimulated by federal R&D expenditures dating back to the Second World War, resulted in industrial growth areas now identified as Silicon Valley near Stanford University, and Route 128 near Harvard and MIT (cf. Lindsey 1988). In Europe the 'Third Italy'[11] served as a key example of an industrial region that was based on a *flexible specialization* economy – small scale, but technologically advanced, and highly productive. Within these districts, companies showed both competitive and cooperative behavior. Firms competed strongly for innovation in style and efficiency, while cooperating in administrative services, raw materials purchases, financing, and research. These networks of small firms thus combined low vertical integration and high horizontal integration.

Among these firms, the high-growth small and medium-seized enterprises were characterized by having an aggressive growth strategy, being strongly market-oriented, being export-market oriented, being 'new product development' oriented, showing a willingness to seek external capital, being extensively externally networked, being strategic in outlook, and having a fairly broad enabling knowledge base (Rothwell and Dodgson 1993: 12). The willingness to seek external capital is an important feature, because it indicates that entrepreneurs following business objectives are willing to sacrifice autonomy and independence to achieve them. Other, less growth-oriented entrepreneurs, might value 'psychic income' – such as personal achievement and job satisfaction – more than the material rewards achieved from high growth (*ibid.* 13).

Piore and Sabel (1984) hypothesize that the flexible industrial districts represented the 'wave of the future' in the world economy, and signaled the end of Fordism. Harvey (1990: 189) accounts that the authors saw "a parallel between the current conjuncture and the missed opportunity of the mid-19th century, when large-scale and eventually monopoly capital ousted the small firm and the innumerable small-scale cooperative ventures that had the potential to solve the problem of industrial organization along decentralized and democratically controlled lines."

Also, government-directed basic research was no longer celebrated, and the autonomy of the scientific community came to be viewed as unfeasible. Policy makers were criticized for the fact that they devoted minimal attention to the economic payoffs of public R&D investments. When in the early 1980s the economy was in a recession, "the ideas undergirding the semi-autonomous science establishment by including science among the state functions in need of privatization, deregulation, and commercialization" became contested (Slaughter 1993: 281). As part of its antistatist ideology, the Reagan Administration wanted to loosen the hold of the federal government on science and technology development. In addition, the relationship between university research and industry was strengthened to unleash entrepreneurial energy, thereby rejuvenating the economy. Business-university partnerships that had been initiated during the Nixon years expanded and National Science Foundation began to subsidize commercially relevant applied science. Thus a combined market-pull and technology-push approach replaced the earlier science-based, government-directed strategy.

8.1.6 The Techno-Economic Roots of the Californian Ideological Coalitions

The different approaches to innovation involve political, management and technological traditions. The different ideological coalitions, which represent different political traditions, have different opinions about which actors should be involved in the development of technology. In view of the concept of the innovation network, the techno-economic roots of the various California ideological coalitions, as described in Part Two, will be studied in this subsection (Figure 8-1 and 8-2).

Harvey (1990: 124) adopts the view that "the long post-war boom, from 1945 to 1973, was built upon a certain set of labor control practices, technological mixes, consumption habits, and configurations of political-economic power, and that this configuration can reasonably be called Fordist-Keynesian." When he talks of a 'Fordist political consensus' (ibid. 167), it is clear that this embraces the liberal consensus between the corporate capitalist and public interest coalition that was described in the previous Chapter. However, by adding Fordist to the political consensus, the strong relationship between the concept of the Keynesian welfare state and the techno-economic power of large hierarchically organized corporations becomes evident.[12]

This liberal consensus came to shape federal science policy (Williams 1961). A strong involvement of the government in technology development was advocated as a means to strengthen the liberal welfare state. Both politicians and corporate managers supported a science-based strategy to facilitate technological progress. Although fundamental research was seen as principally unmanageable, the liberal establishment trusted that scientists would come up with results anyhow.[13]

Within this conception, there is a corporatist approach to setting goals of technology development, and the scientific community solves the 'technical' problem in a normal engineering fashion. In the terms introduced above, one could say that the liberal establishment advocated combined *corporate/civilian innovation networks* in which large corporations played a dominant role in the business pole and the technical pole, and in which the government invested a lot in basic science. This type of innovation network would foster economic growth and profits as well as heighten general welfare.

This creed became tarnished within the public interest coalition during the second half of the 1960s. Pragmatic members maintained their belief in technological progress and corporate America. Its sectarian part started to stress that techno-economic innovation was not automatically in the public interest; scientific results could be misused, and the immense productivity of the United States was often realized at the expense of the environment and consumer safety. The sectarians promoted a more active role for the government to control technological development and regulate the market. In the search for solutions they became inspired by and often supported the views of the Appropriate Technology movement.

The soft side of the Appropriate Technology movement promoted community-directed technology development. This clearly is a perspective on innovation that matches the philosophy of the communitarian coalition. Science and technology are no longer seen as autonomous, but subordinate to local needs. A "self-help technology, or democratic or people's technology – a technology to which everybody can gain admittance and which is not reserved to those already rich and powerful," (Schumacher 1973: 163) was promoted. Technology development becomes a political activity to empower people at the local level. Ideally, political, economic and technological aspects form a unified whole.

The communitarian coalition derided the lack of democratic involvement within corporate/civilian innovation networks. The grassrooters abhorred the rigid hierarchy that existed within the American engineering community between theoretical and practical knowledge, and condemned the dehumanizing nature of mass production systems. Instead, they envisioned *communal innovation networks* that embodied a sociocratic way of designing technology based on the natural and cultural context of its utilization.

With respect to the business pole of such a network, the Appropriate Technology movement's soft side pictured small entrepreneurs who pursued a steady business related to emancipation, democratization, and craftsmanship. The consumer is interpreted in terms of the community.

The hard side of the Appropriate Technology movement incorporated a vision on technological innovation that was clearly linked to libertarianism. It favored *entrepreneurial innovation networks*. As with the views of the liberal establishment, the public was amorphous, but now captured by the 'market'. Large over-organized firms making cheap mass products were no longer the proper actors to fulfil the needs of the market. Instead, small-scale, high-tech adhocracies mastering flexible production technologies were regarded as the key players in renovating America's economy and technological capacity.

Innovation network pole	Corporate innovation tradition	Civilian innovation tradition
Science	Corporate R&D laboratories, and engineering departments	Universities, national laboratories
Technics	High-tech; Academic engineers	High-tech; Academic engineers
Business	Large corporations; Taylorist/Fordist mass production	In principle all types of firms; Taylorist/Fordist mass production
Consumer	Mass consumption	Public interest
Political	Corporate capitalist coalition	Public interest coalition
General	Large corporations have the proper technological resources to drive the process of technology change. They can also organize mass production, which enables mass consumption, and thus general welfare.	Scientists and engineers drive the process of technology development. This can be guided in the interest of the general public by regulation of markets or national civilian R&D projects.

Figure 8-1
The techno-economic roots of the American corporate capitalist and public interest coalition.

Innovation network pole	Communal innovation tradition	Entrepreneurial innovation tradition
Science	Not applicable	Universities
Technics	Appropriate technology; Craftsmen	High-tech; Academic engineers
Business	Small, local entrepreneurs seeking 'psychic income'; production by the masses	Growth-oriented small and medium-sized firms; Taylorist/Fordist mass production
Consumer	Local community needs	Mass consumption
Political	Communitarian coalition	Libertarian coalition
General description	'Appropriate' technological renewal guided by small, local entrepreneurs or community efforts.	Small growth-oriented firms drive the process of technological change. Their entrepreneurial dynamism, and internal flexibility make these firms very responsiveness to the needs of consumers.

Figure 8-2
The techno-economic roots of the American communitarian and libertarian coalition.

The libertarians argued that science had to accommodate industrial problems and industrial motivations, and saw speeding up the product innovation process as the key to economic prosperity. "[B]asic and applied science were no longer treated as dichotomous, but as complementary parts of a whole entity: the three-step innovation process." (*ibid.* 290) Basic science was the first step, development the second, and the third involved creating specific products. Not democratically elected politicians, but small innovative entrepreneurs were regarded as the designated actors to attribute meaning to the technology. Consequently, policy actors who follow a free-market approach willingly give up their ability to come to grips with technological renewal.

Three ideal types of American innovation traditions and/or related innovation networks have been distinguished. The forefront of the corporate and entrepreneurial innovation networks are similar, but differ with respect to the business pole. The libertarian coalition favors dynamic and innovative small and medium-sized firms over 'stagnant' large firms, while the corporate capitalists argue the other way around, enlisting government when it protects corporate investments and markets, and stabilizes profits. Free-market libertarians do not want any government, 'with the exception of stop signs'. The communitarian coalition favored communal innovation networks in which appropriate technology was developed by local not-profit-oriented firms.

In the post-war period, the public interest coalition relied on the power of corporate innovation networks. Basic science was supported, and markets were regulated and created. To indicate the heavy government involvement within the corporate innovation networks, the term corporate/civilian innovation network was coined. During the 1970s, faith in corporate innovation crumbled. Sectarians within the public interest coalition came to support either an entrepreneurial or a communal innovation approach. While the public interest coalition is ambiguous with respect to the proper innovation approach, the other three ideological coalitions all have clear techno-economic roots.

8.2 Danish Innovation Traditions

> Records of the Danish industrial history, told after the traditional ideals of the English or American industrial revolutions, do not offer much material to enlighten our questions, apart from the fact that they all share the disappointment of a very slow process of which it is even difficult to decide on a proper take off period. *Peer Hull Kristensen, 1989*

The Danish path to modernization is very different from the American one.[14] In the last three decades of the 19th century, American railroads had brought the corporate form of organization that dominated the urban industrialization process. In this period, Copenhagen was also passing through a stage of transformation. In the early 1860s, the old guild barriers were eliminated by the state. At the end of the decade, industrial development started, creating a new dependent wage earning class. Newly founded Copenhagen banks[15] envisioned an American type of development in Denmark and stimulated mergers and trusts. In rural Denmark, however, a different type of economic development was already underway. Here, the agricultural cooperative movement took the economic lead.

8.2.1 The Cooperative Innovation Tradition

Two years after the loss of Slesvig in 1864, the first cooperative supply association was set up. Only two decades later, the country had more than a hundred consumer cooperatives that provided the inputs local farmers needed for their households or for production. Producers' associations followed. In 1882, the first cooperative dairy was founded at Hjedding in West Jutland. By 1890 there were almost 700, and in 1914 more than 1,500, about one in every rural district. Other cooperative enterprises, e.g. slaughter houses, were also established one after the other. In addition, a complex of regional and national enterprises was born, to organize international trade and deal with financing. The agro-industrial complex was very successful in developing, producing, and marketing food products of a uniform, high standard. State-financed R&D work and quality control supported these efforts. Edquist and Lundvall (1993: 270) note that cooperative ownership stimulated the rapid diffusion of technological innovations. In addition, the farmer's self-organization effectively cut off the Copenhagen banks from gaining control over Denmark's economically most important sector; the agricultural sector.

Political pressure from groups of local cooperatives resulted in the construction of a dense railroad network. When railway stations were build on unexplored land, a local center of cooperative firms rapidly emerged. Their activity attracted many craftsmen. The railroad towns offered ample opportunities for setting up small shops, working both as dealers and producers. These shops evolved step-by-step "into small manufacturing firms that in the beginning worked as subcontractors in order to compensate for seasonal fluctuations and later on developed into specialized industrial enterprises reaching out for both national and international markets." (Kristensen 1989: 173)

The decentralized structure of the rural production system resulted in a decentralized rural electricity production system. In 1891, Køge and Odense became the first two Danish cities to have so-called electrical lighting systems as introduced by Thomas Edison. Though in the following years electricity was introduced in other provincial towns, most developments in this field took place in Copenhagen, where technical expertise, legislative power, and capital all came together (Wistoft, Petersen, and Hansen 1991: 51). At the time, the electrification of small villages and dispersed farms looked like a utopia, and the electrical revolution seemed to pass by the small cooperative firms in the railroad towns. The Grundtvigian physicist and teacher at the Askov folk high school, Poul la Cour, however, refused to throw in the towel.

The 'Danish Edison' wanted "to establish the technological basis for electrification of the rural areas, securing the small associations their share in the blessings of the new electrical energy" (Rasmussen and Øster 1990: 8). To implement decentralized electricity systems in 1903, La Cour initiated the Danish Wind Power Association (*Dansk Vind Elektricitets Selskab*, DVES). At Askov, DVES instructed rural electricians who assisted farmers and cooperatives in setting up rural power stations. At the end of the First World War, the number of rural power plants was 418. Over a quarter was wind electricity system, while the remainder consisted mainly of small diesel-based electricity plants produced by the Danish firm, Thrige Titan, in Odense.

8.2.2 The Flexible Innovation Tradition

In the cities, the former guilds tried to maintain among themselves the old guild bondage through forming new types of associations, and at one very important point. From 1870 onwards, craft-associations in most of the larger towns initiated local technical schools, educating apprentices in matters of theory, technical drawing, calculus, *etc.* Both liberals and socialists, who were usually opposed to the guild system, supported this educational movement, and by the end of the century 120 technical schools had been established. Specialized technical colleges (*Teknikum*) were set up, where skilled workers further educated themselves. This allowed *teknikum* engineers to compete with the emerging academic engineers from the Danish Technical University (*Danmarks tekniske Højskole*, DtH) in obtaining managerial careers.

Increasingly, the state came to sponsor these schools and to regulate apprentice contracts. The educational system was also strengthened by the fact that the unions in Denmark organized workers according to their craft. The institutional formation of craft education spread all over the country. By the 1930s, "craftsmen whether owners of small shops or firms, or skilled workers from cities as well as from small peripheral railway towns shared the feeling of, and *de facto* belonged to a common system of craftsmanship" (Kristensen 1989: 174).

As a result, the social mobility of the Danish worker was relatively high and workers could easily move from one firm to the other. Skilled workers found themselves in the right place to build up contacts and knowledge to start their own small enterprise. Thus common values and a common work ethos came to exist among small entrepreneurs and craftsmen. The educational system of artisans also shaped the organization within big firms.

At the turn of the century, the largest Danish firms produced ships, freezing equipment for ships, trains and wagons, and machinery for dairies and slaughterhouses. Although these large companies often tried to modernize in accordance with their American competitors, they eventually found their strength in producing small quantities of complex and often huge constructions. Instead of Fordist mass production plants, the organization of these companies came to consist of "a conglomerate of smaller craft shops, working together in a pattern similar to that of a building site" (*ibid.* 175). These workshops were run by foremen, recruited from the skilled workers, who often set up their own independent firms, ending up eventually as subcontractors for the larger ones where they had gained their experience and education.

The craft ethos and flexible production in small batches have constantly been challenged by Taylorist organization principles. Especially after the war, many Danish managers tried to modernize their organization along the ideals of Fordism. Kristensen argues incisively that the reverse can also be said: the craft-organization challenged

Fordist management models. He reveals that at the end of the 1960s, many firms had already abandoned the Fordist mass production strategy to a search for market niches. Through the years, a wealth of small firms had kept dominating the face of the Danish production structure. From the mid-1970s, the average enterprise even started to decrease in size. Larger firms in the larger cities in eastern Denmark closed, while small enterprises in rural areas grew in number and in size. "Consequently, in the 1970s and 1980s, traditional craft dominated industries as the iron and steel industries, and furniture industry, became of increasing importance to the Danish economy, reproducing the structural conditions for the strength of craft principles in the work organization at the micro level." (*ibid.* 184)

The electric industry seems to provide an interesting example of how the craft ethos was preserved despite or maybe because of centralization. As has been described, the Danish electricity system consisted of several hundred small hydro and diesel stations. After the war, the initial structure rapidly evolved into a system of large central power stations. Both this process of technical centralization, as well as the concentration of administrative power, was justified by economic arguments, and widely supported within the Danish establishment. During the 1950s, the electric utility industry came to consist of two large power pools.[16] The style of choosing and pushing technology in Denmark differed from the United States (cf. Hirsh 1989: 75-76). The Danish utilities had a tradition of designing and building power stations themselves. As a result of centralized planning, they were able to build several identical fossil-fuelled power plants, from which experience would be acquired and bugs worked out. Thus, within the engineering community of the Danish electric industry, also after the Second World War, the cautious design-by-experience approach to technological innovation was maintained.

The national government and academic community challenged this attitude. Political concern about the future supply of fossil fuels inspired the foundation of a nuclear research center at Roskilde. In the establishment of Risø National Laboratory – Denmark's largest sector-research institution – the internationally celebrated physicist, Niels Bohr, and the then Minister of Finance, Viggo Kampmann, played a vital part. Like other Social Democrats, Kampmann sensed that a new technological epoch was approaching, and novel means were needed to support the industrialization process in Denmark (Risø 1983).

Risø was founded on the American model, with its strong emphasis on basic research. This made close contacts to universities a natural thing. In particular, the collaboration with the Danish Technical University became very active. Still, the relations with the electrical utilities were disappointing. For fear of getting involved with unproven technology and losing their technological independence, the utilities were not interested in nuclear power. When in the early 1970s, the utilities started to express their wish to build nuclear power plants, the relationship with Risø and the utilities improved dramatically. However, the anti-nuclear movement and the resulting change in public opinion became a traumatic experience for many of Risø's staff (Risø 1983: 27). In 1976, Risø's mandate was broadened to include non-nuclear energy research and paid contract work. The institute successfully diversified its energy research, and greatly increased its contacts with the industry, amongst others the wind turbine industry.

8.2.3 The Techno-Economic Roots of the Danish Ideological Coalitions

The transformation of the Danish economy in the last quarter of the 19th century illustrates the importance of social movements and political visions. According to Edquist and Lundvall (1993: 293), "Grundtvig might be regarded as the most important single Danish innovator, as one who has put his marks on the modern Danish system of production and innovation".

His ideas about self-reliance and the educational system he envisioned prepared the ground for the national movement toward farmers' cooperatives. The cooperative coalition advocated so-called *cooperative innovation networks* (Figure 8-3). It wanted to orient industrial policy "toward support for something akin to Schumacher's intermediate technology".[17] The Danes expected small local firms and democratically organized cooperative enterprises to develop these appropriate technologies. Within the user pole, consumer cooperatives were thought to play an important role.

Innovation network pole	Cooperative innovation tradition	Flexible innovation tradition
Science	Not applicable	Universities, national laboratories
Technics	Appropriate technology, local craft; skilled workers	Combination of craftsmanship and engineering knowledge; Academic engineers, *teknikum* engineers and skilled workers
Business	Collective and democratic forms of entrepreneurship; production based on Danish craft ethos	In principle all types of firms; flexible production system based on Fordism and Danish craft ethos
Consumer	Local community needs	Public interest
Political	Cooperative coalition	Established coalition
General description	Technological innovation to maintain the right of cultural self-determination	Technological progress to achieve economic growth and sustain the welfare society

Figure 8-3
The techno-economic roots of the Danish cooperative and established coalition.

As Kristensen (1989) elegantly shows, the cooperatively organized agroindustrial complex had much wider implications for the Danish industrial structure. It obstructed concentrated ownership and financial control, as had developed in the United States, while it promoted the establishment of many small-scale firms in rural areas. The folk high school movement that inspired the farmers thus eventually reinforced the craft-educational system that was set up by the old guilds in the formerly privileged cities.

The craft-ethos also came to influence the production culture of the larger firms.

Post-war Fordism represented a threat to this work. Although the Social Democrats accepted the prospect of degradation of work content, they compensated workers in two ways: by collective agreements and wage increases, and through expanding the educational system (*ibid.* 179-181). Skilled workers in large numbers tried to become *teknikum* engineers, and unskilled laborers developed into semiskilled workers. In this way, the craft character of the workers and the production culture was reinforced. It is fair to say that the established coalition promoted the compromise between Fordism and the Danish craft ethos. This truce fostered so-called *flexible innovation networks.* In the technology pole of these networks both (academic and *teknikum*), engineers as well as skilled workers played a major role.

This in contrast to the corporate and entrepreneurial innovation networks in America. "Historically, the United States has lacked a system for training craftsmen, while possessing an abundance of higher skilled (white collar) and lower skilled or unskilled workers." (Ergas 1986: 27) While, in the United States, a large and unified market pushed manufacturers in pioneering mass production techniques which made little use of craft labor, Danish firms developed small batch, flexible production systems run by skilled workers. Similar to the public interest coalition in California, its Danish equal sees techno-economic progress as a way to sustain the welfare society. Moreover, the established coalition is also ambiguous with respect to the question of which firm size is conducive to techno-economic progress.

8.3 Politically Desired Wind Energy Innovation Networks

Combined with the case descriptions and the political analysis in Part Two, the material in this chapter suggests that several ideal types of politically desired wind energy innovation networks can be discerned in both the United States and Denmark.

The liberal establishment wanted to develop wind power plants composed of numerous large-scale wind turbines that could contribute to the existing centralized electricity production system. The electric utility companies were regarded as the main future producers of wind power, and thus as the main consumers of large-scale wind turbines. With respect to the science and technical poles of the network, the liberal establishment put trust in expert knowledge that was thought to be available within national laboratories, universities and the R&D departments of large corporations. Finally, with regard to the business pole, large firms were assumed to be uniquely endowed to exploit innovative opportunities; large enterprises were seen as the engine of technological change.[18] In line with the nomenclature used in this chapter, this innovation network could be called the *civilian/corporate wind energy innovation network.*

The libertarian coalition challenged the monopoly power of utilities, and wanted small and medium-sized firms to take advantage of new small-scale alternative production technologies, like cogeneration and wind power plants, to get a foothold within the electricity production sector. In contrast to the liberal establishment, the libertarian coalition argued that the regulated utilities were risk averse and technologically conservative. The utilities were not seen as the appropriate consumers of renewable technology, but instead were thought to represent significant barriers to its widespread commercialization. Independent producers of power were regarded as the key actors within the user pole of the wind energy innovation network that the libertarians favored. Ideally, new technology-based, high-growth small and medium-sized firms would make renewables a competitive energy source.[19] The libertarians reckoned that university research laboratories provided a source of innovation-generating knowledge that these private wind turbine producers could use for commercial exploitation. The network that the libertarian coalition desired is named the *entrepreneurial wind energy innovation network*.

The communitarian coalition saw wind turbines as a way to decentralize control over production of electricity. They expected farmers and small neighborhoods to use small wind turbines for the local production of wind power. This would increase local self-reliance and reinforce community values. Preferably, local craftsmen within small, not-profit-oriented firms would build turbines. The communitarian coalition preferred local innovation networks that will be named *communal wind energy innovation networks*.

In Denmark, the established coalition had similar ideas about the implementation of wind energy innovation as the liberal establishment in California. Wind turbine parks would be owned and operated by utilities and were expected to be built by a consortium of large Danish firms. The knowledge for building large wind turbines was to be developed within the Risø National Laboratory in close cooperation with the Technical University of Denmark. In accordance with the terminology introduced in this Chapter, this politically desired network model is called the *flexible wind energy innovation network*.

Like the communitarian coalition in California, the Danish cooperative coalition also hoped for a network of local innovation networks to arrive. It fancied cooperative firms would produce these turbines, which would preferably be used by communities and/or consumer cooperatives. This ideal-type of network will be named *cooperative wind energy innovation network*.

9 Unraveling Wind Energy Innovation

For a successful technology, reality must take precedence over public relations, because Nature cannot be fooled. *Richard Feynman, 1986*[1]

The objective in this chapter is to describe the influence of public policy on the development of technology and industry in the field of wind energy in California and Denmark. This is a complex task, maybe too complex. But hopefully the analysis in the previous chapters will be sufficient to give the reader some useful insights. We have the benefit of hindsight and can build on the thoughtful views of many other commentators and scholars. It should be stressed that the various actors under study could not possibly profit from these advantages. They had to base their actions on uncertain forecasts about technological, political and economic developments.

This book can be seen as an attempt to examine the statics and dynamics of these processes. On the one hand, we follow Jacques Ellul's (1964: 60) assertion, "What else can history teach us? Only the vanity of believing we can impose our theories on history. Any philosophy which asserts that human experience repeats itself is ineffectual." On the other, policy belief systems were identified that were deeply rooted in national political-cultural history. These beliefs, that guide the actions of policy makers, encompass both an idealized past as well as a utopian future. One could say that beliefs serve as evaluative structures both in a *retrospective* and a *prospective*[2] manner. In addition, national techno-economic traditions operate as structures to evaluate techno-economic achievements of the past, and to speculate on the success of future innovation approaches.

The dynamics and statics of the policy formation process were described in Chapter 5 and 6, respectively. In the previous chapter, we studied the statics of techno-economic developments. Long-standing relationships were uncovered between particular political and techno-economic traditions. The term innovation tradition indicated the unity of the various types of traditions. It encompasses the mutual dependency of techno-economic and political cultures, that is, techno-economic traditions rely on the support of political traditions, and vice versa. To realize its political views, an ideological coalition has to support its desired innovation network, by seeking external and internal legitimization of its desired innovation approach.

Chapter 8 also showed that various ideological coalitions have distinct ideal types of innovation networks in mind. Corporate capitalists, libertarians, and communitarian grassrooters preferred a corporate, entrepreneurial, and communal style of innovation, respectively. The public interest coalition was found to be ambivalent with respect to its preferred way of innovating. As yet, in the 1970s this coalition relied on the innovative power of large corporations. The established coalition in Denmark had somewhat similar ideas about wind energy innovation as the liberal establishment in the United

States. It wanted flexible innovation networks to develop wind energy. The ideal type of network for the cooperative coalition was the cooperative wind energy innovation network.

In this chapter, the dynamics of innovation is scrutinized through investigating the interaction of policy makers with other types of innovation actors within innovation networks. As we saw, the success or failure of a certain innovation policy is related to choices made in other poles, and vice versa. Thus, to understand innovation and the way politics influences techno-economic development the object of analysis should be the whole network, which is defined as the complex interaction between the different actors in the various poles. Besides studying the relations policy makers have with other innovation actors within the same network, the interaction between the various innovation networks will be examined.

In section 9.1, the American federal attempt to set up a corporate wind energy innovation network will be described, since it clearly influenced the policy measures proposed in both California and Denmark. In both states, similar efforts were being made to set up a corporate and flexible wind energy innovation network, respectively. Public policy did not exclusively aim at setting up these types of innovation networks.

In California, a common sense approach was followed in which practically every innovative activity in the field of renewable energy was welcomed and stimulated. Besides promoting a corporate/civilian innovation network, communal and entrepreneurial networks were also supported. The political view behind these two innovation styles disapproves of heavy government intervention. To indicate that policy was actually made to achieve these types of innovation networks, we should attach the label 'civilian' to them. We will only do so if it makes sense for our analysis.

In Denmark, grassroots activities encouraged the government to also support the creation of a cooperative wind energy innovation network. In sections 9.2 and 9.3, we will describe the different Californian and Danish experiences until the mid-1990s. Finally, in section 9.4, the wind energy innovation processes in both countries will be compared.

9.1 The Federal Halfway Measures[3]

In the United States, efforts were made on the federal level to set up a corporate wind energy innovation network. Within this network highly novel and sophisticated large-scale wind turbines, developed using a design-by-extrapolation method, would be built by huge companies and used by electric utilities. It was assumed that only huge firms had the technical and managerial resources needed to participate in the Federal Wind Energy Program (FWEP) and privately develop and produce these turbines.[4] It was also thought that only the first half of the network needed active support. Thus R&D stimulation presented the principal policy instrument for setting up the network. After the wind energy technology had been developed within the FWEP, huge companies were supposed to set up the second half of the network.

In this section, we will describe how the FWEP was set up and received support from political actors at the federal level. We will study whether the large manufacturers and electric utility companies acted upon the forecasted technological availability and economic feasibility of large wind turbines.

9.1.1 Corporate Wind Energy Innovation

Similar to the Manhattan and Apollo Projects – although much smaller in scale and of lesser political importance – the FWEP can be regarded as just another 'national project in civilian technology'.[5] At the beginning the large-scale wind systems partly dominated the whole FWEP. Its mission was to develop large-scale, cost-efficient wind turbines that could be used as fuel savers within a utility-owned, high-tension, centralized electricity production system. Three events in particular determined the design of the program: the first Wind Energy Conversion System (WECS) workshop, NASA's involvement, and the first oil crisis.

The FWEP illustrated the gradual shift in federal policy from supporting basic science and the autonomy of scientists to promoting applied research in which business was supposed to play a major role.[6] In terms of innovation networks, this change in technology policy meant a shift in accent from the science pole to the technical pole. This strategic shift was partly inspired by the idea that it was no longer the shortage of scientific knowledge but underutilization that presented the main bottleneck in techno-economic progress. This attitude underpinned the first WECS workshop organized in 1973 to determine research priorities for a new wind R&D program.

Most of the engineers present showed faith in NASA's expertise in aerodynamics and materials, and the current practice within the aircraft industry. Assuming a sufficient technological capability within NASA and American industry, it was commonly held that developing wind turbines presented no technological barriers. The name of the game was rapid economic utilization. The technological community tried to convince the political establishment that solar energy could swiftly be developed into a commercial technology.

At the time, such a program had little prospects of receiving adequate funding, but, coincidentally, the ensuing oil crisis increased the perceived need for energy R&D. It was thought that the energy dilemma required strong government leadership and a massive technological effort. With respect to wind energy, the political establishment wanted viable, large-scale, wind turbines that would save barrels of oil *within the next decade*. The ready-made wind power program benefited instantly from this almost 'instinctive' signal for help from politicians, who urged the technological community "to assume leadership in society," and wished it "Godspeed, more progress – and federal funding".[7] Political pressure to come up with quick results neatly combined with the assessment of the manageability of a rapid development process.

Thus, the federal wind power program presents a clear example of a breakthrough action that is related to the technical pole. To speed up the development, a design-by-extrapolation approach was chosen. The FWEP started off with a 100-kW 'state-of-the-art' wind turbine based on a supposedly proven technological concept. Next, design optimization computer codes and scaling laws would be applied to generate larger prototypes and eventually an economical megawatt-sized machine. Such an approach emphasized "strictly disciplined dynamic modeling," and imposed a sequential, phased development program.[8]

The design-by-extrapolation approach corresponded with the fact that NASA was given the supervision over the program. NASA subcontracted mainly huge aerospace corporations that had already been active within the space program.[9] As an offspring of the technological community that had successfully built the Apollo spacecrafts, these companies were thought to possess all the theoretical and technological know-how to make building a commercial wind turbine a simple engineering problem. The related involvement of aircraft engineers brought in the basic community values and goals of the engineering profession in America. Among its values was the enjoyment of using advanced technology that promised to push back previous barriers in order to improve performance (Constant 1980).

The strong belief in the power of the American technological community prevented that much attention was given to the business and consumption poles of the preferred innovation network. In other words, policy makers showed little concern for the beliefs of industry and utility managers. It assumed that by determining the economically optimal configuration and size of wind turbines the FWEP would be able to come up with wind turbine technology that could be rapidly commercialized. Cost reduction was to be achieved through product innovation. Subcontractors that had the proper organizational and managerial capacity to set up assembly lines for the production of large multi-megawatt wind turbines would then be quite willing to put this new technology into production. With respect to the future usage of wind turbines, the federal policy makers simply trusted on the technological *statesmanship*[10] of utility managers. In short, federal policy makers anticipated that the business and user poles of the corporate wind energy innovation network were already well developed. They speculated that once a proper wind turbine technology had been delivered, government involvement would no longer be needed.

This raises the question of what the would-be manufacturers and users actually thought of the novel wind turbine technology. In the next subsection we take a short look at the actors that were expected to play a role within the production and user poles of the envisioned corporate wind energy innovation network.

9.1.2 The Lack of Private Sector Involvement

In times of recession, technological change and the search for new product lines and market-niches surge to the fore of corporate survival strategies. Judging by what a growing number of political energy scenarios were saying, large wind turbines seemed to provide a new high-potential product. Several firms became interested in wind energy. Some firms, Boeing, General Electric, and Westinghouse, took an easy ride by diverting federal money to cover their first R&D investments.[11] Others, like the Bendix Corporation and the Hamilton Standard Company, were so confident about the market potential for wind turbines that they privately financed their research efforts.

These corporations held that cost reduction is to be achieved mainly through the use of mass production technologies. Economic studies showed that mass-produced, megawatt-sized wind turbines were capable of supplying wind power at costs competitive with conventional power sources. Its favorable economics would even spur on conservative utilities to use wind power as a fuel saver. As a result, wind turbines were expected to produce about ten percent of the nation's electricity use at the end of the century. This business scenario was based on a commonly held conviction within management circles that the development of technology was analogous to the biological life cycle (cf. Starr and Rudman 1973).

The life cycle model distinguished four phases: a conceptual, a rapid growth, a consolidation, and a maturity phase. Along with the change of technologies, production structures were thought to "evolve from open-ended and unstructured processes toward rigid and elaborately structured processes."[12] Each stage demanded distinct marketing and production skills. In the first stage, technology emerges and has to compete with existing technologies that provide the product. Entrepreneurs have to invest money in R&D so that the technology can move to the next, more profitable stage. In the next two phases the use of the technology grows exponentially. Managers must assure that "rapid innovation continues and benefits from lower production costs accrue to the company when advances in the technology make it a 'winner' in the marketplace due to its superiority over others" (Hirsh 1989: 187). Consumer products change from unstandardized, low-volume items, to mass-produced, standard items. Finally, the technology and production processes reach their maturity. In this phase advances occur at a much lower pace.

According to the life cycle model, early involvement in the new wind turbine market would give a company a competitive advantage for the years to come. Since it was commonly held that an adequate technical base to build wind turbines existed, it was expected that the conceptual phase could be passed through swiftly. This made early involvement even more important. 'Manufacturers' hoped that several utilities would be willing to serve as partners in risk taking. Early involvement of so-called *lead users* (Von Hippel 1976) was thought to be crucial in opening up the market to a scale that would justify setting up production lines. These firms stressed the need to "go out and 'beat-the-bushes' to find people who really want to use it."[13]

It was appreciated that the introduction of new energy technologies had historically required risk-sharing relationships between vendors and utility customers (Hirsh 1989). The utilities had grown accustomed to manufacturers providing turnkey contracts at fixed prices, and guarantees on the performance of installed systems. The government's role was mainly to counteract the risk aversion of the utilities by securing utilities against loss or damage,[14] subsidizing the cost of new technology, and by allowing utilities to recover development costs in their rate settings (cf. March *et al.* 1982: 58).

Federal policy makers and large corporations differed with respect to the proper role of government. Policy makers advocated a technology-push approach to encourage the development of technical capabilities in the field of wind energy. According to the large firms, the government's main role was to initiate vital user-producer relations between manufacturers and electric utilities. The federal government only backed the mission-oriented R&D innovation strategy, and overlooked the market-oriented insights of the innovation actors within the production and user poles of the network. Several problems resulted from this.

First, large corporations scolded the FWEP for neglecting to build large numbers of test turbines to rapidly select the best design and confirm the reliable operation of wind turbines, which was crucial for getting receptive utilities involved in wind power. Federal funding was more than sufficient to follow such a practice-based innovation strategy.[15] Despite this clear criticism only a few very expensive high-tech prototypes were build that were at odds with the idea of creating cheap mass-produced turbines (Serchuk 1995: 87). After a few years, a new program manager was hired who was acquainted with the manufacturing principles of the automotive industry to counteract the high-tech influence of NASA. He later explained "I knew what volume production was, and that it wasn't going to come from the aerospace industry. ... I remember looking at the multi-hundred-dollar rivets that they were using on the blades of some of the early machines".[16] This illustrates joint innovation learning. Namely, government executives came to appreciate the demands of corporate business managers, and accordingly tried to alter the risky R&D strategy. Nevertheless, the program was already in full swing and the struggle to integrate high-volume industrial principles within the technological paradigm of the FWEP engineering community was like fighting a lost battle.

Second, it soon turned out that the utilities were not keen on getting involved in unproven wind energy technology. They stressed the need for operating experience to increase confidence in reliability and choice of technology. Since technical reliability was not proven, and the above policy incentives were not available in the field of wind energy, "the [electric utility's] perceived risks of investment into wind turbines exceeded the potential benefits of a technically successful project". (March *et al.* 1982: 62) The federal government did not act upon the utilities' unwillingness to play a role in the commercialization of large wind systems without good financial incentives. This impeded user-producer relations between potential wind turbine manufacturers and electric utilities.

9.1.3 The Public Utilities Regulatory Policy (Breakthrough) Act

More or less from out of the blue, the PURPA legislation was enacted as part of the National Energy Act of 1978. PURPA, together with the tax incentives available to cogenerators and other small power producers, provided significant financial incentives for investment in wind turbines. Utilities were specifically excluded from taking advantage of these incentives. This further dampened the utilities' interest in investing in wind turbines but created a clientele for large wind turbines that did not exist before. In this sense, PURPA can be seen as a breakthrough action related to the consumption pole of the network for large wind turbines.

Third party power producers, like Windfarms Limited, showed their willingness to assist manufacturers in developing wind technology, and financed large wind projects that allowed manufacturers to set up production lines. In an unforeseen way, PURPA completed the establishment of the large wind turbine innovation network.

For several reasons, potential manufacturers failed to act upon this. In the early 1980s there was a general loss of political support for renewable technologies at the federal level. Guided by a hands-off philosophy of government, the Reagan administration slashed the FWEP's budget. By arguing that the technology was proven, it defined the technical pole within the realm of the private sector, and thus forced industrial firms to invest in the development of wind turbines at their own risk.

This did not happen, because the prototype testing demonstrated that the FWEP designs were not ready for production. In addition, the demonstration projects falsified the idea that the technology would be easy to develop. It turned out that the scientific knowledge that was used to model wind behavior and to predict wind turbine aerodynamics and structural dynamics contained major errors. The idea that all the fundamental knowledge was already available proved to be a fallacy. Large wind turbines had been 'optimized' by using incorrect engineering design codes. It became clear to the manufacturers that the scientific and technical poles of the corporate wind energy innovation network were underdeveloped.

Finally, their market projections had been based on the assumption that oil prices would gradually increase through the years. Falling oil prices in the mid-1980s undermined this theory, and made it implausible that for some medium long period wind turbines would be used as economic fuel savers. As wind energy technology was not perceived as a core-business, most large firms eventually stopped their development efforts.

9.1.4 Looking Back on the Innovation Process

The wind energy policy of the federal government has not resulted in the establishment of a large wind turbine innovation network in the United States. Policy makers were politically constrained to focus on the scientific and technical poles of the innovation network. An enhanced federal presence within the business and consumption poles of the network found little political support. It was commonly held that the establishment of the second part of the innovation network was to take place through private decision making. But oddly, the private sector's rationale for making decisions was disregarded. It was soon found out that the R&D strategy within the FWEP was incompatible with the idea of mass production. It also became evident that the conceived users, the utilities, were risk averse with respect to renewable energy investments. Since policy makers expected existing corporate structures to automatically adapt to the political demand for wind energy technology, they neglected the need for mutual innovation learning.

Coincidentally, PURPA brought new opportunities to private sector initiatives. Wind farm developers tried to fill the gap between manufacturers and utility companies. But while these private enterprises were busy creating convergence between the business and consumption pole, the Reagan administration sharply decreased the support for the FWEP. The private sector was to take care of the commercialization process by itself. It failed because of lack of technology, government and business support.

9.2 California: Common Sense Wind Energy Innovation

> It is common sense to take a method and try it. If it fails, admit it frankly and try another. But above all, try something. *F. D. Roosevelt (1882-1945), 1932*

> It was California's alternative energy policy – the product of a progressive social and economic climate – that was instrumental in drawing the industry westwards. Although the budget of the state wind program has been minuscule compared with that of the Department of Energy in Washington D.C., in its heyday its approach was so practical as to be almost visionary. *Ros Davidson, 1989*

In many American states, more than half of all government jobs is subject to patronage. In California, however, over 98 percent of all state employees hold merit appointments (Harvey 1985: 117). Nevertheless, the influence of the governor and his personal staff on policy is enormous. In no policy area did this become more apparent than in the field of energy during Jerry Brown's governorship. He first appointed Wilson Clark as his special aide for energy policy, established the Office of Appropriate Technology (OAT) in May 1976 as a part of the governor's Office of Planning and Research, and gradually appointed renewable energy advocates to the California Energy Commission (CEC) and the California Public Utilities Commission (CPUC). Maybe more important, Brown's encouragement for new ideas in the field of energy really got people excited and attracted 'whiz kids, PhD progressive types' into government. The OAT and the CEC, which had recently been formed, became important vehicles for innovative ideas.

The great surge of young, creative people into the Brown administration had the negative side effect that the everyday bureaucratic and business side of government was neglected. Jan Hamrin, at the time a young PhD herself, says: "Sometimes we were lucky we didn't mess things up worse because of lack of understanding of how business operates. I mean, I didn't know anything about this and I was setting tax credit policy. I was writing it – you know – 'Well, what if we would just do that? That sounds like a great idea! Put it in and get it passed!' I didn't have a clue. ... A political cartoon in this morning's *Chronicle* carried me back to those days (Figure 9-1). It has Yeltsin talking to Clinton at this UN meeting. Yeltsin says 'I haven't the faintest idea what I am doing.' And Clinton says 'Neither do I. isn't it great!' That is a lot of how it was under the Brown administration. I would ask my colleague 'Well, what will we do next?' He would answer 'I don't know, just let us try something!'"[17]

With respect to the implementation of renewable energy, Brown pursued the type of pragmatic strategy F. D. Roosevelt had followed to form the New Deal. The California solar initiative was a chaos of opportunistic experimentation.[18] In terms of innovation networks, not one sort of network was privileged, as was the case at the federal level, but various types of networks were more or less simultaneously supported: a corporate, an entrepreneurial, and a communal one. In this section, we will analyze the wind energy innovation game in terms of the evolution and interaction of these three networks (see Figure 9-2).

Figure 9-1
Cartoon "Mr. Yeltsin and Mr. Clinton share a laugh." (Source: *San Fransisco Chronicle* October 25, 1995)

Figure 9-2 graphically summarizes the history of wind energy innovation in California. An innovation networks is depicted by a box, in which the various poles of the innovation network are marked by their first letter: that is, S/T stands for Science/Technics; B for Business; C for Consumption; and, P for Politics. On top of every box the type of innovation network is written. It is also shown in which subsection(s) that innovation network is described. Figure 9-2 may be helpful to keep track of the complex analysis in this section.

Subsection 9.2.1 describes how the CEC, out of dissatisfaction with the FWEP, set up or proposed a wind power program to stimulate a corporate innovation network, in which large wind turbines would be developed for the California environment. In the same year, the OAT became responsible for stimulating the residential use of small-scale wind turbines. The effort to set up a communal wind energy innovation network is described in subsection 9.2.2.

While public policy originally aimed at establishing a corporate and a communal wind energy innovation network, private initiatives by small and large firms, aroused by PURPA, encouraged Californian policy makers to pay attention to the entrepreneurial wind energy innovation approach. This is described in subsections 9.2.3 – 9.2.6.

Subsection 9.2.7 explains how a lack of policy in the second half of the 1980s and the belief in rapid technological progress led the wind power industry to shift its focus from generating wind power to developing wind farms for electric utility companies. One

Period	Californian wind energy innovation networks		
	Corporate / civilian WEIN (§§ 9.1-9.2.1)		*Communal / civilian WEIN* (§ 9.2.2)
1973-1982 **Energy crises**	S/T: **FWEP & CEC WPP** B: Aerospace companies C: Windfarms Ltd, SCE, PG&E; **PURPA** P: Pro-utility-involvement alliance		S/T: **Rocky Flats Test Station & OAT** B: Undercapitalized firms C: Local users; **Energy tax credit, PURPA** P: Communitarian & public interest coalition
		Entrepreneurial / civilian WEIN (§§ 9.2.3-9.2.6)	
1982-1986 **Wind boom**		S/T: **WPRS** (after 1985) B: USW, Danish firms C: Wind farmers, individual investors; **Energy tax credit, PURPA, ISO4, Wind Information Center** P: Regulated competition alliance	
	Corporate / entrepreneurial WEIN (§ 9.2.7)		
1986-1995 **Policy vacuum**	S/T: EPRI-Utility-USW, AWT Program; after 1990: **AWT-program** B: USW, Zond, FloWind C: Electric utilities P: Sustainable grow-and-build alliance	"**BRPU-process** in preparation" (Regulated competition alliance)	
1995- now **Wind energy in jeopardy**	**Restructuring process** (§ 9.2.8)		

Figure 9-2
Wind energy innovation networks (WEIN) in California. The boxes represent different types of WEIN's. A dotted-lined box indicates that little market development took place in the related innovation network in the corresponding period. The various poles of the innovation networks are indicated by their first letter. Public policy instruments aimed at strengthening a particular pole are printed in bold.

may argue, in terms of innovation networks, that the concept of corporate wind energy innovation, as promoted in the 1970s, was revisited.

The next subsection describes how the electric utility restructuring process killed this private development, and created a very uncertain future for wind energy in California.

Finally, in subsection 9.2.9, we will look back on the wind energy innovation process.

9.2.1 Corporate Wind Energy Innovation

> The parallel development of wind-electric systems should be conducted in the same manner in which the nation in the past developed military systems to defend our national security. *James Lerner and Mat Ginosar (CEC), 1978*

In California, the CEC criticized FWEP's risky development strategy. It proposed a wind power program with quite a number of turbines to explore the effectiveness of several variations in design detail.[19] The simultaneous testing of different designs was intended to exploit the economies of mass production as soon as possible based on ample experience and dependable design. However, lack of political support from the California Legislature impeded such an intensive broad-experience program. While wind power R&D was not regarded a proper state task, assessing the state's wind resources was. To inform potential users on proper wind sites, a Wind Resource Center was established.

After the CEC technology-push approach had got into a blind alley, the consumer pole within the large wind turbine innovation network remained. The appropriate agency to encourage electric utilities to invest in wind energy was the CPUC. The CEC recommended the CPUC to impose wind power capacity quotas on the utilities (March *et al.* 1982: 84-85). Though the proposal was not adopted, the utilities themselves began to make wind power development plans. This was part of a larger shift in management strategy of the utilities caused by myriad events.

The regulatory climate in California prohibited nuclear power and made it hard to use coal. The second oil crisis and the associated prospect of rising oil prices made the extensive future use of oil questionable. Combined, these events only left open the option of alternative energy sources. In addition, the fact that the utilities' nuclear plants were still not on line both created a need for new capacity and threatened their financial situation. The small scale and the modularity of renewables gave them short lead-times and greater flexibility; an argument that the EDF had stressed for years at CPUC rate hearings. Most important, PURPA allowed utilities to reduce their financial risk by contracting with outside developers to build and operate small-scale plants. As described above, some of them had put their faith in the development of giant wind turbines, and were playing a motor role in engendering the interests of large corporations to move into the wind turbine business. The commitment of these firms strongly increased the trust of the utilities in the possibility that this technology might be aggressively developed within the coming decade. Last but not least, the CPUC was seriously urging utilities to purchase power from small producers.

Thus dramatic changes in the business environment made the California utilities receptive to the benefits of an alternative energy supply strategy as advocated by the pragmatic side of the public interest coalition (in particular, EDF, CEC and CPUC).[20] This is a clear example of learning between policy makers and business managers. David Roe (1984) of the EDF's energy program gives a full account of the whole mutual innovation learning process in his book, *Dynamos and Virgins*. The utilities had always seen conservation and solar as 'kid stuff' that did not fit their grow-and-build strategy.

In the early 1980s, the utilities began to realize the economic advantages in using such alternatives. This created congruency between policy makers and corporate managers concerning large wind turbines, and increased the interactions and convergence within the large wind turbine innovation network. Whereas the federal government failed to get the utilities interested in wind energy, the public interest coalition in California succeeded in *matching* the needs of utility actors within the consumption pole of the corporate network with their own political preferences. This is a clear example of joint innovation learning. As described above, lack of reliable technology arrested the further development of the network. Maybe the CEC wind power initiative could have made the technology work where the FWEP failed.

9.2.2 Communal Solar Energy Innovation

California policy makers became known for supporting the establishment of non-utility wind farms with medium-sized wind turbines. They thus became famous for setting up an entrepreneurial wind energy innovation network. The Brown administration's support for the development of large wind turbines is seldom praised or even recalled. The same goes for the effort of the OAT to encourage the use of small wind turbines.

In many respects, the desired communal small wind turbine innovation network was opposite to the large wind turbine innovation network. With respect to the consumption pole of the communal network it was assumed that many people were looking for new alternatives to the high-energy world. Once people encountered these "common sense methods for reducing our dependence on non-renewable fossil fuel resources," the masses would start to build and buy decentralized energy systems. Ultimately, this would lead to energy self-reliance at the local level. Thus, a network of communal solar energy innovation networks was envisioned.

Contrary to corporate innovation, science, technology and mass production were not regarded as key ingredients. Technology needed to be simple and production by small, local not-profit oriented firms was preferred.[21] People's skills, labor, and commitment were the most basic resources of any communal network. After 1976, OAT promoted the establishment of communal solar energy networks. It favored 'implementation through public participation,' and mass enlightenment was its final goal. To stimulate the revival of self-help ventures and cooperative community efforts, providing information on successful initiatives and managing demonstration projects were seen as the proper policy interventions. For example, the book, *Common Sense Wind Energy* (OAT 1982) provided an introduction to small-scale wind energy.

Small wind energy systems did not receive as much private and governmental support as other solar technologies. The most straightforward reason is that most homes and farms in California are located in low wind areas. Related to this, wind power had little tradition in California. Solar water heaters did. In the first decades of the century, solar water heaters had been a common sight in Southern California. This phenomenon was temporary. As fossil fuel prices dropped, gas heaters became more economical.

At the end of the 1960s the interest in solar technologies revived. Backyard tinkerers and hobbyists started to experiment with solar heaters, and some of them started up small businesses to sell and install solar devices to heat water for domestic use and swimming pools. In 1978, approximately hundred equipment manufacturers and two hundred installers with solar experience were active in California. By the end of the decade more than thirteen thousand homes had solar water heating systems, and there were about fifty thousand solar heated swimming pools (CEC 1981: 11).

The state government helped to strengthen and increase the coherence within this solar water heater innovation network. Its solar program tried to help the industry to move the technology beyond the "tinkering phase" through market stimulation.[22] In 1977, a 55 percent solar tax credit was introduced to strengthen the user pole of the network. To benefit from the tax reductions, potential consumers needed to have high tax bills and thus high incomes. This economic incentive made perfect sense because only wealthy people have swimming pools, and thus are the potential users of solar water heating systems. As a result, their sales annually doubled during the five years after its introduction.

This was not solely caused by the solar tax credit. Between 1975 and 1978, twenty pieces of legislation were enacted to overcome various implementation barriers which were thought to slow the growth of solar energy use. To counteract consumer concerns over solar system reliability, a law was enacted to protect consumers from fraudulent businesses. The CEC also had a voluntary testing and certification program for flat-plate collectors, and was working with private industry on the labeling of solar systems that qualified for the state tax credit. Besides, several measures covered other pressing issues, like job training and financing.

The OAT and the CEC promoted the idea to extend the tax credit to wind turbines. In 1978, the Legislature classified wind turbines as a solar technology, which made them eligible for the 55 percent tax credit. With respect to the solar water heater innovation network, the tax credit aimed to energize an existing innovation network. Strengthening the consumption pole was expected to have a positive effect on every other network pole. The communal wind energy innovation network, however, was nonexistent. Thus the wind energy tax credit can be regarded as a breakthrough action, aimed at creating a market for residential wind turbines, so as to stimulate the production pole.

In spite of the tax credit, the sales of small private wind turbines stayed low and most manufacturers sold only a handful of wind machines a year. Barely surviving, the small firms could not afford to improve their products. Consequently, wind turbines on the market reflected the technology from the 1930s but lacked the reliability and longevity of the old windchargers.[23]

9.2.3 Entrepreneurial Wind Energy Innovation

> First recognize that wind energy is an industry in its infancy. This has several important implications. There are few government regulations providing for consumer protection and public safety. Also, there are new companies and products being introduced as the market expands. Some of these may not be in business five years later. Finally, consistent with the American tradition of free enterprise, there are always some sales and promotion companies, who, knowingly or otherwise, tend to exaggerate what their products will do. All of these suggest that you, as the potential buyer, must be as informed as possible. *Jack Park and John Obermeier, 1982: 34*

The above shows that the small wind turbine innovation network could profit from policy instruments that were developed for implementing another type of solar technology that was close to commercialization. Similarly, wind farmers within the entrepreneurial wind energy innovation network could eventually profit from the political impetus created by, in particular, developers of industrial cogeneration projects.

As with the solar water heater industry, the independent cogeneration industry had flourished in California at the beginning of the 20th century, but had since dwindled to only a few projects. After the oil embargo, this changed dramatically. Already in 1976, the State Legislature enacted the Private Energy Producers Act (PEPA) to enable non-utility power projects. This law was especially meant to allow the industry to set up cogeneration plants that would serve their own electrical and steam process needs. PEPA also allowed individual citizens to generate electricity for private consumption by using renewable technologies, like solar cells or wind turbines. Both technologies, however, were still very expensive.

In the same year, the CPUC also began to direct the investor-owned utilities to pursue the development of cogeneration facilities. Cogeneration was thought to be attractive because of its associated fuel use efficiency, its ability to accelerate conversion from oil to cleaner gas-fired generation, and its potentially significant contribution to electricity supplies. Since utility managers questioned the reliability and economic feasibility of the renewables and cogeneration alternatives, they were slow to commit themselves to such projects, despite strong encouragement from policy makers.

PURPA was not only in line with PEPA but even brought it a step further. Namely, besides allowing self-production, PURPA of 1978 required utilities to purchase excess power from renewable and cogeneration power plant operators at the utility's *avoided cost* – i.e., what it would otherwise cost the utility to provide the power itself. PURPA broke the utilities' monopoly on electricity generation, and together with the energy tax incentives, challenged entrepreneurs to become independent power producers by developing small-scale alternative power projects, like geothermal and cogeneration plants, that were on the verge of becoming competitive with conventional central power stations. Several entrepreneurs stepped into this business.

Unexpectedly, in view of the immaturity of wind technology, some entrepreneurs began to anticipate making a profit out of setting up wind power plants. Some wind farm developers put their faith in the corporate wind energy innovation network to swiftly come up with cost-competitive megawatt-sized wind turbines. Most of the wind farmers, however, became involved in the establishment of large arrays of medium-sized wind machines. In the early 1980s, the Brown administration actively encouraged the financial community to get involved in the development of wind farms. Financiers were informed about the state's excellent wind resources and about the possibility of financing wind farms through selling investments in those projects to rich individuals, who could benefit from the attractive federal and state energy tax credit programs, and from the yearly revenues from the sale of wind power.

It is worthwhile to give these events a moment's thought, because at this point in time a lot of development lines come together. The energy tax credit was originally meant to stimulate the use of solar water heaters. The state tax credit was then extended to wind energy to foster the sales of residential wind turbines. After the coincidental passage of PURPA, the policy makers in California began to use this policy instrument to involve the financial community in the wind farm business. They could also point at the abundant California wind resources that had been accurately assessed within the CEC WPP. Besides, a Wind Information Center was available to disseminate wind data to potential operators.

While these public initiatives had little impact on the innovation networks the government intended to stimulate, they turned out to be very effective in stimulating the consumption pole of the entrepreneurial wind energy innovation network. Firms that were accustomed to finance land and condominiums extended their field of activity to setting up wind power plants. By promising high profits, these firms attracted the interest of up to 50,000 individual investors and raised two billion dollars for California's burgeoning wind power plants (Gipe 1995: 31).

The money that poured into the business of building wind farms and generating electricity created a huge demand for medium-sized wind turbines. The new market situation motivated firms within the residential wind turbine business to shift their attention to developing and producing medium-sized turbines, and also attracted new firms to move into the business of building wind turbines. The joint result of the different policy interventions, most originally meant to stimulate the development of either a corporate or a communal innovation network, was an abrupt and strong rise in the convergence of the entrepreneurial wind energy innovation network.

Increased cooperation within the network was based on the widespread expectation among individual investors, wind farm developers, and wind turbine manufacturers that proper wind turbine technology existed or could be easily developed. The lack of progress the decade before implied that a weak technology base existed for small wind turbines. Most firms considered that technology base inappropriate to base the development of medium-sized turbines on. Most of these engineering-based firms pursued similar turbine types as was done within the large wind turbine innovation network, and tried to push the 'state of the art' technology as far as they could in the direction of high-speed, light-weight wind turbine designs (Heymann 1995). Putting trust in engineering tools, like helicopter design codes, the American engineers felt able to solve any technical problem. Consequently, they pursued a design-by-extrapolation approach to rapidly upscale the size of their wind turbines. Such a development strategy was favored by wind farm developers, who strove for economies of scale within their wind projects.

Maybe more important, it also resulted from the fierce competition that existed between wind turbine producers.

This competition was especially based on technological competence. Karnøe (1995: 256) explains that American manufacturers "typically went for the breakthrough innovation, or at least a design that had radically changed features compared to existing technology, in order to give them a technology-based competitive edge in the market". The wind farmers stirred up this innovation process as they defined the needs of the market. Unfortunately, most of them had little knowledge about wind energy technology, and since reliable independent performance data of wind turbines were nonexistent, wind farmers more or less had to take the manufacturer's information for granted. This situation made the strategy of upscaling even more lucrative, because its short-term commercial success seemed guaranteed.

9.2.4 Destructive Creation

The opposite turned out to be true for two main reasons. First, the constant introduction of innovative, larger and supposedly more economic wind turbine types enabled wind farmers to squeeze other manufacturers by forcing them to sell their wind turbines below their original price. Thus, the same innovation process that was expected to increase profits in fact caused profits to decrease, which made the need to innovate even more pressing.

Second, like their colleagues in the corporate wind energy innovation network, the scientific engineers within the small American firms proved to be wrong in using an aerospace design heuristic. Instead of stimulating a process of *creative destruction* – that is, a process in which superior new technologies continually displace existing technologies – the market conditions in California in the early 1980s aroused a process that could be characterized as *destructive creation* – that is, a process in which poor new technologies continually displace existing unsatisfactory ones.

Soon, wind farmers discovered that American turbines lacked reliability and showed poor performance within the California wind environment. Individual investors found out that investing in wind turbines was less profitable and more risky than their financial advisers had led them to believe. Since the various actors within the network were not able to fulfil their promises to other innovation actors, the level of trust within the entrepreneurial innovation network declined. This decay of trust was strongly enforced by regular fraudulent practices and negative publicity from the conservative media that depicted wind farms as 'tax farms.' Lack of reliable information about both the condition of the wind energy technology as well as the efficiency of the tax credit program played a crucial and dubious role.

As we saw, in Denmark wind turbine owners informed each other about bad technology and fraudulent manufacturers because every owner benefited from this service. In California, access to credible information played a major role in the competition between wind farmers. This caught competing wind farm developers in the firing line. On the one hand, they were eager to benefit from the experience of their competitors because it would make them buy proper wind turbines. On the other, they did not want to give away information that would help their competitors in making the right investment decisions.

Ever since the wind industry started its exponential growth in 1981, the CEC had recognized the need for reliable data on product performance. It wanted to tie the level of subsidy to energy production rather than to capital investment. This simple legislative measure would have forced wind farm developers to make public their production data in order to receive the subsidy. Regardless, this obvious improvement was not implemented. To grasp this example of non-decision making it has to be placed within a whole circle of events.

The poor design of the tax credits enabled their abuse. In addition, lack of information made the real amount of tax abuse a matter of speculation. This generated 'bad press.' This problem signaled by the press was grist to Deukmejian's mill. Encouraged by the public mood that demanded a slowdown of state expenditures, the new Governor seized the opportunity to propose the abolition of the state wind energy tax credit program. He used the difficulties with the wind energy tax credit to expose the general problem of tax abuse.

Deukmejian's smear campaign against the energy tax credit defined the atmosphere of the political debate. Again, the lack of reliable information for evaluating the impact and abuse of the wind energy tax credit,[24] enabled the Governor to do so. Proponents also suffered from a lack of data to document their point of view. Nevertheless, bipartisan support thwarted the Governor's attempt to abolish the wind energy tax. The program was extended, but not improved.

Political rhetoric had prevented a sophisticated debate on how to improve the tax credit program. Nobody was really pleased with this outcome. Hence poor tax credit design led to poor information about their impact and abuse, which caused a speculative debate that left the poor design untouched. In short, power politics thwarted obvious policy measures like changing the energy investment tax credit into an energy production tax credit, publicizing production figures and setting up a test program for wind turbines. Since such policy interventions were ignored, the process of *destructive creation* was maintained by new ignorant and/or fraudulent developers. Thus, while the wind industry was clearly committing suicide, the lawmakers gave them a helping hand.

The CEC wanted to break this vicious circle by asking the wind industry to voluntarily report on the performance of their installed machines. Its effort to set up a Wind Performance Reporting System (WPRS) finally found support within the industry in 1985, when already many American wind turbine manufacturers and developers had gone bankrupt. While the CEC tried to control market development, the CPUC instead added fuel to the flames by introducing the long-term ISO4 contract in October 1983. This initiative hastened this process of self-destruction and thus seemed the worst possible outcome. How to explain this remarkable policy intervention of the CPUC?

It should be realized that the CPUC's policy of promoting renewables was realized only in the form of power purchase agreements. No explicit policy for wind energy existed, or for any other renewable energy technology. The CPUC wanted to let the market decide what type of technology would develop. Thompson explains, "We thought that we were doing enough by making the contracts available and financiable. That was the big push. We knew there was a huge wind industry potential in California. We knew there was a lot of biomass and geothermal in California. ... We thought we were doing a great thing to promote the alternative energy industry."[25] In other words, the CPUC was only interested in creating a market for renewables and cogeneration, and not with the other developments within any innovation network. It did not really care about the business of building wind turbines. It did not even care about the business of building power plants. It was interested in the business of generating electricity.

The CPUC's general support for the independent power industry matched the *regulated competition alliance*'s ambition to introduce a competitive element in the electricity production sector. The way this was done – the standard contract approach – was consistent with the idea of a regulated market. With respect to technology or industrial development a pure free-market approach was championed. Thus, the libertarian attitude of the CPUC implied a myopic view on innovation and explains why the ISO4 contract was introduced despite the fact that this would have a catastrophic effect on the American wind turbine industry. The Commission was not informed about that, and if it had been informed, it would not have been bothered because it did not intend to promote one specific type of technology and had faith in market forces.

It can be concluded that the laissez-faire attitude of the CPUC and Governor Deukmejian clearly prevented a fruitful discussion about the introduction of public policy instruments that could have played a beneficial role in the efficient usage of public money for stimulating the development of wind energy technology and industry. In other words, instead of creating advantageous conditions for joint innovation learning among the actors within the entrepreneurial wind energy innovation network, public policy seemed to frustrate or prevent such learning.

The fact that investor confidence in wind farms gradually declined, and project developers were losing their faith in American wind turbine technology meant that the entrepreneurial wind energy innovation network that had been created almost overnight was heading toward major disaster. The introduction of the ISO4 contract appeared to speed up this scenario. Still, this did not happen. Two strategies were contending within the market to escape the above destructive and mutually reinforcing processes: a strategy of vertical integration as developed and used by U.S. Windpower (see § 9.2.5) and a strategy of horizontal integration as introduced by the Danish wind turbine industry (see § 9.2.6).

9.2.5 U.S. Windpower's Strategy of Vertical Integration

U.S. Windpower is the only American producer of wind turbines that survived the California experience. In the early 1980s, this firm chose a business strategy of vertical integration that circumvented many of the market imperfections that troubled the wind energy business at the time.

U.S. Windpower was founded in 1975 by two engineers who saw a business opportunity in wind energy. The firm developed their turbines with internal financing. Nevertheless, their development work was, in an indirect way, stimulated by federal

R&D expenditures, since the firm built up its technological competence in close contact with a wind energy research group at the University of Massachusetts, which had close ties with experts at MIT in the field of aerodynamics and materials.[26] Technology development was placed entirely in service of making profits and creating opportunities for growth. The firm became renowned for its ability to raise external capital to finance its aggressive growth strategy, setting up commercial wind farms that used hundreds of self-produced wind turbines. The knowledge gained in the field was fed back into the development process, and, in 1983, a solid wind turbine was developed.[27] Strategy then shifted from product development to cost reduction via the mass production of a standardized design. The firm started to exploit its innovation efforts and focused on earning money by producing wind electricity.

U.S. Windpower was thus active in both the technical, business and consumption pole of the entrepreneurial wind energy innovation network, and maintained close contacts with actors within the science pole. The firm succeeded in developing the various poles and enhancing their coherence. Due to its business strategy of *vertical integration,* the company in fact was an innovation network in itself, and was able to avoid many dependency problems that were afflicting other wind energy companies.[28]

For example, while other wind farmers lacked reliable product information, U.S. Windpower knew the poor state of its initial turbines. Still, it did not have to compete with other producers. This isolated position tolerated a design-by-experience approach, while other manufacturers were forced to upscale their designs. U.S. Windpower's vertically integrated income circle, therefore, reduced the problem of raising the level of cooperation between different innovation network actors to the problem of increasing the spirit of partnership and trust between actors (general and R&D managers, and engineers) within one organization.[29] In other words, U.S. Windpower's business strategy of independence created ideal conditions for joint innovation learning.

The lack of solidarity between innovation actors within the same pole and across poles of the entrepreneurial innovation network made vertical integration a rational business strategy for survival. Exactly because of its independent position, U.S. Windpower was able to reap profits from a long-term contract. No wonder U.S. Windpower was the firm most active in lobbying the CPUC to introduce the ISO4 contract.

9.2.6 The Danish Industry's Strategy of Horizontal Integration

A second strategy appeared more or less out of the blue. The 'coincidental'[30] entrance of Danish technology also saved that CPUC policy initiative, because it offered wind farm developers a way out of 'the heap of American wind turbine technology rubbish.' By means of the DV statistics and the Risø certificate, the Danish producers could demonstrate the reliability and performance characteristics of their machines. Trustworthy and high performance data was just what the American machines were missing, and exactly what the 'serious' wind farm developers required. Hence the availability of a long-term power purchase contract was a clear stimulus for wind developers to select wind technology on the basis of its longevity and cost per kilowatt-hour. While trust in

American turbines declined, developers rapidly gained faith in Danish technology. Danish turbines were less infirm than the American ones, and their generous, conservative, and more expensive designs stood the test of time several times better than the 'advanced' lean and cheap American designs. While the influx of superior Danish technology saved a large part of the serious wind farm business, it was the final *coup de grâce* for the American wind turbine industry.[31]

The availability of the ISO4 contracts caused a shift in market conditions. The market was no longer exclusively based on tax credits, but more and more on energy production. Gaining a share of this market required a change from a gold-rush mentality to a more sober business perspective based on long-term continuity, lower profit expectations, and higher credibility. The 'tons of junk' that had been installed in the first years had given the wind industry a bad name and had lowered the interest of rich individuals to invest their money in 'tax farms.' In the 1970s, the lack of public and investor confidence in the electric power industry had given the option of wind power a chance. Ironically, decline in public trust was now threatening the influx of capital into the wind power industry. The presence of the ISO4 contract enabled Danish manufacturers to replace this loss of financial support. Based on their good reputation back home, credible figures and realistic profit expectations, they found Danish institutional lenders and investors prepared to put money into California wind farms, and Danish insurance companies willing to insure these ventures.[32]

The tax credits and unlimited availability of financiable contracts, combined with the close cooperation between the American wind farm developers and Danish capital, technology and production capacity, sustained and changed the character of the California wind boom. From a technological and business point of view, most wind developers were not ready for this rapid growth; siting techniques, maintenance and operating strategies were still poorly developed. The frantic year-end construction schedules to qualify the installed wind turbines for that year's tax relief intensified the disorderly and environmentally destructive way of putting up wind farms. The prospective end of the state and tax credit also stirred up this development.

In April 1985, the CPUC suspended the ISO4 contract. Within the one-and-a-half years that this contract had been available, wind developers had signed enough contracts to keep busy until the early 1990s. Withdrawal of the tax credits endangered the financial feasibility of such projects. The call for economics of government, and the success and abuse of the tax credits had substantially lowered their political support. Although federal tax credits were publicly abolished at the end of 1985, behind the stage the wind developers were successful in getting transitional rules written that allowed those developers who already had a signed contract and a project site to get the credit. Finally, the state energy tax credit expired at the end of 1986, and finished the California Wind Rush.

9.2.7 Corporate Wind Energy Innovation Revisited

The year 1986 represents a clear interlude in the regulatory environment for renewable energy technologies in California. The state energy tax credit is abolished and ISO4 contracts no longer available. A policy review period follows in which it is decided that in the future third-party power contracts should be allocated through bidding. Overcapacity within the electricity production sector enabled the CPUC to develop such a bidding process (the so-called BRPU process) at a very low pace. This bred a 'policy vacuum' whose outcome was highly uncertain to the actors concerned.

U.S. Windpower took a proactive stand and made a virtue out of necessity. It believed that by rapid technological innovation wind power could be made cost-competitive with fossil fuel generated electricity, speculating that utilities would become interested in buying wind power stations, and began to see itself as the future supplier or vendor of wind power plants.[33] This meant a clear break with the company's former strategy of vertical integration. Until then, the relationship between actors within the entrepreneurial network (in particular, the wind power industry) and actors within the corporate network (in particular, the utilities) had been adversarial. The smaller independent entrepreneurs had built and owned the wind projects, and sold power to the utilities under a contract. The latter felt that these contracts had been forced on them.

In the new situation, the innovation activities of U.S. Windpower no longer undermined the monopoly position of the utility industry, but instead came to support it. Theoretically spoken, U.S. Windpower positioned itself within the business pole of the corporate wind energy innovation network, which had not developed in the early 1980s. Now, U.S. Windpower took the private initiative to inject new life into this type of network. In return, the electric utilities supported U.S. Windpower in developing a cost-effective utility-grade wind machine. The idea to develop a breakthrough turbine was supported by the Variable Speed Wind Turbine Development Alliance, founded by the Electric Power Research Institute (EPRI) and joined by, amongst others, PG&E. The assistance of utility partners was necessary for financial and marketing reasons, and to appreciate what 'utility-grade' really indicated. Utility involvement boosted confidence in wind turbine technology in the early 1990s.

From a policy perspective, the innovation activities of U.S. Windpower were helpful in realizing the ideas of the *sustainable grow-and-build alliance*. Its change in business strategy, therefore, serves as a very interesting example. During the wind boom, U.S. Windpower's strategy of vertical integration corresponded perfectly with the political ideas of the *regulated competition alliance*. It also fit the market conditions at the time that were shaped by powerful financial stimulation measures combined with a laissez-faire attitude that left the usage of these policy instruments uncontrolled. After the wind boom, a policy vacuum emerged. Lack of public financial support for wind energy let U.S. Windpower opt for cooperation with investor-owned utilities. Interestingly, a temporary free-market approach to governance caused firms to adopt a business strategy that, if successful, would strengthen the political position of the corporate coalition. Thus the laissez-faire approach of the libertarian coalition was undermining itself.

Meanwhile, worries about urban air pollution and global warming prompted government and utility actors to re-examine alternatives to coal and oil. Wind power received a lot of attention since this option was closest to becoming economically competitive. In the 1980s, current costs had dropped nearly an order of magnitude; at good sites it was estimated at about eight cents per kilowatt-hour (Schaefer 1989).[34] It was held that few of the cost reductions could be traced to the improvement of wind turbine technology, but stemmed "mainly from organizational learning, which involves standardizing procedures. At the factory, manufacturers learned mass-production techniques; in the field, workers learned to site machines more effectively and to schedule maintenance at times of low wind". (Weinberg 1990: 148-149)

The past progress of California's wind farm developers, and the estimate that a lot of technological progress could still be made, caused EPRI and PG&E to argue that with "advanced airfoils, 'smart' electronic controls that adjust system-operating parameters based on wind characteristics, and continued development of advanced materials that yield lighter, stronger components," the cost of wind power could fall to 3.5 ¢ per kilowatt-hour during the next ten to twenty years. Inspired by the remarkable former economic progress of California's wind industry, and the projected economic advantages in the field of wind turbine technology, DOE launched the Advanced Wind Turbine (AWT) program in 1990 as an important part of the Federal wind program, which aimed to create a domestic wind turbine manufacturing base with multiple players (Lamarre 1992: 15). U.S. Windpower condemned the AWT program, which subsidized its competitors and undermined its dominance of the American wind power market.[35]

This firm had pursued a strategy based on the idea that no governmental support for renewables was to be expected. Now the federal government was intervening in the corporate innovation network U.S. Windpower was developing, the company began to stress that wind power did not need any governmental support at all. Its new, privately developed technology would push the price of wind power below the magic 5 ¢ per kilowatt-hour mark and make it cost-effective.[36] This publicity campaign was intended to get utility business seriously interested in wind power and to keep politicians from supporting the wind industry. To keep up the appearance of competitiveness, U.S. Windpower even went so far as to oppose the 1.5 ¢ per kilowatt-hour production tax credit that was introduced in 1992 by the Bush Administration.[37]

Other wind farm developers did favor government involvement, and were bowled over by U.S. Windpower's self-serving and arrogant statements.[38] Nevertheless, they also looked upon U.S. Windpower's publicity campaign with envy. Provoked by U.S. Windpower, and attracted by renewed utility interest and government involvement, they became interested in the utility market for wind power plants and took up the challenge of making them cost-effective. To become competitive on the domestic utility market, wind farm developers understood they needed to have their own wind turbines. Stimulated by the AWT program, Zond stopped using Vestas turbines and began to develop and assemble turbines on its own. In close cooperation with R. Lynette & Associates, FloWind began to market the AWT-25.

9.2.8 Wind Energy Innovation in Jeopardy

Meanwhile, the CPUC had finally established the BRPU bidding rules. The public interest coalition had managed to get a renewable set-aside of about 300 megawatts of the new power for bidding by renewable technologies only. In this respect, the BRPU was expected to create an appropriate initial market for the new advanced wind turbine designs. The BRPU policy was in line with the ideas of the *regulated competition alliance* and stimulated independent wind power development. However, such policy was at odds with the development toward utility-owned wind power stations. In other words, the political actions taken were contrary to the actual innovation activities taking place within the corporate wind energy innovation network. Instead, the BRPU process promoted wind developers to move on independently from the utilities. It gave entrepreneurial innovation a new chance, and seemed to offer wind farmers a very luxurious position.

The flirtation with the utilities had caused them to control both the development of wind projects as well as the production of wind turbines. The BRPU stimulated the strategy of vertical integration that U.S. Windpower had already left behind in the mid-1980s. The outcome of the bidding process was a resounding defeat for the regulated utilities, and seemed to really re-open the possibility of independent wind power production. Things, however, turned out differently. The outcome was genuinely offensive to the investor-owned utilities and regenerated the ancient adversarial relationship between the utilities and the wind power industry. This undermined cooperation between these two innovation actors within the corporate network. Moreover, SCE and SDG&E challenged the BRPU procedures, and forced the CPUC, via FERC, to cancel the power contracts. This was a blow to the ambition to set up wind farms independent from the electric utility companies.

Meanwhile the restructuring debate had become very active, precipitated by the high price of electricity, and more fundamentally, by the gas industry. Over the last decade, the independent power industry, in particular cogeneration plant developers, had become an integrated part of the overall economy of California. Being connected to small, entrepreneurial companies side-by-side with major industrial concerns (e.g., food processors, paper mills, chemical manufacturers, and oil production and refining companies) and present in every legislative district, the independent power industry had gathered a lot of bi-partisan support. In addition, the natural gas industry felt strong enough to compete head to head with the investor-owned utilities. Despite all the rhetoric about the cost-competitiveness of wind power, the wind industry was not able to compete heads-on with the utilities. Wind farmers needed green pricing or a renewable set-aside to continue their business.

As the independent power industry reached its maturation a difference of interest between the large competitive gas industry and the small renewable energy industry became apparent. In the past, the independent power industry had always formed a united front. The marketing director of Zond explains, "The utilities would be just smiling as we were fighting against each other. They would play us off against each other, and we would both lose. So, the gas and wind people realized that we had to work together."[39] From a policy perspective, it is easy to see that this united industrial front corresponded with the *regulated competition alliance*. The low price of natural gas and the strong growth in industrial cogeneration plants – in other words the success of the cogeneration innovation network – made the gas industry want to leave this alliance.[40]

In September 1995, the independent power industry reached an agreement with the investor-owned utilities on how to introduce a competitive market for electric services in California. The *Memorandum of Understanding* symbolized the temporary climax of the rise of the libertarian coalition. The growth of the independent power industry, in particular the gasfire industry, guided and enabled the ascendance of the libertarian beliefs within the energy debate. The same beliefs that had helped to develop the wind power industry now threatened its livelihood. Ironically, U.S. Windpower, the most passionate advocate of deregulation,[41] became its first victim. With the utility market down, the other Californian wind power companies had to muse on new business strategies and wait for better times to come.

9.2.9 Looking back on the Innovation Process

The wind energy innovation process shows an incredible complexity (see Figure 9-2). Three different types of innovation networks were supported in the 1970s. California's effort to set up corporate and communal innovation networks failed because of a lack of political support, support from the business community, lack of demand from users (private users as well as utility companies), and a lack of reliable and efficient technology. In other words, despite the effort of the Brown administration, almost all the poles of these two wind energy innovation networks continued to be strongly underdeveloped.

In contrast, an entrepreneurial wind energy innovation network, rather unexpectedly, came into being almost overnight. It was an unintended consequence of an incoherent set of policy measures (PURPA, tax credit programs, Wind Resource Center) that had been originally introduced to stimulate other types of technologies and innovation networks. Several business people, however, noticed that this set of policy measures offered the opportunity of setting up independent wind power plants to be financed by rich individual investors who would benefit from tax credits. The reinterpretation and creative use of the existing public policies by business people created a strong and unforeseen coherence among the initially disjointed and ineffective set of measures. The coincidental combined result of these measures was an uncontrolled 'Wild West' type of development that led to a technology development process that was typified as *destructive creation*. This was noticed by the public interest coalition, which proposed simple measures to counteract it. However, political rhetoric from especially Governor Deukmejian prevented a sophisticated debate on how to improve the innovation process.

Two strategies were concocted within the market to circumvent or resolve this. Vertical integration was the business strategy that best fitted the policy aim of stimulating independent power and the cut-throat mentality that characterizes mainstream American business culture. A former employer of U.S. Windpower explains, "We never saw ourselves as a cute little windmill company. We always envisioned ourselves as a generating business."[42]

The other strategy was to rely on Danish technology. The Danes exported new investment and insurance structures, and introduced the Danish way of doing business to the innovation network. Theoretically spoken, the technical and business poles of the Danish cooperative/flexible wind energy innovation network became an integral part of the entrepreneurial network. This explains why wind farm developers, at the end of 1984, agreed to cooperate on the Wind Performance Reporting System (WPRS). By that time, wind power projects started to place greater reliance on conventional financing from institutional lenders and foreign investors. Accordingly, verifying performance credibility became a prerequisite for attracting these new sources of financing (CEC 1991: 3). While the strategy of vertical integration presupposed and promoted a winner-takes-it-all attitude to doing business, the imported European-style strategy relied on and fostered cooperation among wind farmers (e.g., WPRS) and between manufacturers and wind farm developers.[43]

The two different strategies also resulted in different approaches to innovation. With respect to technology, U.S. Windpower had a very tough policy on secrecy (Karnøe 1995: 255), and, having developed its basic design, almost entirely focused on the production process. Its basic strategy to achieve substantial cost reductions was to set up a production line that turned out several hundred machines a year. Their USW 56-100 became the 'Model T' of the wind turbine business. In contrast, continuous product innovation and cooperative R&D characterized the innovation strategy of Danish manufacturers. Paradoxically, cooperative behavior that was nurtured and successful on the low-competitive Danish home market led to serious complications once the Danes became active on the high-growth California market, where it overcharged the process of *creative destruction*.[44]

After the expiration of the tax credit programs, the absence of any public policy made U.S. Windpower change its business strategy. The firm started the development of a novel wind turbine for the utility market. In political terms, this implied a shift from the entrepreneurial innovation network as promoted by the regulated competition alliance toward the corporate wind energy innovation network as preferred by the sustainable grow-and-build alliance. At the beginning of the 1990s, the activities within the technical and business poles of this new corporate network were thriving. But as the utilities started to pay serious attention to these wind energy innovation activities the preparations within the CPUC of the so-called BRPU process were finished. With respect to renewables, this bidding process stimulated the development of independent power. Thus, instead of agreeing with the ongoing development in the corporate wind energy innovation network, this policy program tried to breathe new life into the entrepreneurial wind energy innovation network.

Initially, the BRPU process seemed to be rather successful, especially with respect to wind energy. However, this success regenerated the old adversarial relationship between the wind farm business and the utility companies. As a result, the BRPU process was cancelled, which turned initial success into failure. The wind industry was not only forced to cope with the loss of power contracts they had won in the BRPU process, but the restructuring debate also caused the slowly growing interest of the investor-owned utilities in wind power to evaporate. Finally, the political process left the wind power industry in California empty handed.

9.3 Denmark: Living Wind Energy Innovation

> The living word demands that one not only respects his opponent, but also sees him as necessary, yes in the deepest sense as a friend. For the individual person just as for a people it is important that the truth not fossilize into dogma. ... He who wants to be free must let his neighbor be free with him; this is Grundtvig's fundamental perception. Therefore it is not liberation from but *vekselvirkning* with the neighbor, that is the central idea in his view of freedom. *Ove Korsgaard (folk high school principal)*[45]

Basically, the wind energy innovation game in Denmark can be analyzed in terms of two innovation networks that gradually merged in a non-straightforward manner. Figure 9-3 represents a graphical summary of this complex process. As in Figure 9-2, the various innovation networks are represented by boxes, in which the various innovation poles are sketched. Figure 9-3 serves as a guide through this section. On top of each box it is indicated in which subsection the related network or event is discussed.

In the 1970s, the two innovation paths could still be clearly distinguished; one fitting the flexible innovation tradition promoted by the established coalition, the other, taking the cooperative innovation tradition as its point of departure, advocated by the cooperative coalition. The two innovation approaches will be described in subsection 9.3.1 and 9.3.2, respectively. While the innovation activities in the flexible wind energy innovation network stayed limited to a mere scientific enterprise, the activities in the cooperative network enticed lay people to become engaged in technical experiments, and attracted small firms to produce and sell wind turbines. At the end of the 1970s, the myriad activities even encouraged the established coalition to support the cooperative innovation network.

In subsection 9.3.3, it is described how, in the early 1980s, the original cooperative network evolved into a mixed flexible/cooperative wind energy innovation network under the influence of both ideological coalitions. The blending of the two originally distinct ideal types of innovation approaches perseveres throughout the decade. Experiences gained during the California and Danish wind boom (see subsection 9.3.4 and 9.3.5), inspired an ongoing blending. The next subsection discusses how this integration process reaches its temporary completion. The gradually evolved innovation balance is denoted as *flexible ecological innovation*. In the final subsection, we will look back on the wind energy innovation process in Denmark.

9.3.1 Flexible Wind Energy Innovation

During the 1970s, government activity in the field of wind energy was concentrated on the establishment of a flexible wind energy innovation network. To achieve this, the joint Wind Power Program (WPP), of the Ministry of Trade and the electric utilities, was set up in 1977. Since the FWEP in the United States served as an R&D model for the WPP, it will come as no surprise that the scientific knowledge strategy of the two programs resembled each other. Like the Mod-program, the Danish one relied on scientific knowledge and the effort of large industrial firms. The blades of the two

Nibe-turbines were designed at Risø National Laboratory,[46] and the rest of the design work was done at the Department of Fluid Mechanics at the Technical University of Denmark. Still, the Danes chose a different technological wind turbine concept. America and Denmark had cooperated on an extensive test program on the old Gedser wind turbine. This machine had been built in the 1950s by Johannes Juul, who had been trained in the Danish craft tradition at La Cour's course for rural electricians at the Askov folk high school. The Danes basically took over Juul's design.[47] In the United States, engineers were less enthusiastic about Juul's approach to wind turbine technology, and chose to elaborate on the design concept of the German professor Hütter. In the late 1950s, Ulrich Hütter had built a 100-kW prototype wind turbine that "looked more sleek and aerodynamic than the clumsy, cluttered appearance of the Gedser machine" (Gipe 1995: 78). While Juul emphasized simplicity and robustness of design, Hütter stressed that high efficiency and low weight were indispensable to competitive wind turbines. In short, while American engineers more or less fully relied on structural dynamics and aerodynamics, the Danish engineering community used a combined theoretical and practical approach to wind turbine design.

Besides R&D, the Danish program aimed to initiate an industrial platform and set up a so-called *user-producer network* (Lundvall 1988). To stimulate the latter, the utilities were directly involved in the management of the program. The WPP was coordinated by the joint research institute of the Danish utilities, DEFU. With respect to the business pole, the Americans presupposed a ready-made industrial capacity for the production of wind turbines. In Denmark things were totally different. While NASA could hire huge aerospace companies that were expected to be able of building and producing an entire wind turbine, the WPP had to subcontract a whole range of companies with different technological skills. Sometimes, it was even hard to find a Danish firm with the right expertise. It was especially problematic to trace a company that was able to manufacture the 20-m fiberglass reinforced polyester rotor blades.[48]

Some urban industrial companies were subcontracted: the Copenhagen company F. L. Smith, which normally delivers machinery to the cement processing industry, took care of the gear of the Nibe B; the Odense firm Thrige Titan built the generator and the electrical system, and Frichs from Århus constructed most of the machine parts. In contrast to the American subcontractors within the FWEP that were acquainted with mass production, the Danish companies were specialized in producing small quantities of complex and often large constructions. In these firms the old craft ethos had been the motive for accumulation and scale (Kristensen 1989: 175). It can be concluded, that the WPP aimed to simultaneously support the creation and integration of the various poles of the desired flexible wind energy innovation network.

In the early 1980s, the network was imperiled because the consortium of large Danish firms showed little interest in producing large wind turbines. Thus, no mutual innovation learning had taken place between the established coalition and the industry. To counter this lack of interest, the Ministry of Energy initiated the wind turbine company, Danish Wind Technology (DWT), to develop and market large wind turbines based on technology developed within the WPP. This breakthrough action was a, by Danish political standards, very uncommon policy initiative. It can be seen as a last-ditch attempt to inject new life into the flexible wind energy innovation network. By

Period	Danish wind energy innovation networks		
1950-1960 Experiments by Juul	*Flexible* WEIN S/T: Juul's experiments B: – C: **SEAS, DEF** P: Grow-and-build alliance		
1973-1979 Grassroots era	*Flexible* WEIN (§ 9.3.1) S/T: DEFU, DtH; **WPP** (Nibe) B: Consortium of large firms C: Electric utilities P: Established coalition	*Cooperative* WEIN (§ 9.3.2) S/T: Ecological practitioners; **OVE, Tvindmill, Wind Meetings** B: Small local firms (e.g., Riisager) C: **Energy Offices, DV** P: Cooperative coalition	
1980-1984 Industrialization period	*Flexible* WEIN S/T: **WPP** (Masnedø, Tjæreborg) B: **Establishment DWT** C: – P: Established coalition	*Flexible / cooperative* WEIN (§ 9.3.3) S/T: **Test Station** B: SMEs (e.g., Bonus, Vestas, Nordtank) C: Windmill cooperatives; **DV Statistics, Energy subsidy, Connection rules, Type approval system** P: Cooperative & established coalition	
1985 Danish wind boom			*Entrepreneurial* WEIN? (§ 9.3.4) C: Institutional investors; **1984-agreement** P: Little support
1986-1988 Downfall		Downfall of the Danish wind turbine industry (§ 9.3.5)	
1988- now Professionalization phase	*Flexible ecological* WEIN (§ 9.3.6) S/T: **Test Station**, R&D Departments within firms B: Professional medium-sized firms C: Windmill cooperatives & electric utilities; **100-MW Agreements, Wind turbine Law, Modern certification system, Export Guarantee Fund** P: Sustainable grow-and-build alliance		

Figure 9-3

Wind energy innovation networks (WEIN) in Denmark. The boxes represent different types of WEIN's. A dotted-lined box indicates that little market development took place in the related innovation network in the corresponding period. The various poles of the innovation networks are indicated by their first letter. (Sub) policy instruments aimed to strengthen a particular pole are printed bold. The gray color indicates the tentative evolution from flexible WEIN into flexible ecological network.

the time the second generation prototypes – the Masnedø and Tjæreborg turbines – within the WPP began operating, the state had already sold its share in DWT. By that time a flourishing wind turbine industry had come to existence that produced and successfully exported medium-sized turbines.

This industry had its origin within a (from a policy perspective) different kind of innovation network: the so-called cooperative wind energy innovation network. (Its formation will be described in the next subsection). The success of the cooperative innovation approach did not imply a final political victory for the cooperative coalition. Rather, its techno-economic success offered new opportunities for the established coalition. It was realized that the business pole of the cooperative network might also play a role in the flexible innovation network. Hence the government shifted its attention to the consumer pole and forced the utilities to sign the first 100-MW agreement at the end of 1985. This breakthrough action was meant to create a guaranteed market for utility-owned wind turbines and stimulate manufacturers to change their business strategy into developing large turbines for the utility market.

In summary, at the end of the 1970s the focus of governmental policy was on the technical pole of the flexible wind energy innovation network. In the early 1980s, it shifted toward the business pole, and in the mid-1980s toward the user pole.

9.3.2 Cooperative Wind Energy Innovation

> We look for politics in the wrong place, with the wrong terms, on the wrong floors of offices and on the wrong pages of the newspaper. *Ulrich Beck, 1997: 99*

While, in the 1970s, government effort concentrated on the development of large-scale wind turbines owned by utilities, grassrooters, hobbyists, and small entrepreneurs were active in developing wind turbines for private use. Political engagement drove a lot of these initial activities. In the wake of the oil crisis and the antinuclear struggle, the cooperative coalition looked for and experimented with alternative technologies that would empower people at the local level, and make nuclear power redundant.

In Part Two, we described how the cooperative coalition tried to influence the central government's energy policy by parliamentary and non-parliamentary action, and critically observed its implementation. The cooperative coalition was not a mere observer. Besides influencing the parliamentary decision-making process, it also tried to put its political ideas into practice in a more direct manner by trying to set up a cooperative wind energy innovation network. To indicate the actors within the cooperative coalition that were actually involved in this venture the term *ecological practitioner*, as coined by Jamison and Læssøe (1990), seems to be appropriate.

Using Beck's notion of *subpolitics*,[49] the argument here is that the cooperative coalition was both involved in 'normal' state politics as well as in subpolitics. The ecological practitioners in fact implemented an alternative energy *subpolicy*. It is interesting to investigate how the ecological practical organizations supported the creation and integration of the various poles of the cooperative wind energy innovation network. In particular, the Organization of Renewable Energy (OVE) and the Danish

Windmill Owners' Association (DV) played an important role in increasing the level of cooperation within that politically desired innovation network. It should be noted that OVE and DV were both cause and effect of the wind turbine innovation activities in the 1970s.

In contrast, a circle of folk high schools that was engaged in the antinuclear power struggle provided an existing organizational network that could be used to organize both joint local activities as well as coordinate action on the national level. Several folk high schools started to build wind turbines. Of these experiments, the construction of the Tvindmill became the most famous and influential one. On May 28, 1975, the building of the 2 MW wind turbine at the Tvind folk high school, near Ulfborg in West Jutland, started when about three hundred people simultaneously put the 'first' spade into ground, symbolizing the collective nature of the Tvindmill project (Tvind 1976). The project's grandeur and cooperative nature attracted volunteers from all over the country, including engineers and engineering students, and became a national event. During the period of construction, the Tvind folk high school was the place to meet people, work together, share ideas, discuss concepts, and also put them into practice. It was a massive technological experiment that inspired and brought together wind energy enthusiasts from all over Denmark. It was also a grand social experience that created community spirit among wind energy supporters.

The exchange and production of technological knowledge within the cooperative network was institutionalized in quarterly Wind Meetings organized by OVE. Here, individual experiences of different types of actors came together. Producers of (components of) windmills were confronted with practical experiences of self-builders and windmill owners. This enhanced contact between idealistic and critical buyers who formed the initial modest market, and the incipient wind turbine industry. Besides stimulating user-producer relations, it also stimulated cooperation within the business and consumer poles, fostering a continuous joint innovation learning process within the cooperative innovation network.

As an indirect result of the regular meetings, DV and the Association of Danish Wind Turbine Manufacturers (FDV) were formed in 1978, and came to agree on certain minimum design standards. Though serving different interests these two organizations often were of the same mind. Besides being a very serious consumer organization, DV wanted to encourage the private use of wind energy from a political point of view. FDV's objective was to increase the market for small wind turbines, so the two windmill organizations usually combined to lobby public officials and negotiate with utility companies.

Trust and cooperation were two essential characteristics of the grassroots-like cooperative innovation network,[50] whose rise was based on political engagement, and whose actors came to share a kind of community feeling. This was inherent to the cooperative philosophy of the ecological practitioners,[51] but was also reinforced by real events, such as the joint Tvind project and the Wind Meetings. One could say that these grassroots activities created a kind of 'wind energy *Gemeinschaft*,'[52] in which people knew each other on a personal basis, and in which solidarity, engagement, and cooperation were valued. In short, the cooperative ideology and the cooperative way of innovating reinforced each other. The fact that cooperation was a natural thing to strive for created ideal conditions for mutual innovation learning within the cooperative wind energy innovation network.

Grassroots activities were guided by the political conviction of the ecological practitioners, who wanted to show the viability of wind energy technology. User demands were lowered and a small niche-market developed. Sometimes technology was forced to adjust to political visions, such as at Tvind, where the huge Tvindmill was clearly an extrapolated design. This was inconsistent with the cautious design-by-experience method that is normally preferred within the craft-like cooperative innovation tradition. Although from a technological point of view the experiment at Tvind was an exception, it was politically symbolic. Political views took precedence over technological common sense, in other words, the technologists placed themselves in the service of the political community. Although a subservient attitude in the technological community is not without danger, the Tvind project illustrates that the technologists choose a design strategy in collaboration with other actors in the network. Thus, they perfectly followed the notion of joint innovation learning in the cooperative wind energy innovation network.

One could maintain that until the investment subsidy on renewable energy technology was introduced in the Autumn of 1979, the innovation activities within the cooperative innovation network were solely guided by the beliefs of the cooperative coalition. The government observed these developments from the sidelines.[53] At the end of the 1970s, however, the government came to support them, and both political views began to exercise influence, so that policy and subpolicy started to mingle. The introduction of the energy subsidy, the establishment of the Test Station, and the publication of a novel set of siting and consumption rules led to a new regulatory environment. A new innovation situation evolved, with new types of actors in the various poles and new types of relations between these actors. The new type of mixed network, which evolved in the early 1980s, will be denoted as the flexible/cooperative wind energy innovation network, to indicate that the innovation activities within the network were influenced by both coalitions.

9.3.3 Flexible/Cooperative Wind Energy Innovation

> Cooperation is very good among the user and manufacturer organizations, the Ministry of Energy, the public utilities, the National Test Center [at Risø], and the Danish counterpart to your EPRI [i.e, DEFU]. A sound political climate has also contributed considerably to the achievement of practical results. *Ebbe Johansen (Industrial Attaché, Consulate General of Denmark), 1982*[54]

To grasp this blending process, one should realize that, while the cooperative coalition was interested in the whole innovation network, the established coalition was mainly interested in developing the business pole since it favored industrial activity which created jobs and, in principle, rejected decentralized private electricity production. As the first activity implied the second, the established coalition had to choose to support neither or both. Facing major unemployment problems, it considered the economic growth argument more important than the absolute preservation of centralized control over the electricity system by the utilities, and chose to support the development of the cooperative wind energy innovation network. The established coalition's demand to promote industrial activity without endangering the central control principle led to a strategy of controlled development.

Three main policy instruments were used: the establishment of the Test Station for Small Wind Turbines at Risø that both enhanced and regulated the development of wind turbine technology, the modest investment subsidy that both created and limited the market, and the rigid siting and consumption criteria that the utilities imposed to limit the private use of wind energy. The official policy strategy combined with several subpolicy activities meant the change from a pure cooperative innovation network towards a mixed flexible/cooperative network happened in so orderly a fashion that, in all phases, the convergence of the evolving network was maintained. Still, a distinct type of convergence evolved. This subsection describes how a combined flexible/cooperative network emerged, and treats innovation activities within and characteristics of its various poles.

During the 1970s, the multibladed wind rose was the most commonly used wind turbine type. These stand-alone machines were usually equipped with a ten-kilowatt generator that produced electricity for heat generation.[55] After the introduction of the alternative energy subsidy, however, the grid-connected, medium-sized (30-55 kW), electricity producing wind turbine became the dominant design, that is, the most sold and used one. Most of these wind turbines were sold to local cooperatives. Johannes Juul's design of the Gedser mill served as an exemplar for this type of turbine.

Juul's work inspired many wind turbine builders, among whom Christian Riisager, who already in 1975 had connected his self-built wind turbine to the grid. When Riisager tried to get this authorized by the electric utilities he received a lot of support from the ecological practitioners. Through a press campaign, the latter forced the utilities to come up with general rules. Although rules were published, many utilities did not observe these regulations. Moreover, the rules only applied to individually owned wind turbines. At the end of the decade, the efforts of DV and FDV to coerce the utility companies to comply with their own rules and to extend the rules to cooperatively owned wind turbines gained government support. Just after the market subsidy had come into force, the desired new provisional rules were published.

The 1979 rules, in fact, broadened the types of actors within the consumer pole. From then on most wind turbines were sold to local cooperatives. 'Local' because all owners had to live within three kilometers of the windmill.[56] Because the capacity of the turbine was related to the electricity consumption of the owner, and the price per kilowatt-hour of a medium-sized turbine was substantially lower than that of a small one, it was more economic to collectively buy and own a wind turbine. The restrictive connection rules thus both created a market as well as restrained its size by limiting the money individuals and cooperatives could earn by selling excess wind power.

The state investment subsidy had a similar function. It enabled cooperatives to produce electricity for about the same cost as they could buy it from the electricity company, but the state restricted the profit margin of wind turbine owners by carefully adjusting the amount of investment subsidy. As innovations led kilowatt-hour prices of wind power to drop the subsidy would also fall. While the alternative energy subsidy created a market that attracted new firms to the innovation network, this bureaucratic balancing mechanism pinched from these newcomers in the business pole a method to expand the market. In short, the utility rules and subsidy law created a small market that forced the producers and consumers involved not to expect too high profits.

In particular, three small companies in Jutland that manufactured agricultural equipment – Vestas, Bonus, and Nordtank – became interested in this market. Wind turbine cooperatives were normally set up by a group of farmers, and thus were the same kind of customers these firms were used to deal with. Moreover, building a windmill more or less demanded the same kind of design and construction work they were accustomed to. According to Karnøe (1995: 261-262), "Making customized and continuously updated versions of small tanks, electrical trucks, small hydraulic cranes, water equipment, and the like was part of the normal business of these companies. Accordingly, products were produced in small batches or series. ... The early development work was organized by a small team of practical [*teknikum*] engineers, mechanics, fitters, and the like. Redesign, operational experience, and production went on hand in hand, in an organic, integrated process." In short, these companies were characterized by a flat and flexible organization, and adhered to the craftslike and incremental design-by-experience approach. In addition, they were not focused on mass production, but on small series production of mechanical quality products. Like the design of the Nibe turbines, those produced by the small firms were also inspired by Juul's design, which later became known as the Danish standard.

The design-by-experience approach fit the technological tradition within small Danish firms. The complex interplay of several institutional arrangements also stimulated a cautious design approach. First of all, most consumers were well informed about the state-of-the-art in wind turbine technology. A network of Energy Offices (set up by OVE, OOA, and a group of progressive folk high schools) informed people about renewable energy technologies,[57] and DV regularly published the wind turbine production data of all its members. These types of subpolicy initiatives made the wind turbine market in Denmark very transparent; manufacturers selling bad technology could be easily detected and those selling good technology easily selected. One could argue that market transparency produces ideal conditions for technology-based competition between firms, and thus seems to promote more risky product development strategies. For various reasons things turned out otherwise. As we saw, the wind turbine market was a low-growth market. Moreover, the restrictive rules advanced local ownership, which promoted local production since cooperatives preferred to buy turbines from a regional manufacturer whose reputation and products they knew. 'Knowing' in this respect meant actually seeing wind turbines operating and listening to other users' experience.

Thus initially firms mainly depended on word of mouth for sales. They therefore had great interest in building up a steady relationship with their clients, and had to provide physical proof of their reputation by keeping their turbines operating. Geographical proximity enabled them to service their own turbines and regular face-to-face contact between producers and consumers created and sustained relations of trust. Fruitful user-producer relationship came into existence that enabled small firms to learn quickly from their mistakes by making use of consumer experience. The fact that manufacturers operated in localized low-growth markets in which, besides cost-effectiveness, trustworthiness and reliability were crucial competitive elements, to a large extent explains why the companies followed a cautious design strategy,[58] and why technological cooperation via the Test Station could coexist with competition.

Besides the transparency of the wind turbine market, the fact that only turbines certified by the Test Station qualified for the state subsidy prevented poor technology from entering the market. Stipulated as a condition to guarantee that government money would be spent well, it had the side-effect of forcing companies to interact with the Test Station. This gave the latter the possibility to set technical standards to direct the technological development. This road was *not* taken for the simple reason that the engineers at the test facility thought no reliable scientific knowledge base existed that would justify theoretical standards to optimize wind turbine designs.

Instead, they seized the opportunity to become technical consultants for small companies with limited technical experience and financial resources. They considered market technology as their standard, and used normal engineering sense to advise manufacturers on especially the reliability of their designs. Because the engineers at the Test Station identified themselves with the craftsmen in the new industry (Dannemand Andersen 1993: 207) it was able to become 'the common R&D department for the whole industry;' the center where all the early technical experiences could be collected.

As noted above, the market situation, that is, the existence of a local, low-growth market, made technological cooperation a rational thing. To get a reasonable share of the market, producers only needed to follow the technological leaders in the business. The cooperative work of the Test Station helped to keep the leading group together, which enabled each manufacturer to keep control of its own regional market domain. Risø's focus on reliability complied with the need of the consumers for more reliable and safe turbines. In turn, producers tried to comply with the demand for safety and reliability and were happy to be informed by the Test Station about these issues. The advisory work of the Test Station speeded up the selection of a Danish standard turbine design, which was not compulsory, but whose generous design fitted with the safety focus of the engineers at Risø. Still, producers had the final say on this issue. They favored the Juul/Riisager design because of its simplicity, its robustness, its operating experience, and the availability of standard components.

It can be concluded that, in the early industrialization phase, a combined flexible/cooperative wind energy innovation network emerged. In many respects it differed from the pure cooperative innovation network of the grassroots period; the dominant technology differed (the Juul/Riisager design took the place of the wind rose),[59] the political backing differed (the established coalition came to support the network too), the dominant type of users differed (cooperative ownership became the norm), new firms entered the network (in particular, small established and later dominant firms like Vestas, Bonus, and Nordtank), and new institutional arrangements and regulations (alternative energy subsidy, consumption and siting rules) guided development.

Activities, relations between actors, artifacts produced, and rules within the new innovation network both enabled as well as embodied the political cooperation between the cooperative and established coalition. The tension between the core beliefs of the established coalition – the economic growth principle that implied stimulating private energy production versus the principle of centralized control over the electricity system – led to a constrained investment subsidy that both created and limited the wind turbine market in Denmark. The utilities (the sectarian side of the established coalition) imposed rigid siting and consumption criteria to limit the private use of wind power. Interestingly, this complied with the second order belief of the cooperative coalition that such a development should be local. A silent compromise was born that both coalitions could live with.

The fact that in the Danish political situation only a localized, low-growth wind turbine market was tolerated thwarted entrepreneurial opportunism and created ideal conditions for mutual innovation learning. In combination with excellent consumer information and the certification of wind turbine technology, this calm market environment caused the high level of cooperation and trust, which had been produced in the grassroots period, to be reproduced and maintained in the early industrialization stage. In particular, cooperation continued to exist in the technical and user poles. In addition, user-producer relations and the relation between the Test Station and the small firms were excellent. It can be concluded that there is a direct link between the political debate and the level of convergence within the innovation network as it evolved.

9.3.4 Entrepreneurial Wind Energy Innovation in Denmark?

As we saw, as much as the success of the flexible/cooperative network was due to demand stimulation, it was related to control on demand. The balancing took place within the political arena. This is perfectly illustrated by the reaction of the government to the sudden growth of the home market in 1985. The 'Danish boom' was made possible by the 1984 agreement. This private compact between DV, FDV and the utilities forced the latter to buy all surplus power and relaxed the siting criteria. This created a new market situation in which wind turbine parks became profitable investment projects for institutional investors and local municipalities.

Both coalitions disapproved of this speculative development, and recognized the need of rational, long-term regulation. To control the capitalistic development of wind power the government rapidly reduced the investment subsidy and reserved it for private investors who lived within a radius of ten kilometer from their turbine. In this way, early private arrangements that confined the market were now incorporated within the energy subsidy law. As said earlier, from a market stimulation instrument the investment subsidy changed into a policy instrument to control the development of wind turbines with local attachment. This policy intervention ended the short 'Wild West' period in the development of wind energy in Denmark, which was severely damaging the public image of wind power.

Paradoxically, whereas Danish politicians detested uncontrolled development within their own country, they actively supported involvement in the real 'Wild West,' the California market.[60] Moreover, the export success of these companies, resulting from uncontrolled development abroad, strongly improved the image of the wind power industry in Denmark. From a Danish political point of view, double standards are clear. The fact that political principles are forsaken when the border of a nation-state is crossed presents a major barrier to the democratic control of the development of technology.

9.3.5 The Industry's Pride before Its Downfall

While policy makers stimulated exports, they did not genuinely appreciate the big changes this demanded from the industry, nor did they anticipate the consequences. When Danish wind turbine manufacturers entered California at the end of 1982, they experienced a severe shift in market conditions. They moved from a protected, pre-competitive, and familiar home market to a distant, competitive international export-market. In theoretical terms, the undercapitalized Danish companies that had grown up within the protected business pole of the flexible/cooperative wind energy innovation network had to quickly adapt to the new circumstances and codes of conduct within the California entrepreneurial wind energy innovation network.

Nevertheless, this drastic change in market conditions had little effect on the role of the Test Station. A strange situation arose, therefore, in which the Danish companies competed each other on the California market by means of product innovation, while cooperating with each other in the same field back in Denmark. As we saw, the Test Station's role as a common consultant for the whole industry had stimulated the production and diffusion of practical knowledge within the flexible/cooperative wind energy innovation network, which had been beneficial for all involved actors. However, once the Danish firms became active within the highly competitive entrepreneurial innovation network, the (initially beneficial) free availability of technical information had the serious drawback of 'overstimulating' the product innovation process.

From 1983 on, the process of innovation went so fast that several turbine types were barely on the market for one year before other producers introduced new, more effective types. This was made possible because the Danish design lent itself to rapid upgrading, and because of the free availability of the Test Station's certification reports. The rapid techno-economical obsolescence of wind power turbines meant that old turbines could no longer be sold for their original price. Price reductions due to improved technology were not offset by similar cost reductions relating to, e.g., innovations within the production process. Kjær (1988) convincingly argues that the weak balance between progress in product technology and production technology was the essence of the economic problems the Danish producers encountered after 1984.[61] The falling dollar exchange rate aggravated the problems that led to the downfall of the Danish wind turbine industry between 1986 and 1988.

In short, in the early industrialization phase, technical cooperation benefited all actors in the flexible/cooperative innovation network. On the low-growth home market cooperation and competition went hand in glove, since the Danish market was limited through a set of institutional mechanisms that stimulated technical cooperation and spurred technological progress, creating the technological basis for entering the California market. Technical cooperation had made it possible for the Danes to invade the

California market, but this threatened technical cooperation, i.e. the basis of its success. Moreover, technical cooperation itself came to threaten the industry. Because the California market grew exponentially, stakes were high and competition intense. Although strong competition among firms seems to exclude technical cooperation, it still continued. Firms tried to obtain, increase, or maintain their market share through product innovation, which was partially enabled and accelerated through technical cooperation, heightened the competition between wind turbine manufacturers and instead of increasing profit margins, eventually led to a decrease in profits.

9.3.6 Flexible Wind Energy Innovation Revisited

Only after the California wind boom did Danish policy makers, technologists and managers come to appreciate the difference between the gentle and civilized home market and the cut-throat California export market. It became acknowledged that the approval system that was accepted in Denmark as a sufficient safety and quality mark was not suited to doing business internationally. Since 1979, the Test Station had certified wind turbines. The main emphasis was put on the certification of the turbine design. Although, wind machines were certified for Danish weather conditions, the certificate was widely used as a quality mark for export sales. Experiences with Danish wind turbines in California revealed some serious weaknesses of the old certification system. Besides type approval, insurance firms and investors came to demand verification of the manufacturing process, wind turbine installation, operation and maintenance. In the Spring of 1988, the Danish government, encouraged by manufacturers and investors, decided to start working on a new certification system based on internationally accepted standards and codes of practices.

Many Danish firms had been unable to generate the profits promised and to honor service and guarantee agreements when technical problems arose. Beside a professional certification system, an export guarantee fund was established to restore trust in the Danish industry. A barrage of bankruptcies caused a significant change of industry structure. The number of manufacturers was halved and fresh capital was put into firms with well-proven products. As a result, the new firms grew financially stronger and the product know-how and experience gained by the Danish wind power industry during the past decade was not lost, but carried on to the new companies. The more professional firms began to organize product development in a more systematic manner. Within newly established development departments, the use of scientific engineering knowledge gradually increased.

An important condition for the redevelopment of the Danish wind turbine industry was the existence of a home market, or rather, a home wind energy innovation network. Until the mid-1980s, wind energy innovation had mainly taken place in what we designated as the flexible/cooperative innovation network. The user pole of this network, and thus the market, had been dominated by private windmill cooperatives. The Danish boom had led to the first 100-MW agreement that aimed at creating a guaranteed utility market.

Theoretically, one could argue that this policy measure was supposed to inject new life into the incomplete and underdeveloped flexible wind energy innovation network, as originally conceived by the established coalition during the 1970s. In this scenario, the established coalition attempts to attract the 'successful' wind turbine industry to become exclusively involved in the renewed flexible innovation network by creating a guaranteed utility market. Such an explanation has several serious drawbacks because it has its roots in the political, technological and economic setting of the 1970s. It ignores the multitude of policy and innovation learning processes that have taken place since that time. The above explanation is, therefore, anachronistic, false and misleading.

A more fruitful way of analyzing the meaning of the 100-MW agreement takes the ongoing process of integration of flexible and cooperative characteristics of innovation as its point of departure. At the time of the agreement, the political attitude of the established coalition was that the development of wind turbines with local attachment could take place parallel to the development of the utility market for large-scale wind turbines. It found that the user pole of the flexible/cooperative wind energy innovation network offered room for both wind turbine cooperatives and utility companies. The established coalition was no longer only interested in the production pole of this innovation network, which, as a whole, was thought to contribute to its political goals. In this respect, the 100-MW agreement must be regarded as a tool of the established coalition to match the activities of the various actors in the flexible/cooperative innovation network with its own political objectives of economic growth and centralized control over the electricity system.

In contrast to the large 1970s Danish firms that showed no interest in developing utility wind turbines, the current industry was interested in the utility market. As the California market declined, the utility market presented a potentially large growth sector. The 100-MW agreement was regarded as a useful leg up for the high potential (European) utility market. Involvement in this segment of the market created great demand for product development and upscaling of wind turbines. The technological community was afraid that utilities would consider development with existing wind turbines a waste of time and reject proven turbines for utility projects. In other words, technologists feared that involvement of the utilities in the user pole of the flexible/cooperative innovation network would challenge their favored design-by-experience approach.

It was also debated whether the Danish design would still be a proper design if turbines entered the megawatt-range in the mid-1990s. The second 100-MW agreement, as part of the *Energy 2000* plan, made it clear that policy makers demanded a sincere, constructive, and long-term involvement of the utilities in wind energy, not just as users, but as co-developers. Utilities began to purchase the cheapest 'large' proven commercial wind turbine available on the market, and started to tune their R&D effort with the wind industry's agenda for research. In the meantime, as described above, the four to five companies that had grown significantly during the export boom had become large enough to afford their own R&D departments. Firms became able to absorb the engineering knowledge that originated from the work within the science pole of the flexible wind energy innovation network, adopting the moderate design-by-extrapolation approach that characterized the knowledge strategy of the flexible innovation tradition.

9.3.7 Looking Back on the Innovation Process

The various poles of the cooperative and flexible wind energy innovation networks have gradually merged over the past two decades. The ecological practitioners in the cooperative network selected the design of Johannes Juul, who, in the 1950s, had tried to construct a wind turbine that could be connected to the central power grid. Engineers within the WPP did the same. The selection of a similar technological concept provided a good pretext for mutual innovation learning. However, this common ground was only acted upon in the late 1980s. Until that time, the design strategies of the technological communities in the two innovation networks differed too much. Practical engineers within the small firms followed a design-by-experience approach. In the late 1980s, a complex set of factors motivated a shift towards a moderate design-by-extrapolation approach. The establishment of R&D departments and the related influx of scientific engineers encouraged and enabled such a shift. Second, the availability of specialized scientific knowledge (for example, a computer design kit that had been developed within the scientific pole of the flexible innovation network) made such a shift possible and logical from an engineering point of view. Moreover, the new certification system demanded a more scientific approach. Finally, the utilities' wish for large-scale wind turbines stimulated a change in design approach.

With respect to the business pole, wind turbine manufacturers have changed from small under-capitalized firms to professionalized medium-sized firms (Vestas, Micon, Nordtank, Bonus, Wind World, and Nordex) firmly backed by established banks and operating globally. As we saw, the underdeveloped production system represented one of the major drawbacks of the Danish wind industry during the California export boom. Over the last decade, however, most firms have invested a lot in the reconstruction and extension of their production facilities. The Danish companies developed small batch, flexible, up-to-date assembly and production systems run by skilled workers. the wind turbine industry has come connected to the flexible innovation tradition as described in Chapter 8.

It follows that the current wind energy innovation network in Denmark fits well to the flexible innovation tradition as described in Chapter 8. As we recall, exactly such a type of wind energy innovation network was envisioned by the established coalition in the 1970s. From a power perspective, one is, therefore, tempted to argue that over the past decade, the established coalition has managed to recapture its leading role within the development of wind energy. It seems to have succeeded in changing the original cooperative innovation network into its own politically desired flexible innovation network. In other words, it has regained its power position.

Such an image of innovation would be far too static, too dualistic, and embodies too little concern for the mutual dependence of policy, industry, and technology development. Indeed, wind energy innovation has come to fit in the flexible innovation tradition, and has been adjusted to the political vision of the established coalition. Nevertheless, it is essential to make the following two remarks. First, the wind turbine industry builds wind turbines for both centralized as well as decentralized usage. Thus, current turbine technology serves both ideological coalitions. Second, and maybe more important, the political vision of the established coalition has seriously changed over time. Although its preference for a centralized electricity system has not altered, it has adopted many of the policy belief elements that were originally advocated by the cooperative coalition. As a result, the established coalition now sees wind energy as "the environmentally most acceptable way of generating electricity". (ME 1990: 101) Developing and selling wind machines is seen as 'the politically most acceptable way of doing business,' because it fits the widely accepted notion of *sustainable development*; a concept that was already coined by the cooperative coalition in the 1970s. Thus, the political ideas of the cooperative coalition have not been left behind, but, to a large extent, have been accepted by the established coalition.

Our analysis shows that the gradual adoption of new ideas within the policy belief system of the established coalition is not solely a matter of political debate and/or policy-oriented learning. The process of policy learning both influences and is influenced by the convictions and actions of other innovation actors. The new political balance seems to be part of a larger innovation balance that was described above and resulted from continuous, joint innovation learning between policy makers, technologists, managers from the wind turbine industry and users (both windmill cooperatives and, since the mid-1980s, electric utilities).

It follows that the newly evolved ideal of *ecological modernization* does not just represent a new political ideal, but a new model of innovation for the established coalition. Considering the fact that the established coalition agrees with the flexible innovation tradition, we might designate this novel innovation approach as *flexible ecological innovation*, to indicate that the political view behind the current flexible wind energy innovation network is different from the original idea behind it.

9.4 Conclusions

Let us return to this chapter's central question:

> How did the various politically desired wind energy innovation networks in California and Denmark develop over the last two decades, and what role has joint innovation learning (or the lack thereof) played within those developments?

In the previous sections, we described how wind energy innovation developed over the last two decades. Figure 9-2 and 9-3 depicts the analysis of both the Californian and Danish wind energy innovation story, respectively. In this section, we reflect on the role joint innovation learning within those developments. Moreover, it is argued that in the same way policy story lines can be regarded as both cause and effect of long-term integrative policy learning, so-called *innovation stories* are both cause and effect of long-standing joint innovation learning. In this respect, a policy narrative is an integral of a more comprehensive innovation story.

9.4.1 Denmark: The Symbol of Ecological Innovation

The wind energy innovation game in Denmark was played by two innovation styles. With the help of the cooperative way of innovating, the flexible wind energy innovation network that Juul had envisioned in the 1950s evolved into a flexible ecological wind energy innovation network. The indirect manner in which this happened is depicted in Figure 9-3 by a grey color.

During the 1970s, relations between the two innovation styles were adversarial. At first, the flexible innovation strategy failed because the utilities showed little interest in using wind as an energy source, and Danish industry was not interested in developing and producing wind turbines for the utility market. Moreover, technological difficulties with constructing large utility wind turbines thwarted the attempt of the established coalition to set up a flexible wind energy innovation network.

The cooperative innovation style became a success, with cooperation representing a core value within the culture of the cooperative innovation network. Through actual cooperation and face-to-face contacts trust was bred and sustained among the various types of innovation actors within the grassroots cooperative network. Joint innovation learning seemed a natural thing within this subpolitical network, and represents one of its key success factors.

The success of the cooperative innovation activities did not cause the established coalition to oppose those activities, but rather, by the end of the 1970s, to support and control them. Gradually, the initial cooperative network changed into a flexible/cooperative network with both cooperative and competitive features. The consumer pole of this network was dominated by windmill cooperatives. On a very small scale, the cooperative coalition therefore managed to increase the decentralized production of wind power, while the established coalition got itself a new industry. The flexible/cooperative network was sprinkled with compromise which took shape into a set of policy measures that strongly influenced the behavior of and interaction between different innovation network actors. This happened in the following way.

In combination with the renewable energy investment subsidy, a type approval system was initiated to ensure government money was well spent. This forced manufacturers to contact Risø, and had the indirect effect of maintaining and nurturing cooperative R&D. It also protected consumers from buying unsafe and unreliable technology. The established coalition regarded the cooperative network as a kind of niche in which the wind turbine industry could develop and made sure that the consumer pole developed at a low pace. In addition, the association of windmill owners monthly published the track record of almost every installed wind turbine. This set of interventions instituted a low-growth, highly transparent wind turbine market, in which cooperation and competition went hand in hand. As the initial cooperative network changed into a mixed flexible/cooperative one, its cooperative features were maintained. Thus, also in this phase, joint innovation learning played an important role in the gradual and mutual development of wind energy technology, industry and policy.

Two crucial episodes follow in the history of wind energy innovation: the export success and failure to California and the Danish wind boom. While the Danish government had little influence on the first event, it was unpleasantly surprised by the second.

For the sake of the argument we will disregard these two key events for the moment and will come back to them later. The thread of our argument is the gradual integration of the two styles of innovation, which was strengthened by the above two events. Although most of the Danish firms went bankrupt between 1986 and 1988, the remaining poles of the Danish flexible/cooperative innovation network were still in good condition.

The California adventure had rendered the Danish wind turbine technology the world's leader. Danish users were still enthusiastic about wind energy and politicians still supported its development. Moreover, the first 100-MW agreement forced the utilities to buy a fixed quantity of wind power capacity over the coming years. This all gave the battered wind industry new hopes for the future and speeded up restructuring. Since the mid-1980s, the industry has served both the private and utility market, though it is obvious that the main innovation effort has been directed to utility-grade wind turbines. The industry's newly refound creditworthiness, quality control system, and utility-grade products encouraged power companies to become 'seriously' involved in wind energy. In this way, the flexible/cooperative network changed into what we have named the flexible ecological wind energy innovation network.

This label indicates that, although dominated by the flexible innovation style, several important cooperative features were adopted in the course of its evolution. In Chapter 8, it was found that a truce between Fordism and the Danish craft ethos, and a mix of design-by-experience and design-by-extrapolation was typical for the flexible innovation tradition.[62] Thus the story of wind energy innovation seems to be a specific case of a more general phenomenon.

Before we draw some conclusions from this, let us return to the political analysis of part Two. In Chapter 6, we found that the cooperative coalition served as a critical corrective to the established coalition.[63] It represented a type of populism within the Danish political establishment itself. Accordingly, it was concluded that the established and cooperative coalition together formed a long-term learning alliance, or, in other words, an establishment alliance.[64] In a similar way, the cooperative style of innovation seems to serve as a critical corrective to the flexible innovation approach. We therefore coin the term *innovation learning alliance* to indicate this long-standing cooperative innovation learning relationship.

We can even extend on this comparison. In Chapter 4, we introduced the term story line. A policy story or narrative is seen as both cause and effect of a long-term integrative policy-oriented learning process between ideological coalitions. In other words, political story lines enable and stabilize learning alliances, which produce and reproduce story lines. In Chapter 5, we argued that the established and cooperative coalition formed a long-standing learning alliance that was organized around the story line of ecological modernization.[65] In this chapter, we conclude that the competence of Danish policy makers, business managers, and technologists to deal constructively with political and professional differences to a large extent determined the success of wind energy innovation in Denmark. As a result, political learning organized around the story line of *ecological modernization* could go hand in glove with innovation learning in the field of wind energy, in which a grassroots technology was adapted to utility standards. By continuously confronting each other with new options, and (at the same time) by constantly searching for integrative solutions, policy makers, technologists and managers have been pushing each other along in the field of wind energy innovation.

From this one might conclude that the story of ecological modernization involved more than just a policy narrative. Besides enabling members of different ideological coalitions to cooperate, this story line also provided a narrative that allowed policy makers, technologists and business managers to communicate, find common ground, and coordinate their activities.[66] In fact, story lines guided actors from different ideological and professional positions in their search for mutually approved political, technological and economic solutions. In other words, the story line of ecological modernization is both cause and effect of long-term joint innovation learning. Thus, instead of being merely a policy narrative, the story line of ecological modernization might be said to represent a more comprehensive *innovation story* or *innovation narrative*. Consequently, we may speak of the innovation narrative of *ecological innovation*. The Danish story of wind energy innovation represents the number one pioneer of this innovation style. It represents "the symbol of ecological innovation".

9.4.2 Intermezzo: Innovation Stories

Most of the time policy and techno-economic innovation worked in tandem in the Danish case. Since joint innovation learning was omnipresent in Denmark it was concluded that ecological modernization not only represented a policy narrative, but also a more comprehensive innovation narrative.

The complexity of California's energy politics was characterized by four ideological coalitions. The corporate capitalist coalition and the pragmatic side of the public interest coalition formed a long-standing learning alliance that was organized around the story of ecological modernization. Beside the liberal establishment alliance, a competing conservative establishment alliance was discerned, consisting of a part of the public interest coalition and the libertarian coalition, which was organized around the story line of entrepreneurial modernization. This raises the question of whether these narratives just represent policy stories or, like ecological modernization in Denmark, represent inclusive innovation narratives?

Answering this question actually falls outside the scope of this book. We could draw such a conclusion in the Danish case, because in Denmark wind energy played an important symbolic and practical guiding role within the energy debate over the last two decades. Wind power in California clearly did not play an important role. Other technological options, most notably gas cogeneration, assumed such a central position. Although answering the above question goes beyond the scope of this study, we would like to make a tentative assumption that will hopefully help us to put wind energy innovation into a wider innovation perspective and make it easier to structure our comparison of wind energy innovation in both countries.

We would like to assume that the narratives of ecological and entrepreneurial innovation indeed represent two comprehensive and competitive innovation stories. Demand-side management (DSM) was the central tool within the story of ecological modernization. Within the narrative of entrepreneurial modernization, which in the

1990s has come to dominate the energy debate, gas cogeneration played a crucial guiding role. Innovation within the gas industry energized the policy making process, and vice versa. In other words, techno-economic innovation in the field of gas strengthened the story line of entrepreneurial modernization, which stimulated development of the gas technology and industry. We argue that, besides containing a political message, the story of entrepreneurial modernization also includes a related style of innovating. We might, therefore, as well speak of the innovation story of *entrepreneurial innovation*. In this subsection, we will first briefly discuss the rivalry between the two broad innovation stories. In the next one, we will use this discussion to define the position of wind energy innovation and the role of joint innovation therein.

As we saw in Chapter 2, the California electricity quandary was a multi-faceted phenomenon. The utilities saw nuclear power as the way out but this encountered a lot of public opposition and technical problems, and the utilities had great difficulties in getting their nuclear plants on line. Among other things, the threat of capacity shortage offered an innovation space for other innovation scenarios. The two main innovation stories that emerged were ecological and entrepreneurial innovation.

The public interest coalition saw conservation as a structural tool to address the energy crisis. For the utilities, DSM presented a way to save time. The idea behind ecological modernization was that through a greater use of renewables and energy conservation measures, the investor-owned utilities could maintain their monopoly position and profit margin, lower the price of electricity and its environmental impact and increase the reliability of supply. Consequently, all these benefits would help them to regain their political legitimacy. CPUC's ratemaking policy represented the most important mechanism for putting utility investment in DSM at the center of management focus. It was realized that the traditional rate-of-return regulation frustrated goals for energy efficiency and conservation. Consequently, the CPUC adopted the Electric Revenue Adjustment Mechanism (ERAM) in 1982 to decouple the sale of electricity from utility profits (CPUC 1993: 55-56). This complicated set of procedures was thought to motivate utilities to promote conservation as a means to both protect utility financial health and provide consumers with greater rate stability. In this way, ERAM indicated the formation of a new partnership between utilities and regulators. For all that ERAM achieved, the state's utility executives did not fully embrace conservation.[67] Moreover, despite strong pressure by policy makers, the utilities failed to seize the opportunity of gas cogeneration.

Whereas the command-and-control strategy that was central to the narrative of ecological innovation was only partially successful in the first half of the 1980s, a type of policy that emphasized a market-based approach to innovation turned out to be quite successful. The conservative establishment alliance encouraged an entrepreneurial style of innovation that fit in the story of entrepreneurial modernization. Triggered by federal PURPA and California's QF-policy, entrepreneurs of every stripe took their chance. As a result, California became the worldwide leader in geothermal, industrial cogeneration and wind power projects; a novel independent power industry was born.

This created a politico-industrial deadlock, with the utilities on one side and, on the other, the newly founded independent power industry. While the new industry challenged the monopoly power of utilities, the latter tried to undo the independent energy business. The fact that California was confronted with a situation of overcapacity[68] meant this conflict was not in need of an urgent resolution. Regardless, behind the scenes the struggle continued unabated. After an extended period of non-decision making (the second half of the 1980s) the BRPU proceeding was concocted, in which both innovation stories found a place. With respect to ecological modernization, DSM was again perceived as the major tool, though its funding had fallen precipitously between 1986 and 1988. Triggered by the debate on global climate change, a 'state-wide collaborative process' was initiated to renew the utilities' interest.[69] Again the idea was that DSM should benefit both ratepayers as well as shareholders, and needed to be part of the utilities' normal business strategy. The California Collaborative led the sustainable grow-and-build alliance to agree that programs that promoted energy efficiency were the number one priority. Some three-quarters of all future electrical needs would be backed by DSM. Although the regulated competition alliance interpreted this as a way to block the development of the independent power industry, it did not oppose DSM, since it was convinced that the above DSM numbers would never be attained.[70] In 1990, its patience became rewarded, when the CPUC decided that some 1,350 MW of new resources was going to be put out to bid. However, the Californian BRPU was overruled by the Federal Energy Regulatory Commission (FERC). In addition, high electricity rates flared up the electric restructuring debate.

The demand for electricity industry restructuring indicated the political hegemony of the narrative of entrepreneurial modernization. This political dominance was closely related to the techno-economic superiority of gas turbine technology, which at the time had become the most economic electricity generating option.[71] Thus, the economic performance of the independent power industry (in particular, the gas cogeneration industry) strengthened the call of the libertarian coalition for restructuring the electric utility industry. The existence of a competitive cogeneration technology and industry did not merely improved the ideological position of the libertarian coalition, but also advanced its arguments in a very practical way.

As noted before, the *Memorandum of Understanding* (MOU), in which the utilities and independent power industry stated their intent to rapidly move toward an open and competitive electricity generating market, was a compromise between the corporate capitalist and libertarian coalition that to a large extent sidelined the public interest coalition. This implied a political shift; a conservative establishment took over the dominant position of the liberal establishment. The MOU did not merely signal a political take over, but also signified the collapse of the command-and-control style of innovating that had dominated the innovation story of ecological modernization in California. Ecological innovation presented a mixture of the corporate innovation style and the civilian approach to innovation. Although intended to secure the legitimacy of their monopoly position, the utilities forgot to use the opportunities this privileged position and its related favorite regulatory environment offered. They themselves turned out to be significant barriers to ecological innovation. Instead of fighting the independent power industry in the political arena, the utilities would have been better off if they had put more effort into innovating. The techno-economic success of the cogeneration industry made it win the political battle with the investor-owned utilities through the back door, that is, the innovation game. It seems to be just like the saying goes: action speaks louder than words.

One could say that the story of *ecological modernization* was cause and effect of a long-term policy-oriented learning process. This 'successful' political learning process, however, was weakly supported by innovative activity. There was a clear lack of joint innovation learning between policy makers, utility managers and technologists. Ratemaking policy did not entice utilities to adopt new business strategies or experiment strongly with new technological options. In contrast, the story of *entrepreneurial modernization* was full of innovation. The innovation activities of the independent power industry fit perfectly within the political story of entrepreneurial modernization that has come to dominate the energy debate in California. It seems fair to argue, therefore, that if the object of our study had been the gas cogeneration technology and industry, the conclusion had been that the maturation of the gas cogeneration technology and industry and the development of the energy debate in California have gone hand in glove.

In the former subsection we concluded that, in Denmark, the cooperative style of innovation serves as a critical corrective to the flexible innovation approach. A similar conclusion may be drawn for the Californian situation. Entrepreneurial innovation seems to serve as a critical corrective to the corporate innovation approach. A constant, sometimes productive, tension appears to exist within capitalism between monopoly and competition, between centralization and decentralization of economic power (cf. Harvey 1990; Rothwell and Dodgson 1993).

9.4.3 California: The Ugly Duckling of Entrepreneurial Innovation

Here, we will try to describe the development of wind energy in terms of the broader process of innovation within the electric utility sector that was characterized in the former subsection as a competition between ecological and entrepreneurial innovation; a battle that was (temporarily) won by the latter. This means that we will concentrate on variants of corporate and entrepreneurial type of networks (Figure 9-2). Consequently, the Brown administration's attempt to establish a communal wind energy innovation network falls somewhat outside the content of this subsection. Nevertheless, since this development path is very interesting from a comparative point of view we will return to it in the final subsection.

The attempt to set up a civilian/corporate wind energy network became a failure. The CEC initiated a wind power program (WPP) to design utility wind turbines, specially adjusted to Californian weather conditions. The WPP was meant as a supplement to the, from the perspective of the CEC, flawed Federal Wind Energy Program (FWEP). Because most Californian politicians viewed R&D as a federal task, they failed to support the plan. As a result, liberal policy makers were not in complete control over the development of the corporate wind energy innovation network. Though they succeeded in getting the Californian utilities interested in wind power, for technical progress they still depended on the success of the federal research program, and thus had to rely on continuous federal support for the FWEP. The election of Ronald Reagan signaled a drastic change in the political climate in America. As one of its first deeds, the Reagan administration severely cut the budget of the FWEP, which more or less ended the weak development of the corporate wind energy innovation network. It obstructed wind energy innovation as favored by the liberal establishment in California and invalidated, therefore, the process of joint innovation learning that had taken place between liberal policy makers and the utilities.

A competitive type of development did take place; the uncontrolled development of the civilian/entrepreneurial wind energy innovation network. This type of innovation could benefit from policy measures, in particular the energy tax credit program, initially meant to promote the two other innovation networks. In 1983, Deukmejian's rage against the energy tax credit program threatened the viability of the entrepreneurial wind energy innovation network. This crucial event illustrates an innovation paradox. While libertarians promoted the establishment of an independent power industry, they distrusted government interference, which was necessary to give this industry a fair chance. This ambivalence was evident in the discussion around the energy tax credit program. Governor Deukmejian, a representative of the radical side of the libertarian coalition, wanted to abolish the program that originated from attempts to set up communal innovation networks. Since the pragmatic side of this coalition was in favor of the program, its continuation received bi-partisan support.[72] In spite of this, Deukmejian's act hampered joint innovation learning. It created a situation in which politicians became more important than bureaucrats. Since obvious improvements became neglected, the faulty 'old' program continued. Consequently, the wind energy market maintained its murkiness, fraudulent behavior persisted, and the process of destructive creation and related demise of the American wind turbine industry continued.

While California's independent power policy led to the destruction of the technical pole of the entrepreneurial wind energy innovation network, it also led to the over-stimulation of its user pole. In September 1983 the CPUC approved the ISO4 contracts. The long-term contracts for wind only lasted for a year and a half, in which period the wind power industry was strongly subsidized, giving wind farmers the opportunity to set up the civilian/entrepreneurial wind energy innovation network. Somewhat later, the tax credit program was ended, so the California wind industry only received 'sufficient' political support for a very short space of time. In that particular period, support was not only 'sufficient' but truly abundant. Combined with the widespread realization that political support would only be short-lived, this abundance encouraged opportunistic and, even worse, fraudulent behavior. Despite the negative side effects, the California policy led to a boom-and-bust development cycle in the field of wind energy, and quite unexpectedly created an independent wind power industry almost overnight.

As we saw, the sudden arrival of a new independent power industry created a political and industrial deadlock that was characterized by two competing industries and two competing innovation stories. The competitiveness of cogeneration plants increased the assurance of the gas cogeneration industry and engendered a rather militant attitude with an increased crusade against the monopoly structure of the electricity sector. Since the wind industry was far from being cost-competitive, it could not afford such a determined position. Instead of waiting for the outcome of the BRPU process, it tried to adapt its products to the utilities and came to conform to the story line of *ecological modernization*. This was at odds with the overall political climate, but suited the actual situation within the electric industry. So, when after the wind boom California was slowly reconsidering its QF-policy, the wind industry tried to breathe new life into the corporate wind energy innovation network, which seemed to succeed in the early 1990s. The issue of global climate change generated a renewed interest in renewables and it seemed that the utility market for wind power plants would open soon.

Besides, at that time the BRPU process offered new opportunities for wind farm developers to set up independent wind projects. In 1990, some 300 MW had been 'set aside' for bidding by renewable technologies only. Since wind energy represented one of the most competitive renewable sources, the prospects for the industry were very promising. However, the Californian BRPU proceeding was overruled by the Federal Energy Regulatory Commission (FERC), and fragmentation between state and federal policy frustrated the revival of civilian/entrepreneurial wind energy innovation. It cancelled policy measures like the renewables set-aside and renewables portfolio standard policy that would have provided good market opportunities for Californian wind farm developers.

The restructuring process also annulled the policy learning process within the liberal establishment that was organized around the story of ecological modernization. The tidal wave of *entrepreneurial modernization* (temporarily) killed corporate wind energy innovation. It forced the wind power industry to follow a purely entrepreneurial way of innovating. Hence, contrary to the situation in the early 1980s, almost no policy measures were left from attempts to set up other types of networks. This forced the wind power industry to compete heads on with other electricity generating industries. The one question left is: how does wind energy innovation comply with the innovation story of entrepreneurial innovation?

U.S. Windpower assumed these two perfectly agreed. Its former President and Chief Executive Officer, Gerald R. Alderson (1995) stated, "We see the move to privatization and an open, competitive energy economy as reasons for optimism, not pessimism. Rather than becoming victims of change, renewables will benefit from the move to an economically competitive market. Indeed, rather than shrinking from open competition, renewables can actively embrace it as an agent of growth." This point of view was contrary to the prevailing notion in the wind power industry. Most firms feared that wind power would wither away in these new, open markets – unable to compete.

For the time being, the predictions of U.S. Windpower have not yet come true. Worse, the world's largest wind power company became one of the first victims of the shift from a regulated, command-and-control structure to one that features competition and privatization. We think that, during the period under study, wind power was not yet able to comply with the rationale of innovation inherent in the story of entrepreneurial innovation. This leads to an interesting innovation paradox. On the one hand, the wind power industry owes its existence to the story of entrepreneurial innovation. On the other, this innovation narrative, when carried to extremity, forms a huge threat to the livelihood of this industry. Whereas the Danish wind turbine industry was "the symbol of ecological innovation," the Californian wind power industry could be said to be "the ugly duckling of entrepreneurial modernization".

9.4.4 Comparative Conclusions

As with the distinction we made in Chapter 6 between the hard and soft side of populism, we can also distinguish between the hard and soft side of innovation. The Danish cooperative and American communal innovation tradition may be regarded as the soft side of innovation, which is engaged with the social and human aspects of innovation and its implications for democracy. It stresses that innovation should create or maintain egalitarian, convivial, or legitimately hierarchical social relationships. The hard side of innovation, represented by the American entrepreneurial innovation tradition, is in search for innovation that increases economic efficiency and profits.

We found that, while in California, innovation within the electric sector became dominated by the innovation narrative of entrepreneurial innovation, in Denmark it was guided by the story of ecological modernization. While in Denmark, the soft side of innovation inspired the flexible innovation tradition, it had little impact on either the corporate or the entrepreneurial innovation path in California.

In 1970s Denmark, wind turbine technology stood for an alternative energy future. This 'simple' technology gave the cooperative coalition the opportunity to create counter-pictures and emotional symbols of its alternative social vision. In California, solar technology inspired the communitarian coalition. Consequently, even this most obvious proponent of communal innovation initially overlooked the wind power option. Only at the end of the 1970s was wind energy added to the energy tax credit program. Still, a significant market for small private wind turbines did not develop. When Governor Deukmejian came into office in 1983, he immediately abolished the Office of Appropriate Technology, which indicated that the political support for communal innovation, as introduced by the Brown administration, would no longer be continued. In contrast, the Danish cooperative coalition has played an important role in the energy debate throughout the whole period under study, and the cooperative way of innovating had a marked influence on wind energy innovation. While in California the hard side of innovation drove innovation, this innovation type was not tolerated in Denmark. The political reaction on the Danish wind boom exemplifies this. When, in 1985, a speculative wind energy development evolved in Denmark, both coalitions rapidly and clearly expressed their disapproval of such a 'Wild West' type of development. In other words, it became clear to all innovation actors involved that entrepreneurial innovation was *not done* in Denmark.

It follows that the Danish political situation offered room for only one innovation story, that of ecological modernization. More important, innovation in the field of wind energy drove this story. In other words, in Denmark, wind energy innovation fit perfectly in the story of ecological modernization. This can be perceived as both cause and effect of joint innovation learning. In California, two competing innovation narratives could co-exist: ecological and entrepreneurial innovation. Since wind energy innovation was weakly related to both, it tried to be connected to the one that at the moment offered the most opportunities. During the California wind boom, the wind power industry strove to fit the story of entrepreneurial innovation. This was possible only because of favorable public policies. When these policies ended, the wind power industry was forced to change direction and, so to speak, jumped on the wave of ecological modernization. This wave had little impetus though, and in the mid-1990s was flooded by the wave of entrepreneurial innovation, which was stoked by the truly competitive gas cogeneration industry. In the short term, the wind power industry does not seem to comply with the competitive demands of this innovation style. We can conclude that while wind energy innovation in Denmark played the role of pioneer, in California it played only a minor role. It was at the mercy of a competition between two innovation stories that it was only very weakly related to. This was both cause and effect of a lack of joint innovation learning.

From the fact that the Danish political situation only offered room for the innovation story of ecological innovation, it can cot be concluded that the success of the Danish wind turbine industry is solely related to this type of innovation. After 1983, the development of wind power in California and Denmark became both cause and effect of one another's 'success' and 'failure.' Bad technology from American wind turbine manufacturers was irrevocably killing the wind farm business, and many farmers survived because of the entrance of the Danish wind turbine industry. But this entrance finished the American manufacturers off. The large demand for Danish wind turbines generated investment money that could be used to finance technological progress. Without this export success, the Danish government would have probably reduced its support for wind energy, which would have downsized the home market to a level inadequate to fund rapid technological development.[73] In abstract terms, this means that entrepreneurial innovation in California formed one of the breeding grounds on which ecological innovation in Denmark could flourish.

PART FOUR

CONCLUSION

In the introductory Chapter it was mentioned that an examination of wind energy innovation in two countries would not deliver a full answer to the book's central question: How can policy makers guide techno-economic innovation into a democratically legitimate direction? Still, the field of wind energy provided us with a useful test case to develop an interactive network approach to innovation. This strongly rejects technological determinism, where technology is seen as an autonomous force that shapes society. While technology indeed has a major impact on society, historical, political, cultural, organizational and economic factors all influence the path a technology takes. In other words, technology and society coevolve and any inquiry into the process of innovation should take their interdependence as a natural point of departure. According to some scholars, the interdependence of the various aspects of innovation is so strong that making distinctions among them makes little sense (cf. Bijker *et al.* 1987: 3). They insist, therefore, on the need to see the social/technical as a "seamless web."

In the first chapter, we noted that the notion of "seamlessness" leaves little room for the articulation of difference. Moreover, conceptualizing innovation as one complex dynamic carries the danger of disregarding the political character of innovation (Horn 1994: 387). Since the politics of innovation is our main point of interest, we carefully tried to avoid this serious pitfall, by developing an approach that considers both interdependence and difference between the various facets of innovation. Innovation is perceived as an event in which three games – the technical, political, and market game – are played simultaneously, and although the games are interdependent, each game is defined by its unique set of rules and styles of play. This approach, which recognizes the relative autonomy of the political game, as well as considers its interdependence with the technical and market game, is particularly interesting and fruitful from a political intervention perspective.

Appreciating difference and interdependence allows for acknowledging the fact that the three games have different play areas. While national borders are relevant for grasping the political game, the technical and market games have the whole globe as their playing field. Choosing a comparative political perspective and considering the international dimension of business and technology is a necessary condition, therefore, to obtain an accurate insight into the democratization of innovation. In this respect, the choice to study wind energy innovation in states as different as California and Denmark seems to be an astute one. The strong interdependence of the two cases made it possible to incorporate within this inquiry the international scope of business and technology development. More important, it enabled us to study how two entirely distinct societies shaped, or rather, politically guided, a similar type of technology.

This was organized in three steps. In Part One, two thick descriptions illustrated the strong interdependence between the technical, economic and political aspects of wind energy development in Denmark and California. The two stories force upon the reader and analyst the metaphor of a 'seamless web' of society, industry and technology. Still, we decided not to reflect on wind energy innovation in this way. In order to throw light on the politics of wind energy innovation, we analyzed it as consisting of three semi-autonomous games. In Part Two, we focused on the political game, that is, on the policy makers within the energy policy subsystem. We examined how interacting ideological coalitions gave political meaning to and legitimized wind energy innovation. In Part Three, the focus shifted from the political system to its interaction with technological institutes and firms. Within this interaction techno-economic innovation can be said to simultaneously play two different, but interrelated, 'political' roles.

First, techno-economic innovation can be considered an instrument for pre-determined political goals (Sclove 1995: 10). It is a means to generate political ends, and is utilized to materialize, reproduce, strengthen and stabilize political beliefs. Within this view on techno-economic innovation, implementation is considered the main bottleneck. Besides reflecting human intentions, techno-economic innovation also constitutes human intentions. This alternative understanding of innovation can be characterized either as holistic, constitutive, integrative (cf. Follett 1924), interactive (Grin and Van de Graaf 1997; Caracostas and Muldur 1998), or expressivist (Carpenter 1992). Carpenter (1992: 169) explains that "[E]xpressively, technology designs that world and gives form to our most basic priorities and necessities. Expressively, technology actually defines those needs, providing a basis for our self-understanding".

In Part Four, we evaluate our approach to innovation and discuss its implications for the general challenge of the democratic guidance of techno-economic innovation. We also look ahead and discuss the future of wind energy innovation in California and Denmark. In Chapter 10, we address this book's central question. Finally, in Chapter 11, we speculate on the likely future of wind energy in California and Denmark on the basis of the theories and insights developed in this book.

10 The Intelligence of Innovation

> Technologies ... resemble children. Like children, they on the one hand can not forsake parental care and guidance, while on the other, they will constantly surprise their parents, and moreover, will back out of the authority of their parents. Once they are brought into the world, technologies, just like children, have to be brought up into responsible and politically capable fellow citizens. *Martijntje Smits, 1997: 114*

In this chapter we will address the book's central question: How can policy makers guide techno-economic innovation into a democratically legitimate direction? We will argue that policy makers need to consider three interrelated processes. First, a high democratic quality of the political debate is crucial. In section 10.1, we discuss the quality of the decision making process around innovation on the basis of our analysis and findings in Part Two. Political pluralism and policy-oriented learning will play an important role in this discussion.

Besides paying attention to the quality and diversity of the political debate, the process of innovation implementation needs to be optimized and the political and democratic role of techno-economic innovation needs to be monitored, so it can be used as a proper input for the ongoing debate around innovation. In section 10.2, where we refer to the findings of Part Three "The Innovation Game," we touch upon the latter two processes. The mechanism of joint innovation learning and the political role of techno-economic innovation are at the center of this section.

In the final section, we review the implications of our findings with respect to wind energy innovation and the approach to innovation we have developed throughout this book for the democratic guidance of techno-economic developments. The three above-named processes – democratic debating, implementing, and monitoring of innovation – will be dealt with in a more prescriptive manner. In addition, the relationship between political pluralism and techno-economic pluralism is discussed.

10.1 The Intelligence of Politics[1]

> Democracy is the regime in which the majority recognizes the rights of minorities because it recognizes that today's majority may be tomorrow's minority and may have to obey laws that, while they might not represent its interests, will not prevent it from exercising its basic rights. The democratic spirit is based on this awareness of the interdependence of unity and diversity and sustained by never-ending disputes over the ever-shifting border between the two and about the best ways to strengthen their association. *Alain Touraine, 1997: 16*

In Part Two, we examined what kind of instrumental role wind energy technology played for policy makers. A first assumption was that policy makers are guided by hierarchically and laterally ordered policy belief systems. Technology plays a role merely on the lowest level of the first order-discourse, that is, the level of technical verification. "For the policy makers, the artifact is primarily an objective, an end or means for a policy designed to bring closer the preferred social order." (Grin and Van de Graaf 1996: 88) Since the number of belief systems active within a debate is limited, policy actors can be aggregated on the basis of the particular belief system they share. Policy makers were said to form ideological coalitions to indicate that they share hard core values. The 'logic' of policy making was examined in terms of interacting belief systems, or, from an actor perspective, as a confrontation between various ideological coalitions.

10.1.1 Policy Change and Learning

In Chapter 5, we found that the California energy debate was influenced by four ideological coalitions. In Denmark two relevant coalitions were distinguished. For each of these coalitions wind energy technology had a distinctive meaning; it presented 'distinct' means for different political ends. Coalitions may also have different ideas about what their politically preferred wind turbine looks like, and may even talk about a different type of technological object or means. For example, while the cooperative coalition in Denmark was focused on wind turbine technology, the established coalition envisioned the development of wind parks, of which a wind turbine only presents one element. It thus became clear that artifacts can have multiple political meanings, and that different ideological coalitions might have different types of artifacts in mind. For example, while communitarians envisioned small local wind turbines, the corporate capitalist coalition promoted the development of large high-tech wind machines to be owned by the utilities.

Since the viability of wind energy hinges on governmental support,[2] politics has a critical say in its development. Policy makers therefore seem to be in an ideal position to guide wind energy innovation into a democratically legitimate direction. In the 'politics of meaning' model, the democratic fight between policy belief systems defines which political meaning(s) will guide the development of wind energy. By definition, the result of any democratic fight is democratic. Hence, each democratically chosen innovation policy can be seen as a democratically legitimate attempt to guide innovation into a certain direction. This argument would render the central question of this book obsolete.

However, there is more to democracy than just its outcome. The way in which democratic legitimization and support is achieved and/or maintained is also important. When studying the democratic process, the learning-capability of different coalitions within a certain political (sub)system becomes a central issue. Policy-oriented learning then becomes an essential condition for any ideological coalition to become and to remain successful within the democratic battle. A competitive and cooperative form of policy learning was distinguished. While the competitive form is confined *within* the boundaries of the various belief systems, cooperative policy learning requires mutual learning *across* alternative (first-order) policy beliefs. Ideological coalitions that were mutually engaged in such a long-lasting cooperative learning process were said to form a *learning alliance*.

Both in California and Denmark learning alliances were found to play a major role within the energy debate. The dominant presence of learning alliances within the political game puts the process of democratic guidance of technology into an interesting perspective. This can be illuminated by contrasting a democratic model based on the dominance of competitive coalition behavior to one based on integrative behavior. When competitive behavior is the rule, society is seen as made up of warring groups and ideal society as made up of groups in equilibrium. The process of political bargaining would ideally lead to the simultaneous support of different politically preferred type of technologies that each serve different, antagonistic political interests. According to this democratic model, the technological world reflects the political balance of power, or, in other words, it represents the materialization of the bargaining process. Different views and interests will be served by different type of technologies. This model leaves us with a fairly static conception of the process of democratic guidance of technology. Namely, in a situation dominated by competitive behavior the different meanings of technology will not likely change. In contrast, integrative behavior may cause coalitions to perceive technology in novel ways that may not be in conflict with the way another coalition perceives it. Cooperative learning is, therefore, essentially a creative and unpredictable process.

Our analysis in Chapter 5 shows that in the political game around wind energy both competitive and integrative coalitional behavior are present. In California, domination was found within the FWEP, Deukmejian's power politics to abolish the tax credit program, and the way the public interest coalition was practically excluded from the electric restructuring debate. The first failed, the second led to a bad compromise that frustrated both sides, and the outcome of the third is highly speculative at the moment. However, since the case of wind energy shows that domination is often not successful in the long run we should fear the worst. The California and Danish cases illustrate that the process of democratic legitimization needs not to be an equivalent of the terror of the democratic majority. Often minority views are taken into account and attempts are made to give these views a fair chance in the debate or even to integrate them into a collective idea.

In Denmark, the 1970s energy debate was characterized by an ideological conflict. The dominant established coalition promoted the extension of the centralized power system mainly by the use of nuclear power plants. The use of large-scale wind turbines grouped in wind parks that could contribute to the centralized electricity system as fuel savers was also envisioned. In contrast, the marginal cooperative coalition imagined a decentralized low-energy society, which would rely on the efficient use of renewable energy sources. In this view, wind turbine technology was attributed the meaning of locally manageable means to provide environmentally friendly energy.

During the 1980s, the established coalition came to take the ideas concerning energy planning of the cooperative coalition very seriously, and policy learning *across* the two coalitions led to the progressive integration of first order beliefs. The centralization-decentralization controversy played a minor role within this learning process. The two coalitions formed a learning alliance that was organized around the policy narrative of ecological modernization. At the end of the decade, the established coalition came to support the idea of sustainability that had been a guiding principle for the cooperative coalition since the early 1970s. Through *mutual adjustment* (Lindblom 1965) a common political purpose evolved in which wind turbine technology came to be regarded as the "environmentally most acceptable way of generating electricity" (EM 1990: 101) Thus wind energy became democratically legitimized in a constructive manner.

Integrative coalition behavior was also a common feature of the energy debate in California. The corporate capitalist and public interest coalition formed a learning alliance that, like the Danish one, was organized around the story line of ecological modernization. Simultaneously, the libertarian coalition set in motion a learning process that was organized around the concept of entrepreneurial modernization. In the story of ecological modernization, wind energy technology was seen as an environmentally friendly technology. In the policy narrative of entrepreneurial modernization wind energy technology was supposed to live up to the idea that it was able to compete with fossil fuel technologies in an open electricity generating market.

But beside integrative behavior within learning alliances, competitive behavior also existed. In particular, corporate capitalists and libertarians saw each other as arch-enemies. This changed in the mid-1990s, when the corporate capitalist coalition left its *sustainable grow-and-build alliance* with the public interest coalition to join the *deregulation alliance* with the libertarians. This signaled the dominance of the story of entrepreneurial modernization and isolated the public interest coalition.

10.1.2 Policy Tradition and Learning

While the prevailing integrative behavior of ideological coalitions within the Danish energy debate conforms to the general description of the Danish style of doing politics, it seems to be at odds with the common typification of American politics. Danish politics is generally typified by appeasement and a combination of pragmatism and consensualism. As Kickert (1993: 332) remarks, "Danish political culture is saturated with cooperation and compromise." Ejvind Larsen (1983) relates this policy style to the subtle influence of Grundtvig's idea of '*the living word*' that embodies "the program for a type of politics that strongly favors both tolerance of ideological diversity and peaceful conflict resolution". (Borish 1991: 261)[3] In contrast, American political style is often characterized as being both adversarial and dominated by moral absolutism (cf. Hartz 1955). Based on this description one would expect that coalitional behavior in California is mostly driven by the will to dominate or by the search for compromise. However, as we saw, integrative coalition behavior is also a common feature of the political game there. Thus characterizing American policy style as solely adversarial is telling only part of the story. Portraying Danish policy style as solely consensual does not cover the whole picture either. Our study demonstrated this in the following way.

In Chapter 6, we found that at their most general level, the policy beliefs of coalitions are related to distinct ideological traditions. By examining the policy process in terms of interacting beliefs, we, therefore, at the same time, describe it in terms of interacting ideological traditions. At a time in which the end of history and triumph of the capitalist ideology (cf. Fukuyama 1989) are being preached, the discovery of the relevance and constitutive nature of historically shaped ideological stories is an important finding in itself. As stated, from the view of an ideological coalition, these grand political stories represent the principal long-term glue of politics. When we look at the policy process as a whole, however, we might as well say that they form the long-term 'explosive' of politics, that is, ideological difference forms a constant driving force behind long-term policy change. Ideological variety stirs up competition between ideological coalitions.

Ideological traditions do not provide coalitions with clear-cut blueprints for action. First-order beliefs of members of one and the same ideological coalition may be different, while the first-order discourse of members of distinct coalitions may well be similar. This interpretative flexibility enables both competitive forms of political behavior within a coalition (we discerned sectarians from pragmatists), as well as enables cooperation between different ideological coalitions. In the latter case, coalitions are said to form an *alliance.* The type of alliance that embodies long-lasting cooperative forms of learning, and in which core beliefs play second fiddle, was named *learning alliances. Policy narratives* guide actors from different ideological positions in their search for mutually approved policy solutions. We discovered that the national political culture, which was defined as the evolving mix of political traditions, include some type of learning alliances while excluding others.

The political culture in Denmark offered room for only one learning alliance. A kind of libertarian coalition existed, but its ideology and style of doing politics were seen as a threat to the Danish political system. The establishment denied the coalition that embodied the hard side of Danish populism a voice within the political game. The establishment, thus, actively prevented the libertarian story line of entrepreneurial modernization from gaining a firm foothold in Danish politics. This illustrates that the strategy of domination is an integral part of the Danish way of doing politics. Moreover, the rigorous exclusion of 'right-wing extremism' creates a fertile political atmosphere in which the established and the cooperative coalition can peacefully cooperate. It creates a political sphere, populated and guarded by the established and cooperative coalition, which is typified by pragmatism and consensualism. Consequently, throughout the period under study the Danish political culture offered room for only one learning or establishment alliance, organized around the story line of ecological modernization. Within this political cultural setting or learning environment the political meaning of wind energy technology and industry evolved in Denmark.

The learning alliance between the corporate capitalist and the public interest coalition in California resembled the establishment alliance in Denmark. Both covered similar types of policy learning processes, and were organized around the same story line. The shift in American political culture in the 1970s offered room for a second establishment alliance. In contrast to Denmark, right-wing populism was neither ignored nor rejected by other ideological coalitions. Public support for free-market ideology was too big for the kind of exclusion-tactics the Danish establishment had used successfully. Moreover the libertarian tradition was regarded as a typical and worthwhile part of the American political culture. For example, several members of the public interest coalition saw these ideas as a legitimate critical corrective for balancing its own ideas. As a result, a second learning alliance, organized around the story of entrepreneurial modernization, came to exist. The integrative behavior of the public interest coalition towards right-wing populism created an adversarial style of politics between the old liberal establishment alliance and the novel conservative one.

While in Denmark, exclusion of the hard side of populism led to a cooperation within one establishment alliance,[4] in California, inclusion of the hard side of populism led to cooperation *within* the liberal and conservative establishment alliance, but a competitive atmosphere *between* these two alliances. In the 1990s, the narrative of entrepreneurial modernization became dominant when the corporate capitalist coalition abandoned its strategy of integration with the public interest coalition and its strategy of domination toward the libertarian coalition. Its compromise with the libertarians made possible the fast deregulation of the electric industry. The public interest coalition, which had hoped to use some libertarian ideas to strengthen California's energy policy, became sidelined, realizing that its cooperation with the libertarians had benefited the libertarian coalition more.

10.1.3 The Intelligence of Politics

> Only integration really stabilizes. But by stabilization I do not mean anything stationary. Nothing ever stays put. I mean only that that particular conflict is settled and the next occurs on a higher level. *Mary Parker Follett, 1925*[5]

> An alliance can help make up for the defects of a single way of life, but it can never provide a lasting solution. Allies remain competitors; antagonism is always just beneath the surface. *Thompson, Ellis and Wildavsky, 1990: 89*

In Part Two, we addressed the question of how to guide techno-economic innovation into a democratically legitimate direction by looking solely at the political process. As we said before, in Western democracies each democratically chosen innovation policy can in principle be regarded as a democratically legitimate attempt to guide innovation into a certain direction. Consequently, we concentrated on the more interesting question of what way of legitimizing innovation can be considered the most fruitful in the long run.

Domination of one ideological coalition over other(s) seems to offer the easiest way. But it is counterproductive in that it breeds resentment among competing coalitions, and only advances policy learning *within* a coalition, ignoring the notion that policy beliefs of marginal coalitions may have added value. Integration therefore seems to offer a more fruitful way of legitimizing innovation. It treats different beliefs both as a potential threat as well as a potential source of inspiration. This strategy promotes policy-oriented learning *across* coalitions, which aims to satisfy all coalitions. The policy analysis to a large extent confirms that long-term cooperative policy learning across coalitions offers a constructive and intelligent way to democratically legitimize wind energy innovation. More important, our study also illuminates the limitations of such a proposition.

First of all, the national political culture constrains cooperative policy learning across certain ideological coalitions. In Denmark, cooperation of the establishment with the hard side of populism was not done. The establishment defined this perspective as an improper Danish deviance. In California, the soft side of populism disappeared from the political stage in the early 1980s. One might speak, therefore, of the *bounded systems rationality of a democratic system*, or, of the *limited intelligence of democracy*. Our comparative study of the energy debate shows that the limitations of various political systems are different. Moreover, in the same way the mix of political cultures gradually changes, so do the limitations of the system of democracy.

Second, our analysis shows that integrative and dominant coalition behavior often go hand in glove. In Denmark, dominant behavior of the establishment towards right-wing populism created a niche for cooperative policy learning between the established and cooperative coalition. In California, integrative behavior of the public interest coalition towards right-wing populism led to competitive behavior between the liberal and conservative learning alliances. In principle, integrative and dominant behavior are not necessarily interdependent. In practice, however, these two types of behavior often are two sides of the same coin. This means that one might still plead for cooperative learning across two or more coalitions as the best way of democratically legitimizing innovation. Nevertheless, in practice, initiating and preserving a learning process between three ideological coalitions seems to be more trouble than it is worth. Cooperative policy learning seems to be subjected to the *law of diminishing returns*.

Third, integration between ideological coalitions is only possible to a certain level. The use of a well-defined four-layer depiction of policy belief systems offered the possibility to keep track of the degree or level of integration between two ideological coalitions in a precise and verifiable way. In Denmark, the established coalition adopted a lot of ideas of the cooperative coalition. The long-standing cooperative learning process organized around the story of ecological modernization has led to a total integration of policy arguments within the first-order discourse (see Figure 5-6). This was possible only because the two coalitions have carefully avoided their ideological controversy between centralization and decentralization. The same mechanism that enables integration at the same time limits the degree of integration. While integration may lead to the joint support of policy programs, the reasons why various coalitions support these programs remain different. It is this ambivalence (being both attracted to and repelled by rival belief systems) that generates the "switching mechanisms," which continually forge, break apart, and re-form alliances (Thompson, Ellis, and Wildavsky 1990: 88).

The presence of mixed motives means that every alliance is facing several unsolvable inconsistencies. For example, the hope of the cooperative coalition is to achieve a decentralized electricity system that fits its preference about direct democracy. In order to this, however, this coalition has to moderate its suspicion of centralized planning as promoted by the established coalition. In its turn, the established coalition tries to achieve its objectives by using decentralized electricity production technologies as promoted by the cooperative coalition.

In California, the regulated competition alliance may serve as another example. To achieve deregulation, the libertarian coalition had to temper its anti-statist attitude. In its turn, the public interest coalition saw the introduction of competition as a means to increase the efficiency and diversity of the electric system and not as a goal in itself. In the end, the cooperative learning process left the public interest coalition empty handed. This shows that cooperative learning involves both opportunities as well as risks for the coalitions involved.

Our study reveals that policy-oriented learning plays an essential role within the democratic legitimization process around innovation. In particular, cooperative forms of policy learning across ideological coalitions make a valuable contribution to the democratic quality of policy, as they guarantee that other perspectives on the policy problem beside the dominant one get a chance in the political debate. Although cooperative forms of policy learning are more productive than competitive forms they should not be considered a remedy for all ills. Those who are forever forecasting ideological convergence always will be disappointed (*ibid.* 93), that is, cooperative policy learning is not able to get rid of or transcend ideological conflict. But although cooperative policy learning is fallible and limited it seems the best method available to cope with differences in ideology.

Second, as stated above, entering a cooperative policy learning process is a risky and ambivalent venture. A coalition that reaches out for a partner who can compensate for its weaknesses is actually playing the devil's advocate, and one that downplays the ideological differences fits out the wolf with sheep's clothes. Since its outcome is unpredictable, entering a cooperative learning process is a courageous event. Whereas it seems the best way for the establishment to stay in power, it is not a guarantee. While this strategy worked for the Danish established coalition, the opposite was true for the liberal establishment in California.

Maybe some readers will be disappointed by the way policy-oriented learning is put into perspective. The fact that cooperative policy learning faces serious limitations and presents a never-ending, somewhat unpredictable story, however, makes it neither less useful nor less important. On the contrary, cooperative learning presents an essential mechanism both to construct and stabilize as well as to deconstruct and destabilize public policy in a dynamic and complex world. Destruction, creation and stabilization of policy programs often go hand in glove and are essential features of political innovation. We conclude that cooperative forms of policy learning are crucial for the democratic quality of the decision making process around innovation, that is, the intelligence of politics.

10.2 The Politics of Wind Energy Innovation

> The interaction between means and ends is most important. Often ends, both operational and general values (though perhaps not final values), change because of innovation in means. *Yehezkial Dror, 1973: 16*

Coping with a many-sided world that is constantly changing is a highly contingent process. Policy learning depends on, or rather, is enabled and constrained by (amongst other things) national political culture, the dominant policy within a policy subsystem, and techno-economic conditions. Above, it was explained how national political culture enabled and constrained cooperation between some types of coalitions. Our case studies illustrated that the two oil crises in the 1970s confronted the energy policy subsystem in California and Denmark with a situation it was ill equipped to cope with. Hence the existing policy and/or ruling ideological coalition(s) became discredited, and previously excluded alternative policy beliefs were bolstered.

Until that time, the Danish government had had little control over the electricity sector. Since government could not be blamed, increasing command and control over the electric utilities could be perceived as a way out. Since, in California, an elaborate regulatory regime did exist, increasing regulation presented a far less obvious and credible solution than in Denmark. Consequently, policy change became guided by the wish for less regulation and more competition. In Part Three, we examined how the politics of innovation influenced and was influenced by techno-economic conditions and change.

10.2.1 Innovation Networks and Traditions

In Chapter 7, the notion of an *innovation network* was introduced to conceptualize the idea that political innovation and techno-economic development are mutually dependent and continually shape each other. The locus of innovation is the whole innovation network, that is, for an innovation to become successful each pole has to be developed, and their activities have to be tied to each other. Creating a high degree of integration between different network poles requires learning between various innovation actors, which was named *joint innovation learning.*

The political pole can neither be considered the sole locus for legitimizing innovation. Instead of thinking of democracy as simply an interchange of political ideas, taking place within the policy subsystem, it is realized that, outside the political struggle is carried on by other means. Besides by the interaction of political beliefs, the democratic quality of innovation is also determined by technical, managerial and consumer debates around innovation and the practice of techno-economic innovation. The process of democracy is understood, therefore, as the proliferation and interaction of both political and techno-economic ideas and practices. Consequently, we look for the politics of innovation within all poles of the innovation network.

Since technological communities, the management of firms and consumer groups all have *relative autonomy* in perceiving and appreciating innovative situations, different types of means-end chains are always present, continuously evolving in interaction. Danish ecological practitioners neither developed the wind turbine for the industry to become successful, nor for the utilities to set up offshore wind farms. Still, these were indirect outcomes and unintended consequences of their activity. As Follett says, "The purpose in front will always mislead us. ... Activity always does more than embody purpose, it evolves purpose." (Graham 1995: 56-57) For these reasons innovation is seen as both cause and effect of actions of, and interactions between, policy makers, technologists, firm managers and consumers *within* and *across* innovation networks.

In Chapter 8, it was discovered that when policy makers try to set up a particular innovation network, they do not (need to) begin from scratch. Instead of re-inventing the wheel, they are guided by historically developed relationships between political and techno-economic traditions, so-called *innovation traditions*. Normally, different ideological coalitions prefer different types of innovation networks that fit different innovation traditions, each offering 'standard ideal' ways of innovating. In the United States of America, the corporate, civilian, communal and entrepreneurial innovation traditions were discerned. In Denmark, the cooperative and flexible innovation traditions played a major role. The historical macro-analysis was used to structure our micro-analysis, in Chapter 9, of the relationship between politics and techno-economic development in the field of wind energy.

10.2.2 Initial Wind Energy Innovation in California, 1974-1982

In California, the Brown administration played a pioneering role in the development of alternative energy sources. It came to support three types of wind energy innovation networks. The CEC attempted to set up a corporate wind energy innovation network, where large corporations were expected to develop and market megawatt-size wind turbines to be used in large remote wind farms managed by investor-owned utilities. This venture was supported by corporate capitalists and a part of the public interest coalition. The communitarian coalition advocated the decentralized production and use of small wind turbines, which also received government support; institutionally in the form of the OAT, and financially in the form of the wind energy tax credit program. Both attempts failed because anticipated technology producers and users showed little interest. Since the political will to set up these two innovation networks received minimal support within their production and consumption poles, most of these political actions have sunk into oblivion.

Nowadays, California policy makers are mostly credited for backing the establishment of the independent (wind) energy business as supported by the libertarians and a part of the public interest coalition. Their promotion of an entrepreneurial wind energy innovation network resulted in the development of non-utility wind farms in which medium-sized wind turbines were used. Politicians had neither foreseen nor planned this development. The public programs installed had been directed at different innovation objectives; they were aimed at establishing different types of wind energy innovation networks. Business entrepreneurs, however, realized that the existing policies offered them the opportunity to set up independent wind farms. The Brown administration embraced this private initiative and started to encourage these entrepreneurial wind energy innovation activities. This is a typical example of mutual innovation learning. It also illustrates the pragmatic or *common sense* approach of the Brown administration or, in general, the public interest coalition toward energy innovation activities. The public interest coalition came to promote the development and use of renewable energy technologies regardless of whether this was done by families, communities, investor-owned utilities or third-party producers.[6]

The pragmatic stance toward innovation enabled joint innovation learning through variation and selection. The entrepreneurial wind energy innovation approach proved to be the fittest under circumstances that were politically intended to advantage the corporate and communal approach. Does this mean that non-political innovation actors had the final say? In other words, were non-democratic actors deciding about which 'political' direction wind energy innovation was going? The answer is, No! Policy makers still had a choice. As in Denmark, they could have detested this development and changed the rules of the innovation 'selection' game. In California, however, the entrepreneurial approach was tolerated and even embraced by policy makers. The dependence on other innovation actors who have relative autonomy does not directly imply that policy makers are bound hand and foot.

While the activities of the independent wind farmers came as a surprise, its swift political acceptance was not surprising. In particular, in the field of gas cogeneration the entrepreneurial innovation approach was dominant and had led to the enactment of the Private Energy Producer Act in 1976 to facilitate non-utility power projects, and to the determined implementation by the CPUC of the federal Public Utility Regulatory Policies Act (PURPA) of 1978. In this way, wind farmers could benefit both from public policy measures that were originally aimed at guiding wind energy innovation into a direction that was preferred by either the corporate capitalist coalition or the communitarian coalition as well as from policies that stimulated an entrepreneurial innovation style in primarily the field of gas cogeneration. This is exactly what they did. As they imported the entrepreneurial innovation approach to the wind energy field, policy makers approved of this and came to support it; a nice example, therefore, of joint innovation learning.

It can be concluded that the democratic quality of the decision-making process around wind energy innovation during the initial wind energy innovation phase in California was high. The Brown administration did not guide the development of wind energy into one particular political direction. Instead, several innovation approaches, each related to different political beliefs, were supported. Not just the manner in which the liberal establishment wanted to implement wind energy was stimulated, but also approaches advocated by the communitarians and libertarians. At the end of this period, the civilian/entrepreneurial wind energy innovation approach was most successful from a techno-economic perspective.

10.2.3 Initial Wind Energy Innovation in Denmark, 1974-1984

In the 1970s, the Danish government tried to set up a flexible wind energy innovation network. The Danish established coalition and liberal establishment in California had similar ideas about the best approach. In Denmark, a consortium of large Danish firms was supposed to develop and produce large-scale wind turbines to be set up in wind parks owned by the electric utilities. Like in California, this approach failed because of technological difficulties and lack of interest from the industry and utilities. Initially without any direct government support, the cooperative coalition aimed at a cooperative wind energy innovation network. The cooperative innovation approach looked like the one supported by the communitarians in California. While this 'soft' style of innovation did not arouse a lot of practical activity in California, it was successful in

Denmark, where many grassrooters, hobbyists and small entrepreneurs became involved in the development of (small) wind turbines for private use.

Just as several political ideas of the cooperative coalition were gradually adopted by the established coalition, techno-economic development inspired by these ideas was gradually appraised by the established coalition. While its sectarians tried to represent cooperative wind energy innovation as a threat to the existing centralized electricity system, more and more coalition members came to see the new windmill industry as a sign of hope for the badly needed modernization of Danish industry. Consequently, the industry became technologically supported by the Risø Test Station. In addition, from 1979 on, the private use of wind energy was subsidized in such a way that it created a wind turbine market large enough for the industry to slowly develop and make its products ready for export. The Danish home market was regulated by adjusting the amount of investment subsidy in order to restrict the profit margin of wind turbine owners, and by imposing rigid siting and consumption criteria. The set of regulations that was the result of a political settlement between the established and cooperative coalition, which was induced by the initial techno-economic success of the latter, led to a modest and slowly growing home market that provided fertile ground for improving wind turbine technology. A flexible/cooperative wind energy innovation network developed, in which the different actors dealt with each other in a very constructive manner; that is, joint innovation learning was widespread.

It can be concluded that the democratic quality of the decision-making process around wind energy innovation in Denmark rose between 1974 and 1982. First, the government only supported wind energy innovation as advocated by the established coalition. At the end of the 1970s, the cooperative way of wind energy innovation was approved of by the established coalition and a flexible/cooperative wind energy innovation network evolved in the early 1980s. The techno-economic success of the cooperative innovation style helped to open up the political dialogue and increased the democratic quality of the political debate. Whereas techno-economic innovation first functioned as a provocateur, it gradually became a political integrator.

10.2.4 Rapid Wind Energy Innovation in California, 1983-1986

The innovation activities of the wind farmers broadened the scope of the libertarian policy beliefs and the related entrepreneurial style of innovating in the energy field. The libertarian nature of this third line of development in the field of wind energy was not clear to all political actors. Not even all libertarians appreciated this ideological transformation. For example, Governor Deukmejian, condemned the wind farms as symbols of Brown's soft energy policy and tax spendthrift behavior, and failed to see them as an example of the entrepreneurial type of development he favored. Other libertarians did recognize this and supported the extension of the energy tax credit program until the end of 1986. Thus, the beginnings of an independent wind power industry made them coalition temper thier opposition to governmental interference. Innovation encouraged the cooperation between the libertarian coalition and a part of the public interest coalition.

While the innovation activity of the wind farmers helped to prolong the energy tax credit program, the lack of techno-economic action in the communal wind energy innovation network decreased the credibility of the policy beliefs of the communitarian coalition. Lack of concrete techno-economic results made the Office of Appropriate Technology an easy victim for the new conservative administration. The communitarian coalition became the victim of the proven inadequacy of its own communal style of innovating. In contrast, the energy tax credit program, together with California's QF-policy, enabled and democratically legitimized the entrepreneurial style of wind energy innovation, which led to a strong growth of the independent wind farm business, but at the same time to the demise of the American wind turbine industry.

Beside wind farms, California's implementation of PURPA led to the development of cogeneration, solar, biomass, small hydro, and geothermal power plants. The quick advent of the independent power industry and related technologies created a brand new political and techno-economic situation. The market ideology of the libertarians and the ecological ideology of the public interest coalition had materialized within new energy production technologies and industries. The story of entrepreneurial modernization was full of entrepreneurial innovation.

What about the civilian/corporate style of innovating? As we saw, the corporate capitalists and a part of the public interest coalition formed a policy learning alliance that was organized around the story line of ecological modernization. This cooperative political learning process enticed far less innovative activity than the innovation story of entrepreneurial modernization. The Reagan administration had drastically cut back the budget of the FWEP. Early interest by the California investor-owned utilities became bogged down because of reluctant large industries, bad technologies and decreasing oil prices. With respect to other types of promising alternative technologies, similar stories can be told. In particular, despite strong pressure by policy makers, the investor-owned electric utilities failed to seize the opportunity of gas cogeneration. The utilities invested their time and money in nuclear power and demand-side management (DSM), which they saw primarily as a means to save time for the nuclear power plants to come on line. Consequently, a strong independent energy efficiency industry did not evolve.

It can be concluded that the arrival of the Reagan administration at the federal level and the Deukmejian administration in California caused a steep decrease in government support of both the communal as well as civilian/corporate style of wind energy innovation. From a political pluralist point of view, this impoverished the democratic quality of the innovation process. While the liberal establishment had promoted a politically diverse set of innovation styles in the field of wind energy, the libertarian coalition only tolerated its own preferred entrepreneurial innovation style.[7] The libertarians even very nearly terminated their support to the civilian/entrepreneurial wind energy network. This probably would have resulted in a complete stop of all techno-economic activities. Instead, the independent wind power business experienced an uncontrolled and chaotic period of exponential growth. Abundant government stimulation, but a lack of governmental regulation, led to strong libertarian-like techno-economic development that clearly provoked the corporate capitalist coalition.

10.2.5 Rapid Wind Energy Innovation in Denmark, 1983-1986

One can conclude that the Americans profited from the fact that the Danish policy makers had regulated wind energy innovation in such a way that it was characterized by weak market development, weak competitive behavior, integrative behavior with respect to technological innovation, easy availability of technological information, and excellent market transparency. Under these favorable conditions, a reasonably reliable technological wind turbine design was developed that came in handy for the American wind farmers, who were bursting for reliable wind turbine technology. The cautious, incrementalist approach to innovation, however, generated a too modest economic development. While the Americans relied on Danish regulation, the Danes relied on the uncontrolled, speculative development that Californian policy makers had fostered. Profits from the Californian market provided the Danish wind turbine producers the investment money needed for boosting their technology and professionalizing their industry.

In this way, the Danes got the best of both worlds. At home, wind energy innovation was regulated in agreement with the political ideas of the cooperative and established coalition. In California, they found an export market that had stimulated sufficient techno-economic activity to establish an enduring flexible/cooperative wind energy innovation network. From the perspective of democratic steering, it is interesting that the Danes profited from a type of development abroad that they clearly rejected at home. Just as the Danish establishment alliance rejected right-wing populist ideas, it repudiated, as proven by its reaction to the Danish boom, the related entrepreneurial style of innovating. Danish policy makers preferred a more regulated and transparent style of innovating, but while they did not tolerate a Wild West type of development in their backyard, they did not mind profiting from the wind boom in the real Wild West. The democratic debate around wind energy innovation thus stopped at the Danish border. The fact, however, that the wind turbine industry crossed that border in search for export markets facilitated its rapid development and technology, which in turn secured cooperation between the cooperative and established coalition in the field of wind energy. A thriving export market represented an important precondition for joint innovation learning. In other words, successful techno-economic innovation driven by the massive export market functioned as a political integrator.

10.2.6 The Exclusion of Wind Energy Innovation in California, 1987-1995

In Part Three, we found that the two policy narratives distinguished in Part Two presented two comprehensive innovation stories. This meant that, besides containing a political message, the story of entrepreneurial and ecological modernization included a corresponding style of innovating. Because the policy narrative forms an integral part of the innovation narrative, the political struggle cannot be detached from the techno-economic struggle. We saw that the story of entrepreneurial innovation was the result of, and enabled, a long-term cooperative learning process between the libertarian and a part of the public interest coalition. In addition, public policy programs advocated by this alliance inspired a lot of techno-economic activity. New energy production technologies were developed and a new independent energy industry was founded.

The story of ecological modernization guided the learning alliance between a part of the public interest coalition and the corporate capitalists. Although this establishment alliance was successful from a political or policy learning point of view, the narrative of ecological modernization embodied far less innovative activity than the entrepreneurial innovation story. As techno-economic progress induced by the story of entrepreneurial modernization strengthened the cogency of this narrative (with respect to both the political message as well as the associated innovation approach), the relative lack of techno-economic progress induced by the story of ecological modernization weakened its persuasiveness.

As said before, the sudden arrival of the new independent power industry created a political and industrial deadlock. While the conservative establishment dominated the overall political climate, the liberal establishment still dominated at the level of the energy policy subsystem. This was related to the central techno-economic role of the investor-owned utilities, whose monopoly position, however, was no longer viewed as 'natural.' The utilities' attempt to politically undo the independent power industry was like the last convulsion of their hegemonic reign. They failed and were forced to tolerate and compete with the new industry. As a result, the fight between the two innovation stories entered into a new phase.

Gas cogeneration plants represented the trump technology within the story line of entrepreneurial innovation. Strong techno-economic development in this field over the last decade had made cogeneration power plants one of the most reliable and least expensive electricity production technologies in the State. Thus, gas cogeneration was presented as proof of the success of the market-based approach to innovation as promoted by the libertarians. While the cogeneration industry was able to live up to the techno-economic expectations implicit within the story of entrepreneurial innovation and in this way strengthened the plausibility of this innovation narrative, the techno-economic position of the wind power industry was much weaker and did not (yet) agree with the demand for an open, competitive energy economy. The way in which the wind power industry changed its business strategy in response to the absence of government policy in the wind energy field in the second half of the 1980s proved this.

After the suspension of the ISO4 contracts and the expiration of the tax credit program, the California government slashed its activities with respect to wind energy innovation. This fit the laissez-faire ideology of the libertarian coalition, but caused firms to change their strategy into a direction that supported the monopoly structure of the electricity market. Before, extensive government support had stimulated firms to become independent wind power producers. When the support stopped, companies began to present themselves as wind farm developers that wanted to set up wind farms for investor-owned utilities. Thus, contrary to laissez-faire doctrine, government interference stimulated an entrepreneurial approach to wind energy innovation that defied the monopoly position of the IOUs, while the retreat of government led companies to adopt a strategy that agreed with the corporate wind energy innovation approach that supported a dominant market position of the IOUs. In short, the independent wind power industry arose due to a combined entrepreneurial/civilian innovation approach to wind energy, but when the support for wind energy from the government stopped, the industry changed its course and came to advance a corporate wind energy innovation network since it was not ready to compete heads on with other electricity production industries.

At the time the corporate wind energy innovation network started to bear its first fruit, the BRPU-process created a new possibility for the civilian/entrepreneurial innovation approach. As we saw, this process was overruled by FERC. Meanwhile, California was undergoing the worst economic downturn since the Great Depression, and electricity prices in the Golden State were among the highest in the United States. This made the debate about restructuring the electricity production sector flare up. The political debate led to a new alliance between the libertarian and the corporate capitalist coalition. This political truce was an almost inescapable outcome of the technological and industrial changes that had been politically stimulated after the oil embargo and the passage of PURPA.[8] While creating splendid opportunities for the gas cogeneration industry, the new situation seems not to have been directly beneficial to the development of the wind power industry.

This final part of the California wind energy story clarifies some extra insights in the relationship between politics and techno-economic innovation. First, it demonstrates that relevance and credibility of policy beliefs strongly depend on the innovative power of the corresponding innovation approaches. The relative weakness of the civilian/corporate style of innovation in the energy field weakened the position of the corporate capitalists in favor of the libertarians whose entrepreneurial style of innovation was very successful.

Second, it shows that the competition between various techno-economic options takes place on different interacting levels: the technological, market and political level. An advantage on one level will often lead to an advantage on another level. As a result, some techno-economic options will have a larger impact on the political debate than others. As Smutny-Jones,[9] the director of IEP, expresses it, "The gas cogeneration industry was the big dog in the fight." Its techno-economic success drove the electricity restructuring debate, in which the renewables industry had a more marginal position. While the gas cogeneration industry tried to force the political process, the renewable energy industry was moved to the fringes.

Third, and related to the second observation, the development of the wind power industry shows that technology and industry are neither neutral nor deterministic. The political meaning of both the wind power technology and industry changed when the political support for it dropped in the mid-1980s. Techno-economic innovation appears to be versatile. To become political, it has to be given political meaning. In theoretical terms, this suggests that the activities within the scientific, technical, producer and consumer poles of the innovation network are not political in themselves, they are political only in interaction with activities within the political pole. So, while the whole innovation network is political, the separate poles are not.

This does not ignore the fact that certain technologies might tend toward a certain political usage, that is, they may tend to fit in and support specific political cultures better than others. Nevertheless, for such a tendency or 'political intentionality' (Ihde 1990, 1993) to become real, political action has to be taken. It is exactly their flexibility or political multi-interpretability that provides technology and industrial structures a relatively autonomous role within the political debate around innovation. *Techno-economic innovation can be used to materialize political beliefs, and thereby, to reproduce and stabilize them. But while it can guide the political debate around techno-economic innovation, it can not once and for all remove or block the political debate around techno-economic innovation.* We will come back to this important statement when we recapitulate the various conclusions after the description of the Danish story of wind energy innovation in the period after 1986.

10.2.7 The Inclusion of Wind Energy Innovation in Denmark, 1985-1995

The quick response of the Danish government to the wind boom had several goals. First, it was meant to stop an entrepreneurial type of development that was disapproved of by the cooperative and established coalition. Second, the re-introduction of the rigid siting rule by linking it to the state subsidies for wind energy tried to regulate the private use of wind energy. Third, the first 100-MW agreement aimed at bringing the activities of technologists and firm managers within the flexible/cooperative innovation network in line with the political beliefs of the established coalition. Herewith, the established coalition definitely recognized that the innovation network, initiated by the cooperative coalition, was superior in a techno-economic sense to the flexible innovation network it had tried to set up. Having learned from experience, it came to follow the adagio, "If you can't beat them, eat them."

This strategy worked, and wind turbine firms gradually shifted their business and innovation strategy towards the utility market, while at the same time, the private cooperative market for wind turbines remained alive and kicking. More important, the long-term cooperative policy learning process between the established and cooperative coalition around the policy story of ecological modernization had altered the political reason for 'eating.' The 'food,' or the 'process of eating' had changed the 'eater.' So, the German expression 'Der Mensch ist was er ißt' (Man is what he eats) seems to be a more appropriate saying.

While, in the mid-1980s in California, the political meaning of wind energy innovation changed almost instantly from one that supported a libertarian way of life to one that fortified the adversarial policy beliefs of the corporate capitalist coalition, the meaning of wind energy innovation in Denmark changed gradually, as a result of a steady and cooperative policy-oriented learning process between two different but cooperative coalitions. Techno-economic innovation advanced a policy learning process that led to a novel political understanding of wind energy innovation, which in its turn maintained political support that fostered wind energy innovation. The wind energy innovation process in Denmark again demonstrates the interpretive flexibility of techno-economic innovation. However, while in California this was related to the fact that different coalitions were free to use techno-economic innovation to bolster their political views and power, in the Danish case flexibility refers to a truly creative political process, in which, through a continuous, open and constructive debate, new political meaning was given to wind energy innovation.

10.2.8 The Political Role of Techno-Economic Innovation

Both cases show that the relevance and credibility of policy beliefs strongly depend on the innovative power of the corresponding innovation approaches. This finding has major implications for the way we should look at the question of democratic guidance of techno-economic innovation.

In section 10.1, we thought of the democratic process as a political debating process between various ideological coalitions, of which the outcome decided the legitimacy of the innovation process. In that case, we found that cooperative policy learning determines the quality of the decision making process around innovation. The analysis in section 10.2 shows that the cogency of policy arguments relates to expectations that surround techno-economic innovations, their proven validity or irrelevance, and their presence or non-existence.

It follows that the democratic fight is not solely a battle of words and arguments, it is strongly related to technological and industrial practices. This implies that we should look for the essence of democracy in the interaction between these techno-economic practices and political ideas. The advanced perspective in this section (Part Three) does not render the findings of section 10.1 (Part Two) irrelevant. It only recognizes the role played by techno-economic innovation within the debate, and tries to understand its implications for the ideal of democratic guidance of techno-economic innovation.

We found that techno-economic innovations can play different types of political roles. For example, in the 1970s wind energy innovation in Denmark played the role of a *political intriguer*, by trying to arouse the interest of the established coalition in wind energy. In the early 1980s, it came to serve as a *political compromiser*, as it stimulated the established and cooperative coalition to find common ground. In the course of the 1980s, wind energy innovation came to function as a *political integrator*, or even better, as a *political inspirer*, which equipped a creative policy learning process within the establishment learning alliance.

The cases also show that the role played by techno-economic innovation within the political debate is highly contingent. For example, the initial peripheral position of the cooperative coalition within the 1970s Danish energy debate, and the lack of reliable wind energy technology at that time, forced this coalition to use wind energy innovation as a symbolic crowbar to open up the debate around the use of nuclear power, which at that time played the role of a *political dominator*. The different roles that techno-economic innovation may play within the political debate both are shaped by and shape the interactions of the various ideological coalitions.

10.3 The Intelligence of Innovation

De onderstroom, die niemand ziet, bepaalt de richting op elk gebied. *Stef Bos, 1997*[10]

In his famous analysis of modern *Risk Society*, the German sociologist, Ulrich Beck, (1992: 187) maintains that, "[T]he political system is threatened with *disempowerment* while its democratic constitution remains alive. The political institutions become the administrators of a development they neither have planned for nor are able to structure, but must nevertheless somehow justify."[11]

This opinion supposes that, currently, the central political system only has a modest influence on the techno-economic system. Such a powerless position of the political system might be related to the political dominance of free market ideology. In this case, political impotence is self-inflicted and the only thing that remains for the policy sciences is to demonstrate the democratic limitations of this view. As Touraine (1997: 189) argues, "To some extent, the market economy is democracy's antithesis, as the market attempts to prevent political institutions from intervening in its activity, whereas democratic politics attempts to promote intervention so as to protect the weak from the domination of the strong." In case the weak position of policy makers is not self-chosen, a better understanding of the role of politics in techno-economic innovation and vice versa, is a precondition for strengthening their position.

With respect to the political role of techno-economic innovation, Harbers (1997: 160) holds that, "politics is more a question of design than of debating," and thus concludes that "democracy can be better defined as the proliferation of practices than as the proliferation of ideas". Our inquiry shows that while techno-economic innovation can steer the politics of innovation, it cannot remove or block the political debate. In other words, the political system always maintains a relative autonomy against the techno-economic system for purposes of political intervention. Consequently, we arrive at a more moderate position than Harbers and argue that the essence of democracy is found in the interaction between political and techno-economic practices and ideas.

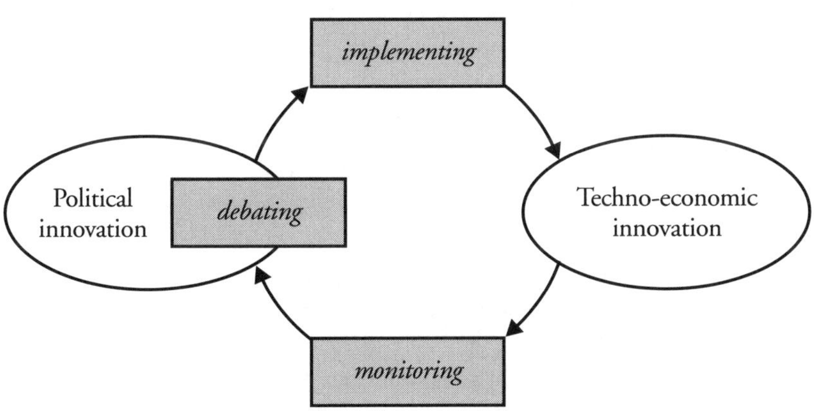

Figure 10-1
The democratic circle of innovation.

Our study shows that if policy makers want to guide techno-economic innovation into a democratic legitimate direction three interrelated processes have to be considered continuously. First, the democratic quality of the debate needs to be high. Second, the process of innovation implementation needs to be optimized. Third, the political and democratic role of techno-economic innovation has to be monitored, so it can be used as an appropriate input for the ongoing debate around innovation. This step completes the democratic circle of innovation (see Figure 10-1).

10.3.1 The Democratic Debating of Innovation

We found that the way politics is actually done in the energy policy field is different from the way in which the contemporary policy sciences usually conceptualize and study the political process. At the moment, it is commonly held that the interaction between coalitions of policy actors is basically driven by competition (cf. Sabatier 1993: 5). From this view, competitive forms of learning within coalitions are dominating the political bargaining process. Politics is characterized by adversarial relations between coalitions, and special type of actors, so-called 'policy brokers', are needed to help coalitions "to find some reasonable compromise that will reduce intense conflict." (*ibid.* 19)

Our findings do not match this competitive image of politics in which dissenters are mainly regarded as enemies. The energy policy field depicts a more sophisticated way of doing politics in which both competitive and integrative coalition behavior is present. Coalitions regularly form learning alliances, in which they downplay their ideological differences in order to search for innovative policy solutions in which both political aspirations find a place. Such learning across coalitions was not driven by a forum "prestigious enough *to force* professionals from different coalitions to participate." (*ibid.* 54)[12] The rationale for cooperation was simply the idea that cooperation might be beneficial (in the long term) for both coalitions.

From a democratic perspective, such an advanced understanding of where a coalition's self-interest lies represents a hopeful and powerful insight. Realizing that a strategy of integration offers a fruitful long-term strategy for staying in power may motivate dominant coalitions to open up towards marginal political views. In the 1970s, the public interest coalition let the ideas of the hard side of populism into the Californian energy debate. In Denmark, the soft side of populism represented the political underdog whose ideas were gradually admitted by the established coalition to enrich the debate. The realization that the ideological 'otherness' of a marginal coalition may embody a source of inspiration for the main coalition safeguards democratic pluralism. Moreover, the strategy of integration allows for the constructive use of political pluralism.

Within the competitive vision on politics, policy learning is bound to stay within the confines of a coalition's policy belief system. Novel situations are perceived as old problems and treated with old solutions. In the complementary cooperative mode, coalitions re-evaluate their perception of the policy problem and look for new common solutions that fit the present situation. Whereas competitive behavior is about following ideals and relating to the past, integration is about creating ideals and relating to the future, and forms the key to political innovation.

We conclude that integrative behavior increases the democratic quality of the debate because it both safeguards political pluralism as well as advances its constructive use. The story of wind energy innovation shows that the demand for and use of political pluralism is not just based on some theoretical democratic ideal, but also on political pragmatism and a sense of reality.

10.3.2 The Democratic Implementing of Innovation

> Good policy analysis always considers resources and objectives, means and ends together, never separately. *Wildavsky, 1979: 10*

For long, basic research was considered as the goose with the golden eggs. The idea was to just put money into the science-and-technology goose and golden eggs would simply roll out in regular pace. In the 1970s, some began to notice that some of the geese were rather slow in laying eggs, or had stopped laying eggs. Others observed that into some of the geese more gold had to be put in than could ever expected to come out. Still, others argued that some of the geese's eggs were rotten and complained about the smell they were spreading. In summary, the blind faith in scientific and technological progress was corroding. Related to this both the effectiveness of 'technology-push' oriented policies as well as the naturalness of the positive societal value of any techno-economic innovation came under attack.

Although technological determinism continues to inspire policy makers (in particular in the United States), an increased understanding of the innovation process has led to a considerable improvement in focus. It was realized that a linear view of innovation that modeled innovation as a number of successive phases (research, development, demonstration, diffusion, and utilization) was no longer appropriate. An increased use of a systems view of innovation helped to open up the 'black box of innovation,' or rather, it helped to visualize the 'seams' of the 'seemingly seamless innovation web.'

Like the morning rime manifests the separate threads of a cobweb, the concept of the innovation network increases the sensitivity of policy makers for the variety of actors and related perspectives involved in the process of innovation. It became acknowledged that, beside scientists and technicians, a wide variety of actors, like policy makers, firm managers and users, commonly play an important role in the shaping of techno-economic innovation. The recognition of the heterogeneous nature of the innovation actors and the related interdisciplinary and interactive nature of innovation has led scholars and policy makers to abandon the linear model of innovation and to adopt a network approach. In this view, innovation is seen as an interactive process, combining scientific, technical, economical, consumer and political aspects and involving a multitude of public and private actors.

Within such an interactive vision of the innovation process, the source of innovation is no longer solely reserved to science and technology, but can, in principle, be located (simultaneously) in each network pole. Accordingly, no part of the network should be neglected in advance to public innovation policies. As Caracostas and Muldur (1998: 148) put it, "From the most basic research to industrial design or social experiment and market research, nothing, in principle, may be excluded from the scope of government intervention."[13] This insight requires policy makers to be intensively engaged with all potential facets of innovation. In short, by exposing the full complexity of innovation, the network approach strongly widens the scope of innovation policies, placing greater demands on policy makers and the content of innovation policies.

As explained in this book, a learning perspective is needed to encourage effective and democratically legitimized innovation. The systems view tells that innovation only works when all innovation activities are geared to one another. The demand for a high degree of integration requires continuous learning between all the players involved in the innovation process; so-called joint innovation learning.[14] With respect to policy makers, joint innovation learning should be aimed at the effective implementation of democratically legitimized innovation. For effective implementation, policy makers should grasp the reciprocal behavior of other innovation actors well enough to anticipate it, in order to stimulate them to find common or convergent interests and views. In the linear model of innovation, policies stimulated technological modernization or market development. Within the interactive network approach the policy maker becomes a public manager of interdependence, an *innovation coordinator* or *integrator*, whose goal is to help innovation networks emerge, to aid in their development and to stimulate interactions between innovation actors in order to increase the network's degree of convergence (cf. Callon et al. 1992).

Next, to democratically implement innovation, policy makers have to safeguard its political meaning, by arranging that all relevant aspects of innovation come to agree or stay in line with the evolving set of political goals. While current literature is much occupied with the question of *how* to boost the effectiveness of innovation policies, little or no attention is given to the question of the democratic quality of innovation. Consequently, many scholars readily translate their insights about the interactive and dynamic nature of innovation into policy recommendations, without considering the interactive role policy makers themselves play. To paraphrase one of Ellul's (1964: 436) concluding remarks in the book, *The Technological Society*, "None of our wise men ever pose the question of the end of all this. The '*wherefore*' is resolutely passed." If the question of societal value is addressed, political goals are often figured to be pre-defined and stationary during the process of innovation.

So, despite all attention given to the diversity of innovation actors, and the dynamic and interactive nature of innovation, current literature refrains from discussing the role policy makers play in an interactive and dynamic fashion. Most Science & Technology scholars treat policy makers as mere innovation outsiders, and make the mistake of considering means and ends, or problem-solving and problem-setting, separately.[15] In this book, we have tried to avoid this typical error and find theoretical concepts to study and take account of the dynamic and interactive role played by policy makers within the innovation process.

A first logical step was set in Part Two, where we analyzed policy change over a period of two decades. This exercise increases the awareness that policy means and ends are constantly on the move. It breaks up the myth of innovation as being guided by a fixed set of political objectives, and shows that, on the contrary, innovation is steered by a gradually evolving, mixed set of objectives, and is directed at continuously moving political targets. Consequently, innovation actors cannot regard politics as a stable beacon in the midst of a changing world. They are like navigators orientating on a misty sea using policy lighthouses that slowly change their position and code of signal. Even worse, the lighthouse keepers are modifying the co-ordinates and coding of their lighthouses partly by anticipating and reacting to the behavior of the sailors whom they are trying to keep away from shallow water and guide into their safe harbor.

Next, the mutual dependence of political and techno-economic activity was conceptualized by including a political pole within the innovation network. Examining this interdependence from an historical perspective clarified that political traditions are often historically linked with techno-economic traditions, and vice versa. This exercise demonstrated that implementing innovation policies is not always about establishing jone particular innovation network. When, like in the case of wind energy, a politically diverse set of innovation networks exists, the policy maker's has to decide on whether and in what way each network is to be supported.

Since techno-economic innovation has political qualities (Winner 1977), the democratic quality of innovation is not only determined by the quality of the political debate. In other words, politics does not solely constitute arguments, but is also embodied in technological and economic development. Democracy has a material side to it (cf. Achterhuis 1997). To examine the democratic features of innovation, the researcher is forced, therefore, to widen the field of investigation. Besides, to safeguard the democratic quality of innovation, policy makers need to widen their scope with respect to the design of innovation policies and the monitoring of the political nature of innovation.[16] *If democracy is the means to safeguard political pluralism, it also has to safeguard techno-economic pluralism.* Innovation policies should, therefore, stimulate and give room to marginal innovation networks. In other words, a democratic politics should attempt to set up innovation policies that prompt the proliferation of as many politically distinct innovation approaches as possible (cf. Harbers 1997: 160). We should add that our case studies show that what is possible is defined by the national political culture.

Does this policy recommendation actually make sense, or is it just a democratic ideal of a daydreaming scholar? For answering this question, we will make use of the, by now, familiar distinction between competitive and integrative political behavior. Within the competitive view on the politics of innovation, innovation is characterized by rival ways of innovating. In this situation, joint innovation learning is bound to stay within the confines of a coalition's preferred innovation network, and accordingly, the political nature of innovation will remain unaltered. Political impulses are expressed by means of techno-economic innovation, which strengthens, maintains or weakens the political position of the related coalition. From a competitive view, it makes no sense for the dominant coalition to support alternative innovation approaches. The success of such an alternative approach may well undermine its political hegemony. Coalitions will try to set up their own politically preferred innovation network. Although the above plea for a pluralist innovation implementation strategy makes little sense from a competitive perspective, both in California and Denmark such a pluralist strategy became adopted in the wind energy field.

In California, the Brown administration simultaneously supported three distinct wind energy innovation networks. Hence, this pragmatic approach caused the pluralist nature of the energy debate to be reflected in innovation policies. In Denmark, things worked the other way around. First, the government only supported the innovation approach as advocated by the dominant established coalition. Yet, the cooperative innovation network prospered. Its techno-economic success drew the attention of the established coalition and caused the government to subsidize networks. Accordingly, the pluralist nature of the innovation game became reflected in the Danish wind energy innovation policy.

Let us present a third example. In, a lot of public money was put into wind power, but the innovation process was barely regulated. In Denmark, wind energy innovation was just sufficiently stimulated and strongly regulated. We discussed the pros and cons of both situations in subsection 10.2.5, and maintained that the relative 'success' of wind energy innovation in Denmark and California crucially depended on the coincidental and mutually beneficial mixture of the two policies or, in general, innovation styles.[17]

These examples demonstrate that, as concluded in subsection 10.3.1, a competitive perspective on the politics of innovation should be completed with a cooperative one. Within the latter view, a strategy of variation and selection is not only dictated by some democratic principle, but also by pragmatism and an advanced understanding of where a coalition's long term self-interest lies. In the cooperative view, innovation approaches promoted by marginal coalitions are not only regarded as threats to the innovation approach promoted by the dominant coalition, but are also seen as potential sources of inspiration that may even help the dominant coalition to stay dominant in the long term. This insight may cause dominant coalitions to open up towards marginal political views and prompt alternative innovation approaches. For example, in Denmark, the soft side of populism not merely inspired the debate, and the related soft side of innovation enriched the innovation process. In the 1970s, the public interest coalition both allowed the hard side of populism to inspire the debate, as well as encouraged the related hard side of innovation to strengthen the innovative power of the electricity sector.

The observation that marginal innovation may embody a source of inspiration for the main coalition's preferred innovation approach pleads in the defense of a pluralist innovation implementation strategy as promoted by Harbers (1997). Such a strategy is expected to safeguard the ideological pluralism of techno-economic innovation. It only makes sense for the dominant coalition, if a coalition can integrate elements of the alternative innovation approach into its own preferred style of innovating. This requires joint innovation learning across networks. Our study illustrates that the integration of networks is not just the annexation of a politically marginal network by a politically dominant one. The techno-economic integration of innovation networks is often accompanied by a cooperative policy learning process across related coalitions, which ensures that, in the process of integrating, the evolving network receives new political meaning.

It follows that political pluralism or the democratic quality of innovation can be ensured by an implementation strategy that either promotes a variety of politically distinct innovation approaches, or fosters an innovation approach that integrates a variety of different political values. Consequently, to safeguard the democratic quality, policy makers should not only consider the initial development of a politically diverse set of innovation networks, but should also be engaged with guiding their interaction, based on the political debate, or rather, the interactions of the various coalitions that define this debate.

10.3.3 The Democratic Monitoring of Innovation

In order to have a refined political debate about innovation policy makers should have a good understanding of the political nature and role of techno-economic innovation. Within current literature there is a rather simplistic understanding of the democratic role of techno-economic innovation. Winner (1977) is among the first to explore this view. Following Mumford (1964), he discerns two democratic types of technologies, "one authoritarian, the other democratic, the first system-centered, immensely powerful, but inherently unstable, the other man-centered, relatively weak, but resourceful and durable." (Winner 1977: 19) Probably, Winner's advice to politicians would be to stimulate the development of democratic technologies, while avoiding technologies that establish authoritarian social relationships (cf. Sclove 1995: 62). We do not adhere to such a reductionist dichotomous vision on the nature of innovation (cf. Smits 1997: 112).

A combined competitive and cooperative perspective provides us with a more realistic and sophisticated approach to the political role of techno-economic innovation. Techno-economic innovation is neither authoritarian nor democratic, but may assume a whole range of political roles. Besides acting as a political dominator, techno-economic innovation can also prompt political compromise or even integration. Whereas the role of political dominator leaves the political meaning or nature of innovation unchanged, the integrative role may change the political meaning of innovation. For example, in Denmark, the success of the cooperative wind energy network paved the way for a long-term cooperative policy learning process with the established coalition, in which new meaning was given to wind energy innovation. This led to a policy that encouraged the integration of the cooperative and flexible approach to wind energy innovation.

In the political struggle trying to be the fittest is often not a matter of proving one's beliefs (means and ends) to be better than the beliefs of others, but a matter of helping one another by finding out how the different views can complement each other. Our case studies have demonstrated that mutual cooperation often provides a more sensible strategy than face-to-face competition. The finding that cooperative political behavior is not just preferable from a moral point of view, but also makes sense in everyday politics presents a hopeful sign for democracy because, as we concluded in subsection 10.3.1, integrative behavior increases the democratic quality of the debate by safeguarding political pluralism. Since political pluralism implies techno-economic pluralism, innovation policies should stimulate the development of a politically diverse (set of) innovation network(s).

Democratic pluralism implies a politically diverse innovation space, in which the various approaches to innovation may freely compete and/or collaborate with each other. Guiding innovation into a democratically legitimate direction thus implies giving room to and stimulating a variety of politically distinct innovation approaches. It also acknowledges the relatively autonomous position of techno-economic actors with respect to the political process. Techno-economic action may create totally new options that policy makers may perceive as either risks or opportunities. The democratic challenge here is to politically interpret these techno-economic developments in order to give them a proper place within the political debate, and adjust them to political needs. This would require a type of joint innovation learning that is not aimed at optimizing the implementation process, but at better understanding the political meaning and role of techno-economic innovation

Again, this goes against contemporary common sense that techno-economic innovation has a dynamics of its own and cannot be adjusted. Such a juggernaut vision on techno-economic innovation would automatically prove Beck's (1992) thesis on the disempowerment of the political system with respect to the techno-economic system. Our study provides us with a far more positive outlook on the prospects for policy makers in bringing techno-economic development in line with political needs. For example, the Danish case showed how the design of the siting rules for wind turbines formed one of the crucial political instruments that shaped the development of wind energy. As is embodied within the interactive network approach on innovation, and as illustrated numerous times within our study, there exist myriad ways for policy makers to guide techno-economic development into a democratically legitimate direction.

Three things have to be ensured for making innovation intelligent from a democratic perspective. The political debate on innovation, as well as the implementation of innovation, should take place in a pluralist manner. Moreover, to detect and assess the unintended and intended political consequences of innovation, policy makers need to continuously monitor and assess the democratic role techno-economic innovation is playing.

11 Wind Energy Futuribles

> I often wonder about whether ideas are modern or not. I think that originators, certain creators who choose for the playful side of their genius are always modern, because modernity always comes from the playful side of the spirit. *Ben Okri, 1995*

In this final chapter, we will speculate on the likely future of wind energy in California and Denmark on the basis of the theories and insights developed earlier. In particular, we will reflect on the likely implications that the principle of innovation pluralism would have for the development of (wind) energy in these states.

The discussion on the future of wind energy innovation in section 11.2 is based on the dubious assumption that the political situation in Denmark will remain the same over the next decade. Over the last two decades, the stable political climate in Denmark created a fertile breeding ground for a long-term policy learning process across the established and cooperative coalition, which led to a pluralist wind energy innovation process. However the future is uncertain; competition is being introduced in the other Scandinavian countries, and is under discussion within the European Council. This means that it is doubtful whether the current social-liberal establishment alliance can further its contemporary cooperative policy learning process. If our assumption proves to be false, the forecast of the California situation in section 11.1 can serve as an insightful comparative example. With respect to Californian politics, it is assumed that the conservative establishment will continue to define the political climate in the near future.

11.1 California's New Wave of Environmental Passion

Despite its current dominance, the future of the new conservative establishment alliance is uncertain, as its political tenability will not only depend on whether the restructured electricity industry will be economically efficient, but also whether the free market is able to realize the goal of a more sustainable economic and social world. Social and environmental values are stressed by the public interest and communitarian coalitions. The arguments of these ideological coalitions are lacking in the current energy debate. From the perspective of political pluralism, these ideas form a necessary input for attaining the above goal. We will start our discussion with the public interest coalition and ask: What kind of strategy would be appropriate for this coalition to regain its sway?

The existence of a sectarian and pragmatic segment presupposes at least two types of coalition behavior. In his perceptive book, *Losing Ground*, Dowie (1995: ch. 5) denotes the pragmatic strategy as *third-wave environmentalism*.[1] The third-wavers believe in the enlightened self-interest of America's private sector, and assert that the environmental argument needs to be translated into profits, productivity, and economic benefits for industry. This ideology, which clearly represents a flirt with libertarianism, has been dominant since the Reagan years. This non-adversarial stance creates a dilemma. While the combined attitude and ideology of the third-wave environmentalists enable them to join the free-market alliance, it also makes it doubtful whether their participation would make a lot of difference. Moreover, the sole existence of a common ground between the two has strengthened the argumentative power of the conservative establishment, and has thus played a part in isolating the public interest coalition.

In principle, the sectarian wing seems to offer a more inspiring and less compromising way out, that is, to form an alliance with the dormant communitarian coalition. Namely, since direct access will bring the energy issue back to the local communities, the public interest coalition could campaign to get customers to buy green electricity. CEERT's Jablonski favors the idea of moving away from intervening at the federal and state level, and states, "I like grassroots stuff, so I look forward to going out into a community and moving the environmental case on a local level".[2] Taking a much broader perspective, Dowie (1995) even predicts such a grassroots revival sometime early in the twenty-first century. He speculates that this wave of environmental passion will be "democratic in origin, populist in style, [and] untrammeled by bureaucracy." (*ibid.* 206-207) Moreover, this fourth wave of American environmentalism will encompass "many traits that characterized the American Revolution – dogged determination, radical inquiry, a rebellion against hegemony, and a quest for civil authority at the grassroots". (*ibid.*) Paradoxically, the grassroots strategy appears to face the same danger as the libertarian one. With its trust in personal choice and disregard of government involvement, it tends to drift away from the basic tenet of the public interest coalition, that is, to protect the public good itself.

Still another form of coalition behavior can be envisioned that has a European social-democratic tinge to it, and would lead to a broad alliance made up of the public interest and the libertarian coalition. This would breathe new life into the dismantled regulated competition alliance. It would support efforts to make the market for electric generation more competitive, but would add the proviso that competition itself is not the goal, backing the approach of the grassroots, but not as the main strategy. It would impose minimum standards, and encourage companies and consumers to go beyond those standards. From the perspective of that alliance, the political defeat of the renewable energy portfolio standard was clearly a lost opportunity.

Our study shows that in the current California situation such an alliance lacks support. The opposite is the case. The libertarian and corporate capitalist alliance favor what Weinberg and Hamrin call, the 'market-machismo' scenario. This says "What we really want is cheap electricity. We care less about the environment. All this global climate change is just a bunch of scientist's pipe dream that want research money. What we really want, and what really is good for everybody is to have cheap electricity. Therefore we ought to move to a very market-driven, price-driven, short-term price-outlook market, which is the spot market."[3]

Weinberg argues that this scenario will usher in a difficult period for renewables and energy efficiency. Together with Hamrin, he also offers a different scenario, the so-called 'eco-quake' scenario. "That is when all of a sudden everyone wakes up and finds out 'Oh my God. All of these effects of fossil fuels that we thought were phony are really true. All these hurricanes all over the world, are really caused by global climate change mechanisms.' Lots of pollutants turn up in the air and the water, and everywhere, and they say 'We have got to go to energy efficiency and renewables, and we don't care what it costs. Just give us all the wind power you can get."[4] This scenario looks at long-term life cycle cost analysis, puts value on risk and diversity. Weinberg predicts that eventually the two scenarios will merge, but wonders "How, in the strange politics of the United States, is all of this going to work out?"

Although the above analysis does not serve as a crystal ball, it might offer a handle to address this question. Three strategies were uncovered that were open to the public interest coalition in its search for a balance between economic, and environmental and social aspects. Since each one of the strategies possesses its own fundamental flaw, a combined diverse strategic approach seems to be necessary. For this, the third-wave environmentalists need to acknowledge the tension that exists between the means they propose and the ends they want to meet. Moreover, a grassroots revival seems to be an indispensable ingredient. This time not, as in the 1970s, to undermine the position of the liberal establishment, but to balance it with that of the conservative establishment. This would restrengthen the power base of the public interest coalition and ultimately help in moving toward an electricity system that is clean, secure and reliable, safe, economic, and robust.[5]

11.2 Denmark's Power of Openness

> Renewable energy was tolerated as long as it did not 'bother' the established system within the energy field. But now we have reached the point where an unconditional: 'Yes please!' is required. *Ejvin Beuse (OVE), 1993*

The analysis reveals that the centralization-decentralization controversy has played a minor role within the above learning process. On the one hand, the pragmatic side of the established coalition tolerated the decentralized development path.[6] The decentralization of control has never been seriously considered, and the decentralized path has been kept carefully in check through imposing siting and consumption criteria, and through creaming off the profitability of private wind turbines. Through close monitoring, the private development was kept large enough to create a home market for the industry, and small enough not to endanger central planning. On the other hand, the pragmatic wing of the cooperative coalition accepted the utilities' involvement in wind energy for environmental reasons. As a result, both paths have encountered broad support. This 'peaceful' coexistence, however, is threatened by the wish to further expand the use of wind power in Denmark.

One option to avoid such a conflict is to release the utilities from the obligation to fulfil the 100-MW agreement. To still fulfil the goal of 1,500 MW of installed wind power capacity in the year 2005 the sectarians within the cooperative coalition suggest replacing old small private windmills at windy sites with larger modern wind turbines

(Energystyrelsen 1993). The short-term potential of this option, however, is only of the magnitude of the 100-MW agreement (Madsen and Bjerre 1993). Although the replacement option could displace the contribution of the utilities, it is obviously not enough to reach the total *Energy 2000* objective. Besides, under the current Wind Turbine Law it would be almost impossible for existing wind turbine cooperatives to find enough members to buy a large (250-1000 kW) wind turbine.

Utility-owned offshore wind turbine parks represent a second option. In that case, the involvement of local governments is not needed, and public resistance against large-scale wind projects can be bypassed by minimizing the visual impact of offshore wind turbine parks as seen from the Danish coast (Energistyrelsen 1994). The political viability of this option depends on the potential techno-economical improvement of wind turbine technology and on achieving a better insight in a range of environmental issues, among these the effects on birds.

Apart from the above two one-sided courses, some proposals try to go beyond the centralization-decentralization problem. These pragmatic approaches are based on the belief that physical planning should be a local affair, and emphasize that the attainment of national goals should depend on the *voluntary* cooperation of local people or governments. Tage Dræbye puts the question like this: "How to organize the large-scale development of wind energy while preserving public participation?"[7]

Establishing projects in cooperation with wind turbine cooperatives is a method that has been pioneered by NESA A/S since 1988 (Mørup-Petersen and Pedersen 1992). The largest electricity distribution company in Denmark is responsible for implementing a part of the 100-MW agreement. The North Zealand company, however, encountered siting problems within its densely populated service area, in which the wind energy resources are below the Danish average. When in 1988, a project from a wind turbine cooperative to install three wind turbines threatened to be reduced to two machines for lack of public interest, NESA joined in to buy the third one. Since then, several successful joint ventures have been initiated.[8] While this strategy postpones the above question, it does not provide an answer to it. Namely, considering that the utilities want to use the most cost-effective wind energy technology in their wind farms, and that the contribution of wind turbine cooperatives to joint projects is limited by the consumption and siting criteria in combination with its organizational capacity to find new members, it becomes evident that public participation will diminish in absolute terms as the use of megawatt-size machines becomes normal.[9]

It follows that the siting and consumption criteria present the principal bottleneck. Their abolishment would imply major changes. It would require the established coalition's support for widespread private decentralized electricity production, thereby renouncing the utilities' monopoly position. Giving up those criteria would also represent a major challenge to the cooperative coalition. Abandoning the consumption limit runs counter to their wish to link the production of electricity to the demand. Letting go of the principle of local attachment flies in the face of their 'criterion of proximity', which implies that "all social tasks should be carried out as closely as possible to the individual citizen". (Meyer, Petersen and Sørensen 1981: 64) Since both coalitions abhor uncontrolled, speculative development, abandoning a connection between consumption and production seems not to be debatable at this moment.

Loosening this criterion as well as relaxing the requirement that cooperative wind turbines should be locally owned present realistic propositions that give room to the development of wind energy in Denmark in the short term.

This route is strongly promoted by the Association of Danish Wind Turbine Owners (DV) and the wind turbine industry (FDV), and has already been taken. Since the beginning of 1997, siting and production criteria have been liberalized to a certain degree.[10] To continue along this road seems to be a logical strategy. A gradual loosening up of the siting criteria will eventually give 'local' wind turbine cooperatives the possibility to recruit members all over Denmark. In case the public is interested in wind energy, such regional, large-scale cooperatives would get enough play to set up large wind projects, even off-shore. This could be done either in cooperation with the utilities, or without their support.

From a techno-economic point of view such a development seems to be logical and effective. Namely, since the early 1970s, techno-economic innovation in the wind energy field has been directed at increasing the size of wind turbines. This is guided by the belief that large wind turbines are more economic than smaller ones. As we saw, this techno-economic development threatens to undermine the role played by wind energy cooperatives. Consequently, it undermines the current pluralist character of wind energy innovation in Denmark. Expanding the size of the current cooperatives may preserve cooperative ownership of wind power plants (both on and off shore). This strategy would level the techno-economic playing field of the two political coalitions, which seems to be an essential precondition for influencing the political debate. In this way, the political pluralism or democratic legitimacy of the development of wind energy may be maintained.

Nevertheless, even such a pluralistic or democratically legitimate development faces several dilemmas. It is questionable that the established coalition could agree on such a development since it thwarts their monopoly position. Paradoxically, increases in scale also impose problems on the cooperative coalition, for they would compel wind turbine cooperatives to transform themselves into more professional consumer organizations. The involvement of technical experts usually means a more passive role for ordinary people, which would be at odds with the coalition's belief in the leading '*folkelig element.*'

Two strategies may cope with this dilemma. The pragmatic side of the cooperative coalition will probably try to maintain the role of cooperatives within the large-scale utilization of wind energy. In this option, wind turbine cooperatives seem to have no other choice than to institutionalize the 'wind energy' cooperative movement, similar to the way the agricultural cooperative movement was institutionalized a hundred years ago.[11] The big challenge in this option is to find the proper organizational form for cooperation between experts and lay people. The cooperative view gives important clues towards productive collaboration: first, the adage "The best weapon against the power of experts is openness," (*ibid.* 174) and second the idea that "the professional classes should be advising, not making the decisions." (*ibid.* 69)

A second possible option is less straightforward and will find most of its adherents among the sectarians of the cooperative coalition. Their philosophy goes back to the original plan of a decentralized electricity system. In this view, local cooperatives possess their own hybrid, decentralized electricity production system, consisting of solar heaters, biomass installations and small wind turbines, totally detached from the central grid. Since such a development might only be possible in a technical sense, but still far too expensive, many people may see this option as far-fetched. Nevertheless, the notion of innovation pluralism means precisely that the development of small-scale stand-alone wind turbines should be boosted in order to open up this political option by making it a techno-economically viable and realistic one.

In fact, on a very small scale, this is happening in Denmark. Since its establishment in 1983, the People's Center for Renewable Energy (*Folkecenter for Vedvarende Energi*) in NorthWest Jutland has tried to keep the option of local electricity production alive by focusing on the technical improvement of small wind turbines. The fact that this work receives funding from the Danish government, and the fact that the Risø Research Center has recently increased its interest in small wind turbines[12] appears to show that innovation pluralism in the field of wind energy is in good Danish hands.

Notes

1 Introduction

1 In this book we will use a broad definition of innovation. Innovation is not simply similar to techno-economic innovation. In our view, innovation involves technical and economic change, the transformation of consumer behaviour, and political change. Moreover, the broad definition embodies the idea that political innovation and techno-economic development are mutually dependent and are constantly shaping each other.

2 The relative simplicity of a wind turbine assembly makes it a surveyable system technology. Basically, there are two different types of wind turbines: the horizontal (HAWT) and vertical axis wind turbine (VAWT). The HAWT has the axis of rotation of its rotor parallel to the wind stream, while the VAWT has the rotor axis perpendicular to the wind stream. The most commonly used type in California and Denmark is the HAWT. A modern HAWT is composed of six basic subsystems: "1. The rotor, which consists of one, two, or three blades mounted on a hub and may include aerodynamic braking systems and pitch controls; 2. The drive train, including gearbox or transmission, hydraulic systems, shafts, braking systems and nacelle, which encases the actual turbine; 3. The yaw system, which positions the rotor perpendicular to the wind stream; 4. Electrical and electronic systems, including the generator, relays, circuit breakers, droop cables, wiring, controls, and electronics and sensors; 5. The tower; and 6. Balance-of-station systems including roads, ground-support and interconnection equipment." (Cavallo, Hock, and Smith 1992: 136)

3 This implies that the two cases are interdependent. Whereas case contamination is normally seen as a problem, in this case it offers us a unique opportunity to study the same technology under two totally different political and economic conditions. Moreover, while a government's area of jurisdiction is bounded to the nation-state, techno-economic innovation has an international orientation. The interdependence of the two cases should be seen as a natural characteristic of a study that focuses on the interaction between politics and techno-economic development.

4 For a comprehensive introduction to the development of renewable energy technologies over the last 20 years, see *Renewable Energy: Sources for Fuels and Electricity* (Burnham 1992).

5 The company concerned is Kenetech Corporation. Before its bankruptcy, this was one of the three largest producers in the world. Before 1993, it was called U.S. Windpower. Since more people recognize the original name, 'U.S. Windpower' will be used throughout the book.

6 In 1996, some 1,200-megawatt (MW) of wind capacity was installed worldwide. About sixty percent of this market was produced by Danish wind turbine manufacturers, representing 9,000 man-years of work (Madsen 1996).

7 In the same article, George B. Leonard (1962: 30) states, "*California is a window into the future.*" In the 1960s, this notion became widely accepted within America. In the early 1970s, Kenneth Lamott (1971: 4) confirmed Leonard's statement, saying "California is our distant warning system for the rest of the United States. California is our window into the future. California is the center of the whirlpool, where all currents come to focus." The myth of California as the fountain of innovation is still vivid (cf. Harvey 1985: 1).

8 Quoted from the interesting introductory book, *Discover Denmark* (Himmelstrup, Hegelund, and Askgaard 1989).

9 The same argument is made by Schwarz and Thompson (1990), who recommend to 'recognize and analyze the inchoate.' See quote at the beginning of this Chapter.
10 Paul Gipe's (1995) book *Wind Energy Comes of Age* is probably the most comprehensive and accessible guide to the technology, economics and politics of wind energy available.
11 See also Strauss and Corbin (1990).
12 During the research the Advocacy Coalition Framework (ACF) has been adjusted to such an extent that its crucial influence can probably only be traced back by genuine ACF-insiders.
13 Both in California and Denmark some twenty key figures have been interviewed; see 'List of Interviews.' Most of the interviews have been audiotaped.

2 California: Common Sense Wind Energy

1 Before the Commission received its current name in 1946, it had been called the Railroad Commission, which had been created by state constitution to protect the consumer against unreasonable rates and see that the railroads and private utilities provide adequate facilities and services. Its duties and powers had been put into effect by the California Legislature with the passage in 1912 of the Public Utilities Act (CPUC 1994b: 1).
2 Utility managers desired two things from new technology: reliability and lower costs. Obtaining economies of scale with larger generating units (steam and water turbines) played an important role in providing good economic returns. Scale economies derived both from geometry and simplified organization. For example, building one large unit instead of two smaller ones required only one foundation for a turbine and one control room. Moreover, since the operation of a large unit did not differ significantly from operation of a smaller unit, the company saved by having staff for only one control room. Together, these capital and operating economies of scale proved attractive to many companies. For this reason, utility managers desired large-scale technology (Hirsh 1989: 40-1).
3 Capital intensity was also influenced by the low cost of money. Furthermore as a natural monopoly the electric industry had no difficulty borrowing money (Hyman 1988: 194-5). As a result, the capital expenditures of investor-owned electric utilities in California more than tripled between 1950 and 1965 (CPUC 1993: 21).
4 For example, PG&E constructed well over twenty fossil fuel units during the 1950s and 60s, ranging in size from 52 to 340 megawatts. In 1948, there were only two power plants larger than 500 megawatt in the United States. By 1972, there were 122 such plants (CPUC 1993: 21).
5 The words "go-go" refer to the hyperactivity of growth and boom psychology that had characterized the California scene since the Gold rush.
6 In the first half of this century, the California electricity system was dominantly based on hydraulic power. When in the early 1920s, California petroleum became available, oil-fired steam plants complemented the hydroelectric power from the Sierras by carrying peak loads when demand rose high and winter freezes or droughts lowered the capacity of the hydroelectric system. By the early 1950s, it became clear that hydro had its limits. Encouraged by Eisenhower's Atoms for Peace program, utilities started to consider the nuclear option. For example, PG&E announced in 1963 with some fanfare its "Super System," which was designed to serve the company through 1980, and put major reliance on nuclear power (Roberts and Bluhm 1981: 127-8). Nevertheless, crude oil became the fuel choice for most new electric generating plants. Its prices were stable and relatively low, and its supply was abundant.
7 See Mooz and Mow 1972, Ball *et al.* 1972, and Doctor 1972.
8 Just the year before, a proposal from conservationists – Proposition 20 – had been passed. This initiative had created the California Coastal Commission, empowered to plan for and license all

oceanfront land use and development. The political themes surrounding the referendum question were similar to those that characterized the power siting controversy. The fighting brought conservationists face to face with developers, and local government face to face with the state. Supporters of the state-wide plan insisted that coastal problems were not simply local concerns. Moreover, they claimed, it was the inability of local governments to counter the pressure of developers, realtors, and construction unions. Along with other industrial interests, and much of organized labor, the utilities campaigned vigorously against the initiative, as it added a major constraint on constructing new generating capacity at coastal locations (see Harvey 1985: 190-1).

9 See NSF/NASA 1973.

10 At the time Alfred J. Eggers, Jr. was the Assistant Director of the NSF's program on Research Applied to National Needs (RANN). The above quotation is taken from Eggers' opening address to the first workshop on Wind Energy Conversion Systems (NSF/NASA 1973: 2-3).

11 The expert on wind energy within the Solar Energy Panel was William Heronemus, a professor of civil engineering at the University of Massachusetts. Heronemus questioned the nuclear dream, and was a bold advocate of the large-scale use of wind power (NSF/NASA 1973: 197-203).

12 Jacobs had been building three-bladed wind generator propellers 4 m in diameter. The windmills were supplied in both 32-volt and 110-volt direct-current models. Jacobs Wind Electric Company of Minneapolis, Minnesota, sold more than 75 million dollar worth of windmills, before the company closed its doors in 1957 (Clark 1975: 539-541).

13 A detailed description of the Smith-Putnam experiment is given by Palmer C. Putnam in his book *Power from the Wind* (reprinted in Koeppl 1982).

14 In those years, the belief in R&D, and the faith in technological capability of the U.S. were widespread, and at their very height as a result of America's successful Apollo program. On July 21, 1969, Neil Armstrong had become the first human being to set foot on the moon. By showing to the world their technological supremacy, the United States of America had aimed to demonstrate the inferiority of the communist ideology (cf. Schulte Nordholt 1963: 191-196). Regardless of whether the space program affected the Soviet Union, it strongly strengthened the technological self-confidence of American engineers and scientists. The act of 'putting a man on the moon' became the symbol of both the nation's technological capability, as well as of the omnipotence of technology itself.

15 A few years later, Louis V. Divone became the branch manager of the Department of Energy's Wind Energy Division.

16 Quoted in Serchuk 1995: 76.

17 After the Apollo program, NASA had been forced to look for a new challenge and among other things became involved in managing the large scale wind systems part of the NSF's Solar Program. Besides belief in R&D, NASA brought into the wind program both the knowledge base of the aircraft industry, as well as the involvement of its leading companies. The agency subcontracted to mainly huge aerospace corporations that had already been active within the space program. It was expected that the offspring of the technological community that had successfully built the Apollo spacecrafts possessed all the theoretical and technological know-how to make building a commercial wind turbine a simple engineering problem.

18 The report is named after its principal author, Professor Norman Rasmussen of MIT. Rasmussen concluded that there was only about one chance in three million per year that an accident at a nuclear plant would cause ten or more deaths (NRC 1975).

19 The initiative is "a form of direct legislation, that allows citizens to qualify a proposal by petition and place it before voters for approval, bypassing the Legislature." (Harvey 1985: 253). The initiative or Proposition stems from the state's Progressive era. In 1910 the Progressives had come to dominate California state politics. Their leader, the liberal Republican Hiram Johnson, had won the general election and Progressive candidates won control of both the state assembly and

state senate. Their legacy has been enormous. Among their social reforms were prohibiting child labor, providing free school texts, extending the eight-hour workday, and expanding conservation programs. The political reforms also involved significant changes in the processes of government. In 1911, voters passed a constitutional amendment, promoted by the Progressives, that introduced some tools of direct democracy, which included the referendum, the recall, and the initiative (cf. Bell and Price 1980: ch. 6; Harvey 1985: 43-45).

20 Jan Hamrin. Interview Oct 25, 1995.

21 *Idem.*

22 This theory was advocated by, among others, Californians for Economic Democracy, led by former student leader and anti-war activist Tom Hayden (D), and by the consumer group TURN (Toward Utility Rate Normalisation), headed by Sylvia Siegel (Leary 1980).

23 The design of the Mod-2 (2.5 MW, 91 m diameter) had started in the autumn of 1977, with a contract award to the Boeing Engineering and Construction Company. Cost reductions were thought to be achieved through reducing capital costs by lowering the weight of the wind turbines. The Mod-2 wind turbine incorporated many 'advanced-design' features (e.g. a teetered rotor and a flexible tower) that were intended to reduce machine weight and therefore system cost. Also automated production and discounts due to quantity purchase of materials and components were expected to contribute to cost savings (cf. Koeppl 1982: ch. 4; March *et al.* 1982: ch. 2; Gipe 1995: 103-107).

24 Governor Brown was so famous for his reversals and compromises on political issues, that the *California Journal* caricatured him as "the man of the thousand faces." Despite this dubious reputation, Brown has always stood by his disavowals of nuclear power and his support for renewables. The consistency of his political ethic in the field of energy seemed to have been the exception, which proved the rule.

25 Allowing for moderate oil price escalation, Ginosar calculated that the achievement of this goal would save the state 17 billion dollar between 2000 and 2005 (Lerner and Ginosar 1979: 133).

26 The basic philosophy was to simultaneously develop six different prototypes: model I (125 ft diameter, 0.5 and 1 MW options); model II (150 ft, 1 and 2 MW); and, model III (180 ft, 1.5 and 3 MW). For each model two units would be developed; one machine for moderately high wind sites and another for very strong wind sites such as the San Gorgonio Pass. In addition, three units would be manufactured for each design to gain sufficiently wide testing experience and to help anticipate production problems. Thus, a total amount of 18 wind turbines were supposed to be tested under the proposed state wind energy program (Lerner and Ginosar 1979).

27 The R&D money had been spent on "mission studies, regional application studies, legal studies, radio frequency interference studies, design studies, environmental impact assessments and statements, etc., etc, etc." (Tennyson 1978: 156)

28 For years, utility companies had desired reliable and low-cost technology. Obtaining economies of scale from more powerful generating units and raising thermal efficiencies had been the two basic strategies to lower generating costs. The Rotor Committee at the first WECS workshop, however, had concluded that it made no sense to look for improved aerodynamic performance of the rotor blades (NSF/NASA 1973: 213). Thus from a pure technology development perspective, the wind program was only left with the economies-of-scale approach.

29 In 1977, the Department of Energy began to respond to the criticism of the soft energy proponents. In fiscal year 1978, the budget for the farm and rural systems program more than doubled compared to the years before, and received somewhat more than 20 percent of the total FWEP funding. President Carter's Sun Day speech of May 3rd, 1978, included an announcement that a 100 million dollar solar initiative would be made. The wind program received one-fifth of this figure. Half of the 20 million-dollar increase went to the small wind systems part of the program, which received about one-third of the total budget in fiscal year 1979 (Tennyson 1979a: 33-34).

The tripling of the budget for small wind turbines was not at the expense of the large wind systems part of the program. Nevertheless, this shift in accent reflected the growing influence of the soft energy perspective within the Carter Administration, and the success of the lobbying efforts of AWEA.

30 Quoted from the original Assembly Bill (AB 2976), introduced on March 20, 1978.

31 Quoted from the Mello Wind Bill (AB 2976) after Committee amendments. The Bill was amended in the Assembly on May 17, 1978.

32 The CEC plan had earmarked 8 million dollar for the task of identifying high-wind sites (Lerner and Ginosar 1979: 136). Thus, the 800,000-dollar authorized by the California Legislature was 10 percent of the amount originally requested. This cutback was probably influenced by Proposition 13 (see above). Moreover, the amount granted equalled only 1 percent of the proposed total program budget. This, however, is not to be explained by the tax revolt. Namely, the Committee amendments had crossed out the program's prototype-development part a month *before* California's voters approved Proposition 13. The strong curtailment reflected a fundamental political disagreement between the state legislators and the Brown Administration. It exposed their dissenting views on the state's role and the size of its involvement, concerning the promotion of wind energy technology in California.

33 The organization SUNRAE – Solar Use Now for Resources and Employment – played a major role in the original legislation creating the 55 percent solar tax credit. The Sacramento-based organization took no part in the anti-nuclear struggle. Instead, it emphasized the job potential of the solar energy industry (Reinke 1980).

34 See Public Utilities Code Section 2801 and 2802.

35 AB 1558.

36 A 1979 SUNRAE analysis showed that of all solar bills introduced by the Legislature, only $26 million (about five percent) was appropriated out of a proposed $516 million in funding (cf. Drager 1979: 393).

37 AB 3623, Hart.

38 Jan Hamrin. Interview Oct 25, 1995.

39 The consumption of imported oil was supposed to be reduced by "replacing oil with abundant domestic fuels; reducing energy demand through improved efficiency; increasing production of conventional sources of domestic energy through more rational pricing policies; and building a base for the development of solar and renewable energy sources" (NEA 1979: 3).

40 During an interview, Thomas Starrs (Oct 26, 1995) revealed that he had talked a lot with Richard Hirsh – author of the book *Technology and transformation in the American electric utility industry* – about this 'mystery' of American energy policy. He explained two points. "One is, that section 210 was one section in one bill out of five laws. There was so much going on, that it was barely noticed by the utilities. No one understood the importance or the relevance of it. If they had known, the resistance would have been far greater. Second, there was no real independent political constituency for it. There was no industry to advocate this law." Starrs' explanation can be brought a step further, by realising that the energy debate took place in a strongly ideological plane. 'Soft' and 'hard' energy, were perceived to depict two fundamentally different approaches to energy development. The two confronting views determined in what way choices were made and options were assessed. Section 210, was clearly a 'soft' proposal. At the time, colleagues in Washington D.C. told Hamrin (Interview Oct 25, 1995), it was envisioned "to allow for small applications where that made sense, and where people wanted to do it on an individual base." Its advocates were not thinking about it in a commercial sense. Nor did its opponents. To them 'soft' was similar to "soft-headedness, undependability, a femininity inappropriate to the masculine rigors of managing an industrial economy" (Roe 1984: 11). As a result, this 'soft' law was perceived beside the point, irrelevant, and thus unworthy of too much attention.

41 Legislative efforts in California to create a market for small-scale alternative energy production, in fact, preceded PURPA. Already in 1976, the State Legislature had enacted the Private Energy Producer Act (PEPA), designed to promote non-utility sources of electricity (see Public Utilities Code, Section 2800 ff.; see also § 2.2.5).

42 Jan Hamrin holds that this was the case, and explains that PURPA's section 210 was originally meant to allow "mom and pop, who live on a farm to put in a wind turbine. Or to put a little hydro facility on the stream that ran through their property. Or for somebody to put a photovoltaic system on the roof." (Interview Oct 25, 1995)

43 Jan Hamrin. Interview Oct 25, 1995.

44 Wayne Van Dyke, president of Windfarms Limited, picked these words to express his confidence in the rapid development of wind technology (quoted in Flavin 1981: 31-32).

45 This observation was made by the president of Hawaiian Electric Company, and quoted in Flavin 1981: 32.

46 At the time, no test results were available from the Mod-2 machine, since the first prototype had only just begun to produce power in late 1981. Test data from its predecessor were already partially available. The Mod-1, however, had never performed as expected, and in January 1981, the machine sustained a failure of the low-speed shaft assembly due to fatigue loads. The Mod-2 program included four prototypes, including a cluster of three machines in Goodnoe Hills, Washington, near the Columbia River. PG&E purchased a fifth Boeing Mod-2 turbine, and started to operate it late in 1982.

47 At the SCE wind test site located near Palm Springs several wind turbine designs were tested. In 1980, SCE purchased a huge three-bladed, horizontal axis, variable speed, 3-MW Bendix wind turbine, and a 500-kW Darrieus vertical-axis machine, manufactured by the Aluminium Company of America (ALCOA). The latter machine was destroyed in a windstorm during commissioning activities early in 1981 (March *et al.* 1982: 19). In September 1983, it was replaced by another machine of the Darrieus design, the two-blade 500 kW DAF Indal wind turbine. Due to reliability problems the Bendix turbine was modified in October 1982 with a constant speed drive train, and a 1.5 MW induction generator. SCE also tested some smaller units (Yinger 1983). By 1982 PG&E began to operate a Boeing MOD-2, a two-bladed, 91-metre diameter wind turbine with a rated power of 2.5 MW. Boeing Engineering and Construction Company had developed the MOD-2 under DOE contract. Carl Weinberg, by then manager of research and development at PG&E, recalls "The machine was a problem. It had gear breakages and all that kind of stuff. But we did a lot of tests on TV interference, noise, and avian impacts. There is a lot of data that came out. It was a tremendously successful R&D experiment, but a commercial failure. Too big, too fast." (Interview Oct 27, 1995)

48 In his book *Dynamos and Virgins*, David Roe (1984) gives a detailed eyewitness account of EDF's role in advancing this issue.

49 The ELFIN model was used to compare the long-term financial and economic effects of utility development of conservation and alternative energy sources, on a large scale, as opposed to the utility's planned central station coal and nuclear generating units. The results showed significant financial advantages to investments in relatively small-sized, short-lead-time projects, as compared to large, long-lead-time projects (Plummer 1985: ch. 17).

50 CPUC Decision 91109, December 19, 1979, p. 28.

51 Quoted in Leary 1980: 219

52 Lovins himself insisted that "the soft path can largely implement itself, if allowed to show its obvious economic and social benefits." (quoted in Frank 1978: 46). Groups sympathetic to the Brown Administration criticized its Lovins-like naive and inactive stance. In November 1979, for example, Alan H. Mirviss of SUNRAE complained that the "Government hasn't been setting an example." He argued that "The time has clearly come to stop paying lip service to these assured

renewable resources and start developing them." (quoted in Drager 1979: 393) David Roe of EDF made the same comment, but also signalled a change. According to Roe (1984:119) "Governor Brown preferred lifting his sights to wringing his hands, but his political advisers were telling him that the time had come when visionary comments to the press ... had to be reinforced with a state legislative package of concrete programs."

53 Actually, the appropriate technology movement did not intend to cure the electric industry. As the saying goes 'one man's death is another man's breath,' the movement anticipated that a widening industrial crisis would compel the public to accept its ideas and act upon them. Ultimately, they wanted to get rid of the monopoly. Their rationale was that "The roots of many of our problems are ... in concentrated control by few over natural, technological, political, and economic resources." (OAT 1981: 4)

54 In the pragmatic approach, the words 'soft' and 'appropriate' loose their original cultural meaning and ideological intent. These adjectives pointed at the oppositional culture of Appropriate Technology that challenged the hegemonic culture and technology.

55 For PG&E, for example, reserve margins fell below 10 percent during the late 1970s, reaching 6 percent in 1981 (CPUC 1993: 51).

56 In 1980, the CEC sponsored a conference in Palm Springs to attract financial interest in commercial wind development (Gipe 1995: 32; Jan Hamrin. Interview Oct 25, 1995).

57 Carl Weinberg. Interview Oct 30, 1995.

58 Afraid that their competitors would benefit from their R&D efforts, USW manufactured wind turbines solely for its own use. Other wind companies were less frantic about protecting their technology. For example, two other major wind project developers, Fayette Manufacturing Corp. and FloWind Corp. (starting in 1983) both fabricated turbines for their own projects and offered their machines to other wind farm developers.

59 Paul Gipe. Interview Nov 2, 1995.

60 John Obermeier, at the time wind energy specialist at the Office of Appropriate Technology, reveals that Ron Robie's involvement was strongly encouraged, and USW's plans were enthusiastically received by both his boss, the Secretary for Resources Huey D. Johnson, and Governor Jerry Brown. (Interview Jul 26, 1993.)

61 In the early eighties, there still was a big difference between the two developments. While at the project-side of the wind business, for every two or three new wind farm developers, one project developer left the business, in the manufacturing business newcomers just added to the existing group of manufacturers (cf. CEC 1985). The fact that the 'turnover' in wind developers was higher seems to suppose that in the beginning of the eighties developing projects was more risky, than developing and selling wind turbines.

62 In several ways economies of scale lower the energy cost per kWh. First of all, larger turbines are more cost effective from an operating view. Philip J. DiVirgilio, Director Wind Turbine Marketing of Zond Systems Inc. (Interview Nov 6, 1995) explains that nowadays a two-man crew can operate and maintain a hundred turbines annually. It takes them the same amount of time, the same number of inspections they have to do, whether the turbines are 100 kW or 500 kW. Secondly, the control equipment is also more or less independent of the size of the turbine. The same is to a certain extent true for installation costs, like high-voltage transmission lines, building roads, and groundwork. It should be noticed that economies of scale have an optimum that is site-specific, and depends on the technological state of the art.

63 Paul Gipe. Interview Nov 2, 1955; and Jan Hamrin. Interview Oct 25, 1995.

64 Some of them, like the Texas-based firm Carter Wind Systems, and Enertech Corporation had been selling to residential applications during the 1970s. Their machines had been tested at the small turbine center at Rocky Flats. Moreover, Enertech had got some grants from the test center to improve and enlarge its turbines. The company had started off with a 1.8 kW machine, and

under the federal small wind systems program developed the 20 kW version of the Enertech E44, which was used in the wind farms in California in the early 1980s. Energy Sciences Incorporated (ESI) stands out as a bizarre example of a new firm that wanted to profit from the new market. The fact that the company was founded by five former employees of the Rocky Flats wind test center left a bad taste in the mouth of the tight wind community. Well informed about the state of the art wind turbine technology and acquainted with the market demand for larger turbines, the former public servants developed the biggest machine of the day, the 50 kW version of the ESI-54. This was followed by a 60 kW, and later on by a 80 kW version. Eventually around 1984, they came out with the ESI-80, which was a 250 kW teetered machine, with two wood-epoxy blades, and a passive yaw system, and was designed by Sandy Butterfield. (Interview Paul Gipe. Nov 2, 1995.)

65 To illustrate the state of the technology, this anecdote from Paul Gipe, who recalls what happened when Zond Systems, Inc. installed its first four (40 kW) turbines in 1981 "They all failed immediately. ... The turbines had a coupler between the gearbox and the generator. When they energised them, the coupler flew off the turbine. That's fast! There might be business reasons why these companies failed, but if you look at this kind of equipment, my God, you turn the electricity on and the windmill breaks." (Interview Nov 2, 1995.)

66 Jan Hamrin. Interview Oct 25, 1995.

67 Many manufacturers played 'liar's poker' by exaggerating their turbines' capacity (Serchuk 1995: 250). Paul Gipe explains that a turbine, 10 m in diameter, is normally rated at 25 kW. He recalls that at the time, Fayette Manufacturing Corp. rated their 10 m diameter machine at 95 kW (Interview Nov 2, 1995).

68 Instead, reminded of the energy crisis by yet another oil supply shortage and another oil price increase, Congress enacted the Crude Oil Windfall Profits Tax Act. Among other things it increased the federal energy tax credit from 10 to 15 percent and extended the expiration date of the credits from December 31, 1982 to December 31, 1985. These measures provided both an additional financial incentive and added security for the investment community (Starrs 1988: 116-117).

69 These are the financial benefits from the credits alone. Lamar (1986: 23) stated that a $150,000 investment would yield $75,000 in federal and state tax credits, $138,000 over five years from depreciation deductions, and $12,000 per year in revenues from the sale of electricity (Starrs 1988: note 246).

70 This second problem had motivated Deukmejian's move. The eight-year Brown governorship had been a golden era for the special interests. Critics argued that the plethora of special provisions added to the tax system over the years had "increased the opportunities for tax evasion and cheating and at the same time given the many taxpayers the impression that the tax system is inequitable" (Walters 1983: 256). The new Governor thought it was time to begin looking seriously at tax loopholes as a way to gain revenue without further burdening all of the taxpayers. He singled out the solar energy tax credit, the fourth biggest special-interest tax break, as a test case (cf. Mirviss 1983).

71 Subsequent efforts of the CEC eventually led to the adoption of Wind Performance Reporting System (WPRS) regulations in 1984. These regulations require all Californian operators with projects rated greater than 100 kW who sell electricity to a power purchaser to report quarterly performance and related project data to the Energy Commission (CEC 1992: 1).

72 The CPUC decision describes SCE's methodical efforts to manipulate independent power developers into contracts which systematically priced the power substantially below the utility's avoided cost (Decision 82-12-055, December 13, 1982, p. 255).

73 Standard Offers 1 and 3 pay for energy and capacity on as-available basis; Standard Offer 2 pays for energy on as available basis, but unlike 1 and 3, the contract pays fixed prices for capacity for

up to 30 years. Standard Offer #1 was designed for intermittent technologies with relatively high delivery uncertainty, such as wind, run-of-river hydro, and solar. For a detailed description see for example, IEP 1993: ch. 3.

74 Jan Hamrin, Executive Director of IEP at the time. Interview Oct 25, 1995.

75 *Idem.*

76 See introductory quotes.

77 First of all, the Supreme Court's support for PURPA was central to its success (Starrs 1988: 133). Those in the energy and utility industries who were hostile to PURPA had turned to the courts for relief as soon as the NEA had been passed in 1978. Two of these cases reached the United States Supreme Court: *FERC v. Mississippi* in 1982, and *American Paper Institute, Inc. V. American Electric Power Service Corp.* in 1983. Both decisions upheld the challenged provisions of PURPA (*ibid.* 129-134). The interim long-term contract eliminated much of the financial uncertainty that had held back initiative and investment by energy entrepreneurs. At the same time, the favourable court decisions eliminated the remaining legal uncertainty, that had also stalled development of cogeneration and small power facilities. Secondly, a vibrant political atmosphere of lingering promise had triggered a part of the Californian temperament: the habit of speculation (cf. Starr 1973: 67). The dream of independent power was a call to action. Hamrin (Interview Oct 25, 1995) remembers the excitement, "But boy, right in 81, 82, 83, you just had people walking up and down the streets. This really excited. The political context is so important, because if people believe that there is something good they can do and they can get money out of it, they will just flock in. That was sort of the basis of a lot of the success." In short, expecting to make their fortune, hundreds of people had been bursting with impatience to set up small power projects. Furthermore, independent power producers felt that this was a good time to set up projects, and realized that times could change quickly. The temporary character of both ISO4 and the energy tax credits were uncertain factors. Finally, technological advances in the cogeneration market, coupled with decreasing fuel prices, increased both cost-effectiveness and project profitability under ISO4 (CPUC 1993: 69).

78 Jan Hamrin. Interview Oct 25, 1995.

79 Between 1983 and 1985, the utilities brought on line about 5,200 megawatts of new plant capacity. More than 3,500 megawatts of those additions came from the nuclear units. SDG&E added roughly 600 megawatts of nuclear capacity to its system during this period, while SCE added approximately 1,800 megawatts of additional nuclear capacity. Also PG&E's 1,100 megawatt first nuclear unit at Diablo Canyon came on line (CPUC 1993: 73-74). It had encountered a farcical situation in late 1981 that had an effect on the credibility of nuclear power in the US second to that of the Three Miles Island accident. As preparations were being made to load fuel into unit one, thirteen years after construction began, it was noticed that diagrams showing the location of major pipework were incorrect – unit 1 had been built to the unit 2 design and vice-versa! (Evans and Hope 1984: 18-19)

80 Jan Hamrin (Interview Oct 25, 1995) recalls the reaction of the CPUC when she informed them about IEP's position, "They said 'Oh no, we could not do that. People would think we don't support these technologies'."

81 About a year later, the Commission also suspended Standard Offer 2. This contract offered levelled fixed capacity payments for up to thirty years in return for a specified level of performance. For gas generators ISO4 and SO_2 looked very much the same.

82 Paul Gipe earnestly declares that "There were so many suckers back then," and proves that the credits were subject to abuse by recalling the following amazing anecdote: "I went to one windmill in Tehachapi. I climbed at the top and there was nothing inside it! It had a big bolt that came out in front that had a big peace of wood on it. ... It was advertised in all the alternative energy magazines. They had people. You could call them up and talk to the engineers, and they

would tell you how great this windmill was!" (Interview Nov 2, 1995)

83 Two real-world examples illustrate this point. First of all, just like their American colleagues, the Danish manufacturers had a hell of a job to get their turbines out in time. Wind turbine producers essentially assemble turbines using 'standard' components delivered by supply companies. In the early days, manufacturers had problems building up a reliable network of suppliers. John Obermeier remembers that he was trying to repair the control system of six supposedly identical BONUS turbines. He was surprised to find six distinct control systems, delivered by six different companies (Interview Jul 26, 1993). Secondly, similar to the U.S. turbines, the Danish machines did not even get close to their performance projections. The main reason for this was that the wind regime in California differed from the one in Denmark, where the turbines had been tested (Paul Gipe. Interview Nov 2, 1995).

84 Paul Gipe. Interview Nov 2, 1995.

85 The product choices made by the California wind project developer Zond Systems, Inc. serve as a perfect illustration of this general trend (cf. CEC 1985). In the first two years, 1981 and 1982, Zond installed 118 turbines from two different American manufacturers, i.e. Wind Power Systems (78), and Carter (10). In the second half of 1982, they also tried out 10 turbines from the Dutch company Polenko, 30 turbines from the Danish company Windmatic, and 2 Danish Vestas-turbines. The next year, Zond bought 15 Carter turbines, but stopped buying the Stormmaster turbines from Wind Power Systems. Instead, they bought one Dutch, 20 Windmatic turbines, and 155 turbines from the Vestas company. Finally, in 1984 the people at Zond had apparently made up their mind about which turbines suited them the most: they purchased 520 Vestas turbines for use in their wind farms, and installed 4 Polenko machines to check them out.

86 Hap Boyd, Judicial Advisor, Lobbyist, at Zond Systems Inc. Telephonic interview. Nov 7, 1995.

87 The notion of entrepreneurial modernization is a very comprehensive concept. It will show up several times in this book when we try to understand the development of wind energy in California on the most general level (cf. §§ 5.3, 9.4). Here, it suffices to say that in the context of this study 'entrepreneurial modernization,' with respect to the role of the state, refers to the fact that market-oriented ideas challenge the idea that an interventionist, planning state is the most suitable means for promoting the welfare of citizens. Privatization together with the introduction of competition are being taken seriously, and the belief grows that innovation is best guided by the 'coercive laws' of a deregulated market.

88 Hulett made this statement in testimony before the FERC, in San Francisco, Mar 27, 1987 (quoted in Starrs 1988: 135, note 221).

89 Thomas W. Thompson, CPUC. Interview Nov 9, 1995.

90 *Idem.*

91 The Assembly Member sponsoring AB 3995 was the Law Professor Byron Sher (D). Sher has been the intellectual, and legislative leader on environmental laws in California.

92 AB 1090, adding Public Utilities Code Section 701.3, was sponsored by Assemblyman Tom Hayden (D).

93 The title of this section is based on an analogy made by Paul Gipe (1995: 226) in his book *Wind Energy Comes of Age* between USW's claim that "its new technology would push wind below the magic $0.05 per kilowatt-hour mark," and vice president Thomas Riley Marshall's famous statement "that what the country needs is a good 5-cent cigar." (See also quotation at the beginning of this section)

94 Carl Weinberg. Interview Oct 27, 1995.

95 Earl Davis, at the time development director of USW, and later EPRI's manager of wind power integration, points out that "Utility ownership is cheaper because utilities, which have access to much more capital, can get lower financing rates than the smaller, independent developers and entrepreneurs, who are viewed as a riskier investment. In addition, because more middlemen are

involved in financing the projects of independent developers, their up-front costs are greater and add to the total installed cost." (Lamarre 1992: 9)

96 Carl Weinberg. Interview Oct 27, 1995.

97 Eric Miller. Interview Oct 26, 1995.

98 Carl Weinberg. Interview Oct 27, 1995.

99 With constant-speed wind turbines the extra torque generated by wind gusts are absorbed by the drivetrains. The gusty nature of the wind in the Altamont Pass caused tremendous stresses on the gearboxes. Variable-speed models were expected to limit drivetrain torque, and therefore to increase the machine's fatigue life.

100 Quoted in Davidson 1989b.

101 The initial R&D budget was about $20 million. The utility partners put in about $13 million. In total the project cost some $60 million (Eric Miller. Interview Oct 26, 1995).

102 Despite the fact, that at the time NREL was known as Solar Energy Research Institute (SERI), only the modern name is used to prevent confusion. The institute received its current name around 1992.

103 See note 64.

104 A clear reduction in costs of energy can be found from 1982 to 1988. In this period, the price was reduced to 45 percent of the 1982 figure (Petersen and Andersen 1995: 13).

105 To the chagrin of Vestas, Zond hired its former managing director, Finn Hansen, to lead Zond's design program in 1993 (Gipe 1995: 82).

106 Philip J. DiVirgilio. Interview Nov 3, 1995.

107 Eric Miller. Interview Oct 26, 1995.

108 *Idem.*

109 USW won 140 MW of firm capacity. Since a capacity factor of 25 percent for wind was used in the bidding, this meant 560 MW of installed wind capacity. Zond won about 160 MW of wind power.

110 Philip J. DiVirgilio. Interview Nov 3, 1995.

111 Eric Miller. Interview Nov 3, 1995.

112 Jan Hamrin. Interview Oct 25, 1995.

113 In the CPUC ruling concerning this matter, a footnote was created that said that the Commission would reject any buy-out settlement that went beyond recovering the cost QF-developers had incurred, i.e. bid-preparation costs. Thomas Thompson. Interview Nov 9, 1995.

114 Electric rates in California were up to 57 percent higher than the national average (cf. Savage 1993: 35).

115 Jeff Dasovich, CPUC. Interview Nov 17, 1995.

116 Besides SCE and IEP, the California Manufacturers Association, and California Large Energy Consumers Association were involved in the MOU.

117 The parties that signed the *Framework* are Utilities Consumer Action Network, UCS, TURN, Sierra Club California, Public Citizen, NRDC, EDF, CEERT, California Public Interest Group, California/Nevada Community Action, and AWEA.

118 The Commission suggested that the standard only be in place for a period of two years, and was silent on whether the portfolio standard should apply to all retail sellers of electricity, or the regulated utilities (Lori Jablonski from CEERT. Personal correspondence. Apr 9, 1996).

119 Nancy Rader, who as a consultant for AWEA designed and lobbied for the RPS policy. (Personal correspondence. Aug 30, 1996.)

120 Philip J. DiVirgilio. Director Wind Turbine Marketing of Zond Systems Inc. Interview Nov 3, 1995.

3 Denmark: Living Wind Energy

1 In 1940 there were about five hundred distribution companies, primarily owned by municipal enterprises or consumer cooperatives. Thirty years later, this number of electric utilities was reduced to 150 (DEF 1972a: 109). The distribution companies had their own area of supply, and were owners of the regional power companies (Olsen 1994: 2-3).

2 The development in fuel consumption for the electricity production was guided by the principle of lowest possible cost. The price-relationship between oil and coal on the world market caused a rise of the use of oil to 80 percent in the early 1970s.

3 Despite the increasing recognition of pollution in Denmark there was little public debate in the 1960s over environmental deterioration. Therefore, Jamison and Læssøe (1990: 67, 73) characterize this period in terms of an increasing 'cognitive dissonance.' For example, see quotes at the top of this section.

4 The electric utility companies were also severely criticized for the visual impact the strongly growing number of pylons had on the Danish landscape (DEF 1971-1972a: 5).

5 During the interbellum period, a cultural radicalism (*kulturradikalisme*) had developed in social circles of architects, artists, teachers *et cetera* that "was engaged in functionalism in architecture, sexual liberation, a new pedagogy based on non-authoritarion learning methods, an agitative form of art directed against militarism, capitalism, Nazism (and the bourgeoisie), and a broad, aggressive, but straightforward (and at the time important) propaganda for, what was perceived as the 'democratization' of the society." (Fonsmark 1990: 143) They were not merely critical, but played an important role in the development of Danish cultural life. The best-known representative of this tradition, the architect Poul Henningsen, provided a steady stream of original contributions to the Danish public debate. After the early 1950s, the fight against consumerism – and its corresponding system of mass production which is typified by aggressive marketing techniques and the search for profit instead of quality – became the main theme in his critical writings (*ibid.* 229). A second dominant theme of the radically oriented intellectuals was the protection of "the community against state power, and the rights of the individual against collectivism." (Jamison and Læssøe 1990: 69) They argued that since within the democratic practice people were powerless, they should organize within a network of grassroots democratic subcultures to fight against collective control over individual affairs (cf. Fonsmark 1990: 228).

6 Between 1966 and 1968, Denmark had a socialist majority in the parliament for the first time in history. The so-called 'Red Cabinet' was supported by the Social Democratic Party and the Socialist People's Party (*Socialistik Folkeparti*, SF). From 1968 to 1971, the three main liberal parties formed a majority government. This *VKR*-government consisted of the two conservative parties *Venstre*, the *Konservative Folkeparti*, plus the social liberal party the *Radikale Venstre*. In 1971, again the Social Democratic Party took over the government (cf. Fonsmark 1990: ch. 4).

7 According to the American anthropologist Steven M. Borish (1991: 214-215) four sets of "deeply cherished" related principles can be distinguished within the Danish cultural tradition: "(1) democracy and egalitarianism; (2) balance and moderation; (3) the Danish art of sociability as reflected in the concept of *hygge*; and (4) welfare and social responsibility." Concerning the Danish orientation towards the latter, he treats three components: "(1) the function of the inclusion principle as it is demonstrated in informal social interaction, formal welfare policies, alternative grassroots welfare efforts, and the Danish political party system; (2) the Danish social conscience with its quick readiness to criticize perceived social injustice; and (3) the notion of *folkelighed*, which can be traced back to the writings of N. F. S. Grundtvig." (*ibid.* 285) The term *folkelighed* is a multidimensional one and cannot simply be translated. Max Lawson, an Australian expert on Grundtvig, argues that "perhaps Peter Manniche's translation of *folkelighed* as 'community life that embraces everyone' is helpful." (Lawson 1992: 118) This corresponds with

the way the Danish Cultural Institute defines *folkelig* cultural activities as "informal, amateur, and community based." (Himmelstrup *et al.* 1992: 142) In a broader sense the term "is concerned with the preservation of identity, of a people's literature, poetry and a way of life." (Lawson 1992: 118).

8 Denmark delivers 175 members of parliament, Greenland and the Faroe Islands each have two representatives.

9 After 1973, the leader of *Venstre*, Poul Hartling headed a one-party government, which was called the 'small'-Hartling government, because *Venstre* had only won 22 mandates. The Social Democratic Party, lead by Anker Jørgensen, governed by itself with 53 and 68 seats in the *Folketing* between 1975 and Autumn 1978, and between November 1979 and September 1982, respectively. The period in between, the social democrats formed a minority government with *Venstre*, the so-called *SV*-government. After September 1982, four successive minority governments under the premiership of the conservative Poul Schlüter came. Between 1982 and 1988 the so-called 'Four-Leaf Clover' (*Firkløver*) government reigned. This minority government consisted of the four bourgeois parties: the Conservative People's Party, Venstre, the Center Democratic Party, and the Christian People's Party. From 1988 to 1990 prime minister Schlüter headed the *KVR*-government, which was a coalition between his own *Konservative Folkeparti*, *Venstre*, and *Radikale Venstre*. Finally, the Conservative Party governed together with *Venstre* until 1993.

10 The name of the *Radikale Venstre*, which directly translated means Radical Left Party, is misleading. Like the liberal *Venstre* Party (the farmer's party) it broke away from in 1905, its name testifies more to its history than to any recent political positions. During the last few decades, the *Radikale Venstre* can be typified as a social liberal party, which tends to be economically conservative, while remaining progressive on ecological and peace issues (cf. Borish 1991: 293).

11 In English: Nuclear power? No, thank you; and: We can say no thanks, without coming to lack anything.

12 The Ministry of Trade threatened to interfere with electricity rationing, if the utilities failed to reduce electricity consumption on their own (DEF 1973-1974a: 5).

13 From 1975 until the second oil crisis, the proportion of the use of coal and oil within Danish power stations did not change. The sudden hikes in oil prices at the end of the 1970s, however, led to another increase in the use of coal, and the Danish electricity generating system became almost totally coal-based. The drastic fall of oil prices in the mid-1980s did not alter this situation (DEF 1971-1995a).

14 OOA originated from a study circle that was organized by Jens Brøndum at his homeplace. The student priest had just returned from a visit to the United States, where he had seen that the new grassroots movements denounce problems concerning technology choice by setting up goal-directed information campaigns (Sørensen 1989: 49).

15 Unlike the situation in California and the Netherlands, the Danish opposition to nuclear power was organized by only one organization: OOA.

16 In 1995, five nuclear plants with a total capacity of 4900 megawatts were foreseen. The first nuclear plant was believed to be ready in 1985, and would generate about 15 percent of the Danish electricity production (HM 1976: 39).

17 Folketingstidende 1976-1977, 1. Samling, Forhandlingerne, p. 342.

18 Bent Sørensen. Interview Nov 11, 1993.

19 The efforts on energy saving were the same in both plans until 1985. After 1985, the HM-plan aimed to stop these efforts.

20 Niels I. Meyer had been president of the ATV since the mid-1960s. In this position, he had become an admirer of Jay W. Forrester (witness the fact that in 1968 Forrester received the Valdemar Poulsen Gold Medal from the Danish Academy of Technical Sciences), and friends with Forrester's associate Dennis L. Meadows (Niels Meyer. Interview Nov 9, 1993). Forrester, a

professor at the Alfred P. Sloan School of Management of Massachusetts Institute of Technology (MIT), is the founding father of the system dynamics approach. In the late 1950s, he had developed systems dynamics as a means of helping to solve management problems in industrial firms, but soon came convinced that system dynamics could be applied to any kind of system, industrial, political or social. In the late 1960s, Forrester wrote a simulation model of the urban system, which was published in *Urban Dynamics*. In the meantime, a group of 30 scientists, educators, economists, industrialists, and government officials had formed the Club of Rome, which was looking for a suitable methodology for a project on 'the predicament of mankind.' The Club became charmed by Forrester's method, and funded his assistant Dennis Meadows to work on a world dynamics model. Meadows' project led to the publication in 1972 of the very influential book *The Limits To Growth* (Simmons 1973). In this report, a computer simulation model, termed the World Model, examined five main subsystems: population, industrial capital, food, nonrenewable resources, and pollution. Interrelationships were defined and then quantified using global data. As an indication of its complexity, the program had nearly 300 variables, and more than 300 feedback loops (Nalder 1993: 24). According to Forrester, people's mental models are inadequate when faced with the dynamics of such complex systems. Instead, the leader of the system dynamics group of the 1970s believed that the convergence of engineering techniques (e.g. feedback theory) and computers could be used to indicate both the source of our problems and some possible solutions (Simmons 1973: 193). The World Model was run many times for varying growth rates and resource levels, and showed remarkable consistent behavior – one of growth and catastrophic collapse. This doomsday forecast both provoked considerable criticism as well as it served as an eye opener for many people. Niels Meyer clearly belonged to the latter group, and the outlook of a global catastrophe made him decide in 1973 to take his sabbatical leave to study system dynamics and economics at Dartmouth College in New Hampshire, NE (Niels Meyer. Interview Nov 9, 1993). He also became politically active, and started to think about a scenario for a different society (Bent Sørensen. Interview Nov 11, 1993). This ended up in 1978 with the publication *Revolt from the Center* (*Oprør fra Midten*), which he wrote together with two other prominent culture radicals – the philosopher Villy Sørensen, and the politician K. Helveg Petersen, a member of the *Radikale Venstre* (see Meyer, Petersen and Sørensen 1981).

21 B. Maribo Pedersen, one of the members of the ATV Wind Energy Committee. Interview Apr 6, 1995.

22 It was clear at the beginning that a majority within the committee was in favor of wind energy. Meyer himself participated in the WEC. The chairman, Jean Fischer, had worked for Danalith A/S, a subsidiary company of F. L. Smidth & Co., a firm that had built and sold propeller-type wind turbines during the second World War. Furthermore, as a member of the Danish resistance, Fischer had developed a secret love for wind energy. He also favored a low-energy society, and was experimenting with designing a zero-energy house (Helge Petersen. Interview Mar 27, 1995; Fischer 1979). Martin Jensen had provided the wind statistics during Juul's experiments in the late 1950s. Frede Hvelplund was a progressive economist from Aalborg University. Bent Sørensen, an engaged anti-nuclear physicists, was interested in wind energy as an alternative energy source (Bent Sørensen. Interview Nov 11, 1993). Finally, B. Maribo Pedersen – a senior lecturer at the Danish Technical University, who had been involved in the research of wind generator rotors since 1974 – also thought that wind energy was a good idea (B. Maribo Pedersen. Interview Apr 6, 1995).

23 During a two-week study trip to the United States and Canada in October 1975, the representative of the utilities' joint research institute (*Danske Elværkers Forenings Udredningsafdeling*, DEFU) within the ATV WEC, Mogens Johansson, became very impressed by the self-confidence of the American researchers concerning both their research program, as well as the future construction costs and the ideal size of wind power turbines (Johansson 1975).

24 The idea of putting up a station to test small wind turbine designs came from the only new member of the WEC, Helge Petersen, who worked as a civil engineer at the Risø nuclear research center. His background in the aircraft industry (he had previously worked for Skandinavisk Aero Industri A/S, a company that designed and manufactured small airplanes, and that in the 1940s was also involved in construction of the blades for the F.L. Smidth wind turbines), and his 'love for new things' had incited Petersen to construct some wind turbines after work. He assures me that these activities were supported by the director of Risø. In 1978, he became the first project leader of the Test Station for Small Wind Turbines (*Prøvestation for mindre vindmøller*) at Risø. Driven by "the fun of cooperating," he also helped in the construction of the windmill at the Tvind folk high school by designing the rotor blades (Helge Petersen. Interview Mar 27, 1995).
25 B. Maribo Pedersen. Interview Apr 6, 1995.
26 The original Danish text is quoted at the beginning of this section.
27 Helge Petersen. Interview Mar 27, 1995. See also note 24.
28 B. Maribo Pedersen. Interview Apr 6, 1995.
29 See DEF 1976b.
30 The independent newspaper *Information* had been a central medium for the alternative political culture since the second World War, when it was started by the resistance movement (Jamison and Læssøe 1990: 79-80).
31 Folketingstidende 1979-1980, Forhandlingerne, p. 10559.
32 Folketingstidende 1979-1980, Forhandlingerne, p. 10570.
33 Erling Olsen saw quality control and system approval as "a matter of finding the right balance between consumer and societal interests, and the fact that it should not absorb too much time nor money." (Folketingstidende 1979-1980, Forhandlingerne, p. 10573).
34 Folketingstidende 1979-1980, Forhandlingerne, p. 10557.
35 The Ministry of Housing initially became charged with implementing this law. Only, the next year the responsibility for the subsidy program was taken over by the newly established Ministry of Energy (*Energiministeriet*, EM).
36 Folketingstidende 1979-1980, Spm. 1384 til handelsmin.: 13114 (Waldorff, VS). August 24, 1979.
37 Folketingstidende 1979-1980, Spm. 1238 til handelsmin.: 12903 (Berrnhard Baunsgaard, RV). July 26, 1979.
38 See EM 1981.
39 Finally in the Spring of 1985 it was decided by the *Folketing* to wipe the nuclear power option out of the Danish energy plans (cf. Wistoft, Thorndahl, and Petersen 1992: 191).
40 In particular, the door-to-door distribution of the information brochure 'Denmark without Atomic Power' (Brunsted *et al.* 1979) made the AE-plan well known to the public.
41 The nuclear accident at Three Miles Island in Pennsylvania, USA (March 30, 1976) 'helped' to strengthen opposition against the introduction of nuclear power in Denmark, and actualized the demand to close the Swedish nuclear power plant in Barsebäck. This claim was enforced by means of mass demonstrations in Copenhagen, which is situated some 20 kilometers from the nuclear reactor site (cf. Wistoft, Thorndahl, and Petersen 1992: 190).
42 Bent Sørensen. Interview Nov 11, 1993.
43 *Idem.*
44 A detailed analysis after the first energy crisis in the winter of 1973/74, had motivated the Danish utilities to abandon the idea of exponential growth, and replace it by the somewhat moderate notion of linear growth. Therefore, "The forecast from November 1973 was the last forecast prepared on the basis of 'the natural law of power supply,' operating with an annual increase of 7-8 percent." (Elkraft 1993a: 24). The second oil crisis also proved the assumption of linear growth to be wrong (*ibid.*). In short, the convential belief in the linkage between growth in energy consumption and economic growth was not only contested for ideological reasons by 'limits-to-

growth' advocates, but also challenged in a practical sense. It was exactly the broad acceptance of the undetermined character of the relationship that made it a policy object.

45 In addition, the sense of isolation was enforced by the fact that the Test Station was placed at an outlying parcel of the large area that was covered by Risø (Dannemand Andersen 1993: 206).

46 In those years the construction work for the new atomic research center in Risø had already started. The principal motives that led to the foundation of Risø were concern about future energy supplies, and the trend towards industrialization (Schmidt 1983: 4-5). Risø represented a prestige project from the Danish government, and although, or maybe just because, it was largely a symbol of high technology the Danish utilities showed absolutely no interest in its research activities during the 1950s and 1960s (*ibid.*).

47 See quotes at the top of this section (SEAS 1987: 76).

48 The project was financially supported by the U.S. Department of Energy.

49 Although the initial development of GRP-blades proved to be too costly, it was expected that on the long run these blades would actually be cheaper, because glass fiber was thought to offer better possibilities for series production than other materials (DEFU 1979: 7).

50 Folketingstidende 1985-1986, Spm. S 128 til energimin.: 1310-1311.

51 This ad-hoc committee was established under DEF's standing Committee for Administrative and Judicial Affairs in 1982.

52 Established in 1980 by DEF to clear up the disputed settling rules (DEF 1980a: 12).

53 See DEF 1984b.

54 Folketingstidende 1983-1984, 2. Samling, Beslutningsforslag nr. B 43, pp. 1792-1797. Proposed on February 24, 1984 by Lone Dybkjær (RV), Bernhard Baunsgaard (RV), Larsen-Ledet (RV), and Helge Petersen (RV).

55 It is interesting to note that the same man who wrote the marketing report, *Danish Wind Power in the U.S.: Rapport om mulighederne for salg af mindre danske vindkraftanlæg*, that triggered the export to California, Claus Nybroe, had also published (together with Carl Herforth, one of the founders of OVE, and earlier of NOAH) *Sun and Wind* in 1976, which had become a surprisingly popular book and had prompted many do-it-yourself activities in the field of wind energy in the second half of the 1970s.

56 The Electricity Tax Law was submitted for the first time in 1974 by the liberal Hartling government, as it was thought to lead to large energy savings. The proposal was taken over in March 1975 by the new SD government, and the Ministry of Trade proposed an electricity tax of 5 øre/kWh, but the *Folketing* withdrew the bill. Two years later, the *el*-tax again showed up on the political stage. A somewhat lower electricity tax of 2 øre/kWh proved acceptable to Parliament and became law in April 1977.

57 Folketingstidende 1982-1983, Spm. S 40 til energimin.: 866-869; Spm. S 659 til industrimin.: 870-871 (Gade, SF).

58 Folketingstidende 1982-1983, Spm. S 659 til min. f. skatter of afg.: 6720; Spm. S 1165 til min. f. skatter og afg.: 11132 (Kalnæs, SF).

59 Particularly the wind power activities of communities could easily lead to a conflict of interests. For example, the city of Ebeltoft put up sixteen 55-kW wind turbines in June 1985, of which the surplus production would be delivered to the power company Midtkraft. At the same time, the community owned the local electricity distribution company, and in this capacity was an interested party within Midtkraft. This raised questions about whether Ebeltoft was entitled to receive the 50 percent investment subsidy for wind parks, which in principle was not available to power companies, and whether Ebeltoft could claim the tax exemption from Midtkraft (cf. Folketingstidende 1982-1983, Spm. S 1162 til energimin.: 11128-11132 (Kalnæs, SF)).

60 See EM 1985.

61 See EM 1985: 1.

62 The utilities had started the negotiations with the Ministry of Energy at 5 km (Karnøe 1991: 23).

63 Paul Gipe. Interview Nov 2, 1995.

64 In 1984, the profit rate before taxation of the Danish wind industry was 11.7 percent. The average number for the total industry was 6.5 percent (Kjær 1988: 19).

65 Jan Svenson. Interview Mar 29, 1995.

66 *Idem.*

67 Philip J. DiVirgilio, Director Wind Turbine Marketing of Zond Systems. Inc. Interview Nov 3, 1995.

68 Between 1985 and 1987, the total export value in dollars fell by 12 percent. Converted into Danish kroner it fell by 44 percent (Kjær 1988: 24).

69 Paul Gipe. Interview Nov 3, 1995.

70 With respect to this point, overoptimistic estimation of wind energy content was the main culprit. The wind resource estimation and the siting of turbines were the responsibility of the wind farm developer, so the Danish manufacturers were not completely to blame for the optimistic performance projections (Stiesdal 1990: 28).

71 Between 1979 and 1989 subsidies to the tune of 39 million dollar were granted to more than 2,500 turbines. Well over 80 percent of these wind turbines were owned by investor groups. Since on average between 50 and 80 families make up a wind plant cooperative, the number of turbine-owning Danish households at the time was well over 100,000. This comprises close on a quarter of a million Danes – one twentieth of the total population (see *WindPower Monthly* 1 (1990) 11).

72 In 1990, Difko managed a total of 63 investor-owned companies, into which nearly 60,000 Danish individuals had entrusted a cash sum. At the time, four out of the 9 wind farm projects in California that were operated by this finance institution made a profit, four made losses and one broke even (*WindPower Monthly* 2 (1990) 15).

73 Folketingstidende 1989-1990, Lovforslag nr. L. 80, pp. 2200-2207. Proposed on November 29, 1989 by the Energy Minister Jens Bilgrav-Nielsen (RV).

74 Lov om garanti til vindmøllebranches garantiselskab. Lovtidende 1989, Lov nr. 813, December 19, 1989.

75 Folketingstidende 1989-1990, Forhandlingerne, Første behandling af lovforslag til Vindmøllebranchens Garantiselskab, December 6, 1989. p. 2923.

76 Forslag til Lov om ændring al lov om statstilskud til udnyttelse af vedvarende energikilder m.v. (Godkendelse af vindkraftanlæg m.v.). Folketingstidende 1989-1990, Lovforslag nr. L. 71, pp. 1762-1782. Proposed on November 16, 1989 by the Energy Minister Jens Bilgrav-Nielsen (RV).

77 At the time, Bonus and Vestas were designing a 750-kW and 1.5 MW turbine with a rotor diameter of 50 m, and 60 m, respectively, and Nordtank was also developing a megawatt-size wind turbine (Østergaard 1993: 14).

78 Vestas was asked to build most of the moving parts (the hub, main shaft, and the yaw system). Bonus took care of the hydraulic system. K.K. Electronic delivered the control system, and the blade manufacturer LM Glasfiber constructed the three-bladed rotor (Østergaard 1993: 13).

79 Søren Krohn. Interview Apr 7, 1995.

80 Industrialized countries were supposed to halve the energy use per citizen in 40 years, and underdeveloped countries could raise their consumption per inhabitant by 30 percent.

81 See EM 1990.

82 Jens Bilgrav-Nielsen in his foreword to Energy 2000 (EM 1990).

83 The technical options for reducing CO_2 are energy saving (end-use savings, and improvement of conversion efficiency), fuel switching, renewable energy, nuclear energy, and CO_2 removal. Denmark had abandoned the nuclear option in 1985, and the option of CO_2 removal and storage (e.g. in exhausted gas fields) did not enter the debate. Therefore, only three options were left.

84 Niels I. Meyer. At a lecture entitled 'Environmental implications of energy supply,' on March 14,

1995, as part of the Euroconference *Sustainability Dimensions of Energy Policy*, held at the Lyngby Landbrugsskole (cf. Meyer, Benestad, Emborg, and Selvig 1993).

85 Søren S. Nørgård. Private conversation after a lecture given by himself entitled 'Central problems of energy policy,' on March 13, 1995, as part of the Euroconference *Sustainability Dimensions of Energy Policy*, held at the Lyngby Landbrugsskole (cf. Nørgard 1993, 1994).

86 Full title of the 20th March agreement: *Aftale mellem regeringen og Socialdemokratiet om øget anvendelse af kraftvarme, naturgas og andre miljøvenlige brændsler samt om eludbygningen i 90'erne.*

87 To a great extent, the March 1990 agreement was an economic rescue-plan for the natural gas project that had started at the end of the 1970s but had become a large financial failure for the government. By both commanding the utilities to rebuild their coal-fired cogeneration plants to natural gas, and demanding an increased use of natural gas in their central plants, the government hoped to save the environment and to make the natural gas project profitable again (cf. EM 1990: 99).

88 Already at the end of 1988, the Ministry of Energy had put down a request for a follow-up agreement. In its reply, DEF had pleaded to interrupt the development of wind turbine parks (and cogeneration plants), and first study the technological and economic aspects of such a development (DEF 1989a: 11). As a result, the Ministry had established a working group to evaluate the first agreement, which published their conclusions in March 1990 (EM 1990). On October 15, 1990, the utilities entered into the second 100-MW agreement.

89 Jan Svenson. Interview Mar 29, 1995.

90 Energy Minister Jens Bilgrav-Nielsen quoted in *WindPower Monthly* 2 (1990) 15.

91 *Idem.*

92 The Wind Turbine Siting Committee (*Vindmølleplaceringsudvalget*) was established in January 1991.

93 Lundholt had a bad personal relationship with her predecessor Bilgrav-Nielsen. Jørgen Lemming reveals that 'Energy 2000' was taboo at the Ministry of Energy during her ministry (Interview Apr 6, 1995).

94 Cf. Ny vindmøllelovgivning. *VEinformation* 49 (December 1992) 20.

Part Two The Political Game

1 Peter Karnøe. Personal correspondence Sep 9, 1996.

4 A Theoretical Framework of Policy Change, Tradition and Learning

1 According to Sabatier (1993: 26), in most policy subsystems, there will be two to four politically significant policy belief systems that are advanced by so-called advocacy coalitions (see below). Moreover, cultural theory's assertion that there are "Five and only five ways of life" (Thompson, Ellis and Wildavsky 1990: 3) – known as the 'impossibility theorem' – carries the same message.

2 In other words, it is assumed that "public policies and programs incorporate implicit theories about how to achieve their objectives (Pressman and Wildavsky 1973; Majone 1980) and thus can be conceptualized in much the same way as belief systems." (Sabatier 1993: 17)

3 Within Fischer's argumentative approach to policy belief systems, at the level of rational social choice, the dominant societal value system and social order are compared to philosophical alternatives in ways akin to Rawls' (1971) 'original position' or, for example, Habermas' (1970) 'ideal speech situation.' Here, a more cognitive approach is followed (cf. Hoppe and Peterse 1993: 34-36). Within the approach taken in this book, arguments at the rational social choice level will be re-

garded as being deeply rooted within a nation's specific political cultural development (see below).

4 Quoted in Graham 1995: 70.

5 The approach to public policy analysis taken in this part of the book is strongly influenced by and builds on the work of Paul A. Sabatier (1987) and Hank C. Jenkins-Smith (1993, 1997), who have jointly developed the Advocacy Coalition Framework (ACF). Many concepts, e.g., the notion of policy-oriented learning, and ways of conceptualizing the policy process, e.g., the focus on policy subsystems, are borrowed from the ACF. Nevertheless, in this study an essentially different way of structuring beliefs and aggregating policy actors is used. The ACF proposes a belief structure that consists of three hierarchical layers that are arranged in order of resistance to change, degree of abstraction and scope (Sabatier and Jenkins-Smith 1997: 19-20). A *deep core* of fundamental normative and ontological axioms is introduced that defines a person's underlying personal philosophy, and which applies to all policy subsystems. Then there is a *near policy core* of basic strategies and policy positions for achieving deep core beliefs in the policy area in question. Policy core beliefs concern a specific policy subsystem. Finally, there is a set of *secondary aspects* including a multitude of instrumental decisions and information searches necessary to implement the policy core in the specific policy area. These secondary aspects only concern a part of a subsystem. The ACF aggregates policy actors on the basis of policy core beliefs, and speaks of *advocacy coalitions*. Although Fischer's method of organizing beliefs is used in this book, one could argue that in this study deep core values are considered as the principal glue of politics. To avoid confusion about this manner of conceptualizing, we prefer to speak of ideological coalitions, instead of advocacy coalitions.

6 This can be compared with the distinction Hoppe and Van de Graaf (1989: 96-99) make between two ways or logics of approaching policy problems: a *reflective* and a *practical* order. Policy makers who follow the reflective order view problems from an ideological and systems perspective. Those who follow the practical order are guided by interpretative schemes that stem from processes of policy evaluation and monitoring, and experiences with implementation.

7 If the outcome of the policy-making process were solely determined by interactive and integrative policy-oriented learning, political discourse would resemble what Jürgen Habermas (1970) calls an *ideal speech* situation in which "there is a genuine symmetry among participants involved, allowing a universal interchangeability of dialogue roles." (Bernstein 1976: 212)

8 The words 'story line' (Hajer 1993), 'story' and 'policy narrative' (cf. Mitchell 1981; Kaplan 1986; Roe 1994) are used interchangeably in this book.

9 In a more or less similar fashion, Roe (1994: 3) uses the term 'policy narratives,' which he defines as "those stories – scenarios and arguments – that are taken by one or more parties to the controversy as underwriting (that is, establishing or certifying) and stabilizing (that is, fixing or making steady) the assumptions for policy making in the face of the issue's uncertainty, complexity or polarization."

10 As we saw, policy belief systems, by definition, offer a set of criteria to judge a certain political situation and to guide political action. Policy narratives do not directly depend for their explanatory power on one or more consistent sets of criteria. Story-lines may, therefore, comprise the power to (temporarily) rise above the various entrenched ideological positions. This argument is in line with Thomas Kaplan's (1986: 770) claim that "One important virtue of stories in policy analysis is that they provide another way – without having to agree on explicit criteria – to explain how we arrived at a current dilemma, what we should do to address the dilemma, and why a particular proposal about what to do is a good one."

5 Policy Change and Learning

1 Thomas W. Thompson, CPUC. Interview Nov 9, 1995.

2 Quoted in Hill 1968: 4.

3 This is a prime example of what Thompson, Ellis and Wildavsky in their book *Cultural Theory* (1990: 263) call *stolen rhetoric.*

4 Interview Gigi Coe. Nov 17, 1995.

5 These quotes originate from John McClaughry's preface to the 1989 edition of Schumacher's book *Small is Beautiful* (xiii). As President of the Institute for Liberty and Community in Concord, Vermont, as board member of the E.F. Schumacher Society in the U.S.), and as senior policy advisor for Ronald Reagan during his 1980 campaign and (what became terminal despair) through the first year of the Reagan White House, McClaughry serves as a prototype of an environmentally aware pro-market Republican (Schumacher 1989: xvi).

6 Howard Allen of SCE, quoted in Roe 1984: 215.

7 The utilities have always taken a half-hearted position within this alliance. In particular, NRDC played an important role in the second half of the 1980s to change this, and achieve the full cooperation of the private utilities. NRDC used the arguments as EDF in the 1970s. The environmentalists wanted the utilities to embrace conservation for bottom-line reasons. This implied that any new energy conservation program needed to be more profitable for the utilities than producing new electricity supply. The utilities joined a dialogue, facilitated by the CPUC from August 1989 through January 1990. The outcome of this so-called *State-wide Collaborative Process* was the report *An Energy Efficiency Blueprint for California* (CEC 1990), which spelled out the central principles and operating framework for a renewed drive for conservation in California (Mazmanian 1992: 208-209).

8 The collaborative process involved fifteen stakeholders, including the utilities, environmentalists, consumers, industry associations and state agencies. Their report "An Energy Efficiency Blueprint for California" (CEC 1990) spelled out the central principles and operating framework for a renewed drive for conservation in California. Also included were specific programs that each of the major utilities pledged to carry out on its own. With only minor changes, the utilities took their conservation programs to the CPUC for approval, which was readily approved (Mazmanian 1992: 208).

9 CEERT represents the renewables businesses and EDF, NRDC, Sierra Club, and Union of Concerned Scientists.

10 The parties to this CPUC filing were the special interest group AWEA, the environmental groups UCS, Sierra Club California, NRDC, EDF, CEERT, and the consumer organizations Utilities Consumer Action Network, TURN, Public Citizen, California Public Interest Group, and California/Nevada Community Action Association.

11 Dan Kirshner, EDF. Interview Oct 11, 1995.

12 The primary interest of the renewables industry was to be sure that it would get paid for existing generation. This short-term interest was especially promoted by CEERT.

13 After a good deal of thought, Steven Borish (1991: 169) suggests the following translation of Grundtvig's notion of *vekselvirkning*: "a balance between two things that remain different, but that should fertilize each other in their differentness."

14 See Meyer, Petersen, and Sørensen 1981: 64.

15 An interesting description about the development of the Danish model is given by Henning Fonsmark (1990) in his book, *The Story of the Danish Utopia* (Danish title: *Historien om den danske utopi*).

16 J.S. Nielsen, one of the pioneers behind OOA, quoted in Jamison and Læssøe 1990: 99.

17 This long-term goal, to be established through a peaceful evolution, is described in the book,

Revolt from the Center (Danish title: *Oprør fra midten*), one of the main political pamphlets of the cooperative coalition. It was first published in Denmark in February 1978 and gave rise to a heated political debate in the Danish mass media. The writers were physicist, Niels I. Meyer, Radical leader, K. Helveg Petersen, and cultural radical philosopher, Villy Sørensen (1981).

18 In theoretical terms, the consistency of first and second order beliefs is the issue here. More specifically, the logical coherence between arguments on the level of situational validation and on the system vindication level is at stake.

19 To a large extent this is owed to careful, thoughtful, and astute work by several researchers who were engaged with sustainable energy systems and alternative energy scenarios and planning systems. The Energy Group of the Physics department at the Technical University of Denmark at Lyngby (led by Jørgen Stig Nørgård and Niels I. Meyer) is one of the best research groups in the world on efficient and sustainable energy futures. At Aalborg University (Ecology and Energy Group, Department of Development and Planning), an interdisciplinary group of scientists contributed to the energy debate, amongst whom Klaus Illum, and Frede Hvelplund. Bent Sørensen from the Institute of Mathematics and Physics at Roskilde University stood out..

20 See subsection 6.2.3.

21 In a similar way, the free town of Christiania – a complete alternative village near the center of Copenhagen, which symbolized the struggle for autonomy and self-determination of the alternative political culture – was officially accepted by the political establishment as a 'social experiment' (Jamison and Læssøe 1990: 93).

22 Søren Riishøj, former energy specialist of the Socialist People's party. Interview Apr 4, 1995.

23 Niels I. Meyer. At a lecture entitled 'Environmental implications of energy supply,' on March 14, 1995, as part of the Euroconference *Sustainability Dimensions of Energy Policy*, held from March 13-24, 1995 at the Lyngby Landbrugsskole.

24 See Schumacher 1989: 259.

25 A third alliance between the public interest and communitarian coalition can be recognized. This can be named the *democratic modernization alliance*. It was short-lived.

26 The fact that in both learning alliances the public interest coalition played a role affirms the flexibility of coalition behavior.

6 Policy Tradition and Learning

1 See quotation at the beginning of section 5.1.

2 The national political culture is often regarded as a single political culture (Thompson, Ellis, and Wildavsky 1990: part 3). For example, American political culture is habitually related to liberalism in its 18th and 19th century meaning. Deregulation of the electric utilities would then be explained by the fact that Americans value individualism and equality of opportunity above all else and that they therefore distrust collective efforts promoted by government. For at least two reasons such an explanation is insufficient. First, it lacks an explanation of why there was and still is an extensive public effort within the electricity sector. Second, and as Robertson and Judd (1989: 2) argue, "American political culture has never been uniform in all times and places." Thompson, Ellis, and Wildavsky (1990) agree with this statement. They maintain that only by acknowledging that countries comprise competing political cultures, political culture can account for political change. The idea of a mix of competing political cultures in which certain political cultures lose adherents while others gain support makes it possible that the overall character of the national political culture changes.

3 Some date the crisis of feudalism as early as 1300 (Hall and Gieben 1992: 81).

4 In the area of the current United States, the Jamestown colony of 1607 in Virginia became

England's first permanent New World foothold. In 1620, fifty Pilgrims aboard the *Mayflower* decided to cross the North Altlantic Ocean in order to seek religious freedom in Virginia. After a sixty-six day voyage, however, the Pilgrims dropped anchor off Cape Cod, far to the north-east of the Jamestown, their intended destination, and founded the Plymouth colony in Massachusetts. Four years later, the Dutch bought Manhattan Island from the Indians that soon "harbored the drunken and profane of many nations, in sharp contrast to the piety of Plymouth." (Snell 1974: 101)

5 The Lockian (and Jeffersonian) idea that land could be appropriated by occupying and cultivating it, also served to legitimize the white colonization of America (Achterhuis 1988: 98).

6 Griswold (1948: 45-46) concludes that Jefferson's view of the small farmer as "the most precious part of a state," is "the classic American statement of the political theory of the family farm."

7 In his philosophical novel *Lila. An inquiry into morals*, Robert M. Pirsig (1991) argues that the Indians were the real inventors of the American way of life.

8 By the early 1870s, one five-hundredth of the Californian population held half of California's land (Starr 1973: 134-41).

9 The Southern Pacific Railroad had brought in cheap Chinese labor to work on the transcontinental railroad. Its completion in 1869 forced thousands of Chinese people to look for other kinds of work. Jobless whites came to blame the Chinese immigrants for their unemployment, and unfortunately, their demand for reforms went with hatred of Chinese workers (Harvey 1985: 15).

10 Both in Los Angeles as well as in San Francisco, where labor for a time dominated the local government, Progressivism took on a somewhat anti-labor tinge (Hofstadter 1955: 241).

11 Henry Ford personified this political belief. Harvey (1990: 125-126) explains that "What was special about Ford (and what ultimately separates Fordism from Taylorism), was his vision, his explicit recognition that mass production meant mass consumption, a new system of the reproduction of labor power, a new politics of labor control and management, a new aesthetics and psychology, in short, a new kind of rationalized, modernist, and populist society."

12 To an extraordinary degree the work of the Progressive movement rested upon its journalism. A central role was played by the muckraker, a journalist who searched out and publicly exposed misconduct of politicians and business. Upton Sinclair was one of the outstanding figures of the muckrake era. In 1904, this socialist reporter-reformer published *The Jungle*, an exposé of the Chicago meatpacking industry (Robertson and Judd 1989: 51). After publication, sales of meat and meat products fell 50 percent. The meatpackers lobbied heavily for federal legislation to restore public confidence and to bolster exports.

13 For example, Amory Lovins referred to himself as a Jeffersonian. At an antinuclear rally in Seabrook, New Hampshire, he once said "I have a *déjà vu* sense that these thousands of faces could very well be eighteenth-century Americans, fighting pretty much the same battles, perhaps in a different technical context. So many different forms of personal and community rights issues come together here as a common grievance with a common remedy. The energy and the ability of these people is very impressive. Of course my long-standing bias favors people like this who have taken the trouble to inform themselves. They are more sensitive, more capable of making wise decisions than a technocratic elite." (Frank 1978: 38)

14 Carroll Pursell (1993: 636) puts it aptly when she concludes that, "Indeed, the culture of Appropriate Technology, as it was expressed from the mid-1960s to the mid-1980s, was more than a little reminiscent of two constructions of masculinity which were widely adhered in the United States at the beginning of the 19th century: that attached to the republican gentleman, with its ideal of proportion and self-restraint, discipline of self and generosity toward others, and that associated with the independent producer, drawing pride of manliness from work, skill, the ownership of tools, self-reliance, and technical competence."

15 The major areas of land reform can be broken down into a fourfold classification (Borish 1991:

125): (1) in 1787, the provision of secure rights to tenant farmers, (2) in 1788, the liberation of the tenant farmer from forced residence on the estate of birth, (3) the formalization and control of the forced work requirement, and (4) the fundamental reorganization of agricultural land use and residence patterns.

16 The most prominent Danish scientist was the physicist H. C. Ørsted (1777-1851), who, in 1828, discovered electromagnetism.

17 Dam (1983: 34) illustrates that Grundtvig could have employed the famous words of Abraham Lincoln (1809-1865) in the Gettysburg speech in 1863 about 'government of the people, by the people, and for the people.'

18 The development of the consensus on the Danish model shows the Grundtvigian 'balance principle' at work. Borish (1991: 240) explains that, "One can observe in Danish society a set of attitudes and ways of acting that constitute a balance principle. ... Its influence leads Danes to view with tolerance the expression of a wide range of competing points of view regardless of the particular issue or context. It tends as well to bring potentially disruptive and divisive social events back into a less contested middle ground." In America exactly the opposite seems to be the case. Political commentator Oscar Garschagen remarks "The political culture in the United States is not destined for sensitive souls. The stake is high, and the game is played extremely rough." Louis Hartz's (1955: 59) comment on the glacierlike nature of American pragmatism also speaks volumes. "It has rested on miles of submerged conviction, and the conformitarian ethos which that conviction generates has always been infuriating because it has refused to pay its critics the compliment of the argument."

19 It is important to note that, while these political arrangements, institutions and ideas genuinely struck root in various Western European countries, they didn't reach their full growth in the United States (see further).

20 Statement by Edward A. Myers, Jr. (Vice President of SCE) at the IEEE Power Engineering Society (1974: 10-13).

21 "The term 'populism' is a notoriously difficult term." (Taggart 1995: 36) For simplicity, the description in *Webster's Ninth New Collegiate Dictionary* (1990) – populism as the belief in the rights, wisdom, or virtues of the common people – is used here as a natural starting point for the discussion.

22 In his book, *American Exceptionalism*, Seymour Martin Lipset (1996: 31) describes the American Creed in five terms: liberty, egalitarianism, individualism, populism, and laissez-faire.

23 Fonsmark (1991: 339-341) explains that the honorary title of being *folkelig* was reserved to the grassroots movement (*græsrodsbevægelse*) – thus, to the 'soft' side of Danish populism – but that the Progress party in fact was "exactly as *folkelig* as every other group." (*ibid.* 340)

24 Already in 1820, Grundtvig wrote these lines: *Og da har i rigdom vi drevet det vidt, når få har for meget og færre for lidt.*

25 See Note 18.

26 Quoted from March and Olsen (1989: 103). Original reference: Finansministeriet (1983). Redegørelsetil Folketinget om Moderniseringsarbejdet i den Offentlige Sektor. Copenhagen: Danish Government.

27 President Bill Clinton's move toward the conservative side that took form after the electorate, for the first time since 1955, had given control of both houses of Congress to the GOP in November 1994, can be seen as a recognition that a shift in political setting has indeed taken place. Clinton's switch toward the current American political center was made explicit in his State of the Union speech at the end of January 1996, as he declared that "the era of big government is over." The 'Neanderthal' Republican and House Speaker Newt Gingrich commented on it as follows "It struck me as a remarkable Republican speech. See how many things you can find in it that Bush and Reagan wouldn't have said." (Judi Hasson. Clinton: 'Finish the job.' He calls for balanced

budget, no shutdown. *USA Today* January 25, 1996.) The fact that nowadays left-wing Democrats and right-wing Republicans describe the President as the first *Republicrat* also speaks volumes. (Oscar Garschagen. De eerste Republicraat. *de Volkskrant* 24 Augustus 1996.)

28 Quoted in Meyer, Petersen, and Sørensen 1981: 186.

29 It should come as no surprise that the Progress Party has continuously expressed its strong antipathy against government involvement within the field of wind energy. Already in 1979, the *Fremskridtparti* characterized the investment subsidy as 'an employment therapy,' and typified the development of wind energy as 'a socialistic development that undermines society.' Ten years later, the extreme right-wing party considered the Export Guarantee Fund as a 'pretext for doing nothing,' and argued that the wind industry should take care of itself. Finally, in 1992 it was the only party that opposed the Wind Turbine Law.

30 A survey around the time of the 1973 election found 78 percent agreeing that "politicians generally care too little about what voters think." (Miller 1996: 143)

31 See the statement of the Swiss Herman Broch at the top of this section. The quotation is taken from the book *Revolt from the Center* (Meyer, Petersen, and Sørensen 1981: 186).

32 In particular, cultural radicalism is rooted within the Socialist People's party and the *Radikale Venstre*. The SF had helped to establish the welfare system during the 1960s. The Radical party is one of the four 'old parties' that constitute the political establishment in Denmark. Through the years, the *Radikale* has tried "to hold the balance of power in the party system and to exercise a significant influence on the formation of Governments and on their politics." (Miller 1996: 62)

33 In Chapter Two of *The Cycles of American History* (1986), the historian Arthur M. Schlesinger, Jr. puts forward the position that American politics passes back and forth between times of absorption in private affairs and times of preoccupation with public issues. Schlesinger's conception of 'the American political cycle' as "a continuing shift in national involvement between public purpose and private interest" (*ibid.* 27) seems in accordance with the fourth conclusion. Schlesinger (1986: 31-31) illustrates how his model fits the political history of the United States in the 20th century. "As the private interest of the 1920s had led to the public action of the 1930s, the 1950s now led into the 1960s and a new rush of commitment: Kennedy and the New Frontier; Johnson and the Great Society; the racial revolution, the war on poverty. ... By the later 1970s Americans were once more ... fed up with public action and disenchanted by its consequences. The compass needle now swung toward private interest and the fulfillment of self. The time received its appropriate names – the 'me' decade; the 'culture of narcissism.' The reaction reached its culmination in the age of Reagan in the 1980s."

34 The above showed that sometimes such a cooperation is also beneficial for the established coalition for legitimacy and pragmatic reasons. Moreover, coalition politics within the fragmented multiparty system of Denmark often imposes a compromising attitude toward the cooperative coalition. With respect to the field of wind energy, a former Radical MP explains that "as long as our requirements were moderate, they offered no problem within the big play." (Tage Dræbye. Interview Apr 7, 1995)

35 In this sense, labels like Reagan and/or Republican Revolution are no exaggerations, but indeed point at a real fundamental change in political culture within the United States.

7 A Theoretical Framework of Innovation, Tradition and Learning

1 Callon *et al.* (1992) put a development pole between the technical resources and the market. What we define as the business pole is more or less the same as what Callon *et al.* mark as the development-commercialization pole.

2 See quotation at the top of subsection 7.1.4.

3 Grin *et al.* (1997: 40) argue that at the lowest second order level, consumer preferences consist of opinions about hygiene, health, upbringing, *et cetera*. At the highest second order level, the preferred level of well-being and comfort for the family or household consumers are part of play a role.
4 See also Grin and Van de Graaf (1994), and Eberg, Van Est, and Van de Graaf (1996).
5 The authors talk of broader concept of *feasibility testing*. In its original sense, feasibility testing urged policy designers to look at the policy design from the point of view of members of the target group (cf. Hoppe, Van de Graaf, and Van Dijk 1987). It was a kind of *backward mapping* (Elmore 1980).
6 According to Callon *et al.* (1992: 227), the introduction of the concept of the *desired network* is fundamental to their approach, since "it underlines the voluntarist nature of any programming and the irreducible uncertainty surrounding it." See quote at the top of this section.
7 See the quote at the beginning of this section taken from Pressman and Wildavsky's classic *Implementation* (1973).
8 Christopher Freeman, Bengt-Åke Lundvall, and Richard Nelson more or less independently began to use the term (cf. Dosi *et al.* 1988: Part V). The research work on NSIs has culminated in the publication of the book *National Innovation Systems*, edited by Richard Nelson (1993). It provides detailed descriptions of the NSI of fifteen industrial market-oriented nations, including the United States (Mowery and Rosenberg 1993) and Denmark (Edquist and Lundvall 1993).

8 National Innovation Traditions

1 Ellul (1964: 47) explains the development of technical consciousness by the conjunction in time of five phenomena: "the fruition of a long technical experience; population expansion; the suitability of the economic environment; the plasticity of the social milieu; and the appearance of a clear technical intention." Ellul (1964: 59) sees the American accomplishment as the result of the exceptional flexibility of the American milieu. In other words, it was due to America's individualism.
2 The international electricity exhibition in 1881 at Paris meant a milestone in the history of electrical lighting (Heerding 1980). At that time, no actual practical application of electricity existed, apart from telegraphy and electro-chemistry. The emphasis was on recent developments, like mechanical generation, telephony, and in particular lighting. For the incandescent lamp, Paris meant its first commercial promotion. The availability of the electric bulb was the main impulse to built power stations. After 1881, bulbs could be purchased at every place where a demand existed. In Paris, almost every electrical lighting system was exhibited. It was exactly the fact that Edison mastered all parts of the system that made such a big impression on the visitors (Lintsen *et al.* 1993: 138).
3 Edison saw this as his greatest achievement. He recalled the venture in these words "The Pearl Street station was the biggest and most responsible thing I had ever undertaken. ... There was no parallel in the world ... all our apparatus, devices and parts were home-devised and home-made. What might happen on turning a big current into the conductors under the streets of New York no one could say. ... All I can remember of the events of that day is that I had been up most of the night rehearsing my men and going over every part of the system. ... If I ever did any thinking in my life, it was on that day." (Kingsford 1969: 132) The total capacity installed at the Pearl Street station was about 895 kW. In 1884, the power station counted 500 clients and 10,000 light bulbs (Lintsen *et al.* 1993: 138). Edison had a lot of success with his lighting system. At the end of 1882, already 153 small installations had been sold. In the same year, Edison also opened a station in London, and a year later, Milan became the first city on the European continent to have a power station.

4 The concept of the modern German university served as an example for the universities in America. In Prussia, the state reforms of 1806 had brought a new type of university. The starting point of this new type of university was its concentration on research and education, dissociated from practical application. In an ideal Socratic environment, curiosity and interest were to be the main motives (Van Dijk 1991: 25).

5 Quoted in Hofstadter 1955: 309. Already in 1936, Keynes had published his controversial book *The General Theory of Employment, Interest and Money.*

6 Total federal R&D expenditures (in 1930 dollars) rose from $83.2 million in 1940 to a peak of $1,313.6 million in 1945 (Mowery and Rosenberg 1993: 39).

7 The advisory report concerning the postwar science and technology policy of Vannevar Bush (1945) to President Truman (1945-1953) was titled *Science: The Endless Frontier.*

8 Hirsh (1989: 116) reports that Electric and Westinghouse were eager to develop commercial nuclear power, but the utility industry hesitated at first, unwilling to become partners in an untested technology. Regardless, manufacturers continued performing R&D. "The final pushes came when the federal government subsidized R&D and fuel costs, when Congress passed legislation that limited liabilities in case of accidents (the Price-Anderson Act of 1957), and when the government permitted utilities to own nuclear fuel in 1964." (*ibid.*) The federal government provided subsidies totaling about $37 billion to the nuclear industry in the three decades before 1980.

9 E. F. Schumacher and D. Dickson are commonly regarded as the two godfathers of appropriate technology (Ravesteijn 1989: 4). Their definition of appropriate technology differed. Dickson (1974) argues that technology plays a political role. He relates modern technology to the class society of industrial capitalism and to the rational thinking of science. According to Dickson, Schumacher (1973) views technological innovation as a neutral process and his so-called intermediate technologies serve as a way to reinforce industrial capitalism. Therefore, Dickson warned that, "intermediate technology carried within its content the danger of small-scale capitalism, with the same inevitable negative consequences as does the current industrial technology." (Riedijk 1986: 117) Here, Dickson's views are tentatively associated with the soft side of the Appropriate Technology movement, and Schumacher's ideas with its hard side. It should be stressed, however, that this distinction is not clear-cut. Besides inspiring many libertarians, Ernst Schumacher's *Small Is Beautiful* was also the bible of the communitarian coalition.

10 Reagan's quote can be found in *Working Together* (1981), a book about the ideology of the Appropriate Technology movement, which was published by the Community Assistance Group of the California OAT in 1981. It again indicates the often ideologically confusing thin line between the hard and the soft side of the movement.

11 The 'Third Italy' denotes north central Italy. The other familiar 'two Italies' are the northern Turin, Milan, Genoa industrial triangle and the backward Mezzogiorno.

12 In his book *The Affluent Society*, John Kenneth Galbraith (1963) distinguishes between two sectors in the economy. First, the *market system* that includes smaller firms and farms, small independent craftsmen *et cetera*. Second, the *planning system*, which is dominated by big firms that control an ever-greater part of an increasingly restricted market. The production of these big firms is based on advanced technology, basic research and product development. These big firms often operate in concert with the state, which in many cases – and in particular in the military industrial sector – is the companies' largest customer. It is fair to say, that the techno-economic structure advocated by the liberal establishment corresponds to what Galbraith calls the planning system.

13 This is related to Schumpeter's (1934) model of the innovation processes. Here, technological invention is seen as independent from the economic system. Thus, invention is conceived as an external parameter. It only becomes an innovation, when an entrepreneur picks up the invention,

and introduces it in the economic system. According to Smits and Leyten (1991: 284), Schumpeter's model, in combination with a supply-oriented view on innovation, and the idea that technology is ready-made, have constrained the technology policy of most Western industrialized nations in the last decades, and even up to now.

14 This section is primarily based on an article by Kristensen (1989; see also 1995). This article is a result of a research project founded on the hypothesis that Denmark is an ideal social formation for studying the transformation from Fordism to flexible specialization, as its industrial sector is strongly dominated by small firms. Charles F. Sabel, who introduced the concept of flexible specialization, cooperated.

15 These banks were led by the big Danish banker C. F. Tietgen (1829-1901), who, in 1857, had established *Privatbanken.*

16 A huge number of distribution companies of widely different size own nine power companies that own the two regional power associations, Elkraft and Elsam. Apart from one, all distribution companies are municipal companies or consumer cooperatives. Thus, ultimately, the consumers own the electric utilities, and can influence the decision making process through representatives, who are either members of the municipal council or elected members of the board in the cooperatives.

17 Cf. Meyer, Petersen, and Sørensen 1978: 162.

18 This view dates back at least to Schumpeter (1950: 101), who in *Capitalism, Socialism and Democracy* argued that "The monopolist firm will generate a larger supply of innovations because there are advantages, which though not strictly unattainable on the competitive level of enterprise, are as a matter of fact secured only on the monopoly level." Five factors favoring the innovative advantage of large enterprises have been identified in literature (Acs and Audretsch 1993: 25). First, large firms have more technological and financial resources. Second, only firms that are large enough to attain at least temporary market power will choose innovation as a means for maximization. Third, the larger firm is more likely to find an economic application of the uncertain outcomes from innovative activity. Fourth, scale economies in production may also provide scope economies for R&D. Finally, innovation yielding cost reductions of a given percentage results in higher profit margins for larger firms than for smaller firms.

19 Acs and Audretsch (1993: 26-27) mention several reasons why smaller enterprises would be more innovative than large-size companies. First, small companies lack the institutional inertia of many larger firms that blocks daring ventures and thwarts innovative activities. Second, some smaller companies put innovative activity at the center of their competitive strategy.

9 Unraveling Wind Energy Innovation

1 This is the concluding sentence of the report of the late physicist Feynman to the Commission investigating the explosion of the Challenger rocket in January 1986 (Feynman 1988: 220-237).

2 Harro van Lente (1993) coined the term *prospective structure* to emphasize that expectations play a major role in the development of technology, in the way that they help to interlock and coordinate activities and build up agendas.

3 The examination in this chapter is inspired by, elaborates on, and relates to the life work of Peter Karnøe (1991, 1995, 1996). He discerns two polar development strategies to wind turbine technology: a top-down and a bottom-up approach. Throughout Karnøe's work the terms 'top-down' and 'bottom-up' refer to different aspects of the innovation process. The small entrepreneurial wind firms that arrived in California and Denmark present the bottom-up strategy. The top-down strategy is exemplified by the Federal Wind Energy Program (FWEP) and its Danish image, the joint Wind Power Program (WPP) of the Ministry of Trade and the utilities. The use of the

terms 'top-down' and 'bottom-up' seems to be the most discerning when applied to the knowledge strategies of technological communities. "The bottom-up strategy is based on principles of learning-by-experience and learning-by-using in which scaling-up is a slow and step-wise process (Rosenberg 1982). Scale effects are assured on each level by producing turbines in series for the market. In contrast the top-down strategy is based on planned research and development programs from which results can be transformed into principles for technical construction. In the top-down strategy the fundamental problem was defined how to determine the 'optimum' design and size, heading directly towards the final wind turbine." (Jørgensen and Karnøe 1995: 68) Used in this way, Karnøe's distinction between the top-down and bottom-up approach to innovation corresponds to the distinction Hirsh (1989) makes between the design-by-extrapolation and the design-by-experience approach. When the terms 'top-down' and 'bottom-up' are used to characterize political and managerial aspects of innovation approaches they seem to become less appropriate.

4 See quote from Rothwell and Dodgson (1993: 13) at the top of this section.

5 Within the energy field, federal R&D programs to commercialize nuclear power, and – after the Arab oil boycott – synfuels received the highest political priority, and consequently were the most expensive national projects in civilian technology (cf. Lambright, Crow, and Shangraw 1988).

6 During the 1960s, American policy makers had become worried about the fact that the nation's scientific institutions, in allowing theoretical interest to determine research priorities, had ignored issues of social and economic significance. In response to this concern, the mission of the National Science Foundation (NSF) was broadened from support of pure science to subsidy for science with direct commercial and social relevance. In 1969, the NSF launched the program Research Applied to National Needs (RANN) to provide a counterpoint to its traditional focus on pure, scientist-managed science. Previously, NSF research had taken place almost exclusively in universities, but RANN attempted to integrate academic efforts with (large) industry and the Federal government's National Laboratories. By 1973, RANN had disbursed more that 200 million dollars to over 1,200 projects, including the wind energy component of the NSF/NASA solar program (Serchuk 1995: 67). In the field of wind energy the funding of the WECS workshop in 1973 was of major importance.

7 Mike McGormack, U.S. House of Representatives during the second WECS Workshop (NSF/NASA 1975: 77).

8 Quoted from the Report of the Committee on Rotor Characteristics (NSF/NASA 1973: 214).

9 The end of the Apollo program had forced NASA to look for a new challenge. Among other things it became involved in managing the large-scale wind systems part of the FWEP.

10 The term *statesmanship* is frequently used to describe "that handful of industry leaders typically possessing strong technical backgrounds who have introduced much of the innovation into the electric utility industry." (March *et al.* 1982: 59)

11 To some players the wind energy development program was very convenient. For example, Boeing suffered from President Carter's closing of the B-1 bomber project and faced lay off of engineers (Karnøe 1993: 54).

12 William J. Abernathy, Kim B. Clark, and Alan M. Kantrow (1983). Industrial Renaissance: Producing a Competitive Future for America. New York: Basic Books. Quoted in Hirsh 1989: 187.

13 John Mockovciak, Jr., manager Energy Systems of Grumman Aerospace Corporation during the panel discussion of the first WECS Workshop (NSF/NASA 1973: 233).

14 The federal government had been particularly active within the field of nuclear power (see Note 315).

15 The CEC planned to use $80 million to construct eighteen large wind turbines. NASA/DOE spend about four times that amount on a handful of multi-megawatt machines (Gipe 1995: 72).

16 Daniel Ancona. Quoted in Serchuk 1995: 78.
17 Jan Hamrin. Interview October 25, 1995.
18 See F.D.R.'s quote at the beginning of this section (Hofstadter 1955: 307).
19 See quote at the beginning of this section (Lerner and Ginosar 1978: 139).
20 EDF talks of 'out-businessing the businessmen.' (Roe 1984)
21 The antinuclear activist Ed Salter, who build wind turbines for residential applications, fit the type of small entrepreneur that was thought to play a major role within the business pole of the communal innovation networks. Paul Gipe describes him as "not a real aggressive businessman, but a kind of an inventor type. ... He is dedicated. He is doing it because he thinks the world needs it." Alike other more growth-oriented entrepreneurs, the creator of the Storm Master, favored lightweight, flexible rotor systems (Gipe 1995: 180). The State of California bought a couple of his turbines for demonstration projects near state park visitor centers.
22 The focal point within state government for assisting the solar industry in the maximum feasible commercialization was the SolarCal Office in the Business and Transportation Agency.
23 This explains the relative popularity in the 1970s of refurbished second-hand windchargers.
24 When in 1986 the debate about termination of the wind energy tax credits started again, the CEC (1986) had done its homework and came up with the report *Solar and Wind Technology Tax Incentive Impact Analysis.*
25 Thomas W. Thompson, CPUC. Interview November 9, 1995.
26 Head of the research program became Forrest (Woody) Stoddard, who had previously worked as part of a helicopter design team. In the mid-1970s, he developed a graduate engineering course specialized in wind turbine technology. About thirty engineers came out of that program and were hired in many of the first companies (Karnøe 1993: 58).
27 This can be considered as a forced learning-by-experience approach that started of at a high-tech level.
28 Besides backward and forward vertical integration, horizontal expansion, and product diversification are common business strategies to control the dependency of industrial firms (cf. Pfeffer and Salancik 1978).
29 This is commonly described in technology management literature as the goal of *third generation R&D management* (cf. Roussel, Saad, and Erickson 1991; Drejer 1996)
30 From an American perspective this was coincidental. The Danish manufacturers were consciously looking for new markets. This was directly related to the controlled and small size of the Danish home market. The quest for export markets was stimulated and sponsored by the Danish government.
31 The wind farm developer and manufacturer, Fayette, serves as a good example. In 1987, the company operated more than 1,400 self-made turbines with an actual capacity factor that was just one-sixth of the best performing American wind farm (Ros Davidson. Shaping up for a new future. *Windpower Monthly* 5 (1989) 14-16.)
32 Most likely these actors had got acquainted during the 'Danish boom' in 1985 that had interestingly been inspired by the development of wind farms in California.
33 This implied designing, developing, building, financing, operating, and maintaining utility-scale wind power projects with a minimum of 50 MW of generating capacity, but preferably not owning them. The California model that evolved out of the wind boom was the so-called wind power plant, which consists of hundreds of machines, all in one place, operated as if it was a single power plant. Because the wind power plant model looks like one power plant to a utility, it fitted neatly into the utility structure. In other words, the concept of a wind power plant was congruent with the beliefs of both the libertarians as well as the liberal establishment.
34 For comparison, at the time electricity from a new coal-fired power plant cost about five cents per kilowatt-hour in the United States (Weinberg 1990: 148).

35 Carl Weinberg explains that "USW's innovation efforts stimulated DOE to do a cooperative program with other wind developers to develop the next generation machine. USW fought this very hard. To some extent I don't blame them. Why should the government give their competitors money to compete gainst them?" (Interview October 27, 1995)

36 This implied that wind-generated electricity would be cost-competitive with electricity from coal-fired plants (see note above).

37 On hindsight one might say that USW was blinded by its own free-market rhetoric. Only in 1995 after the firm had experienced the financial benefits of the production tax credit, the company fought its imminent repeal. Interview Nancy Rader. November 11, 1995.

38 Repeatedly USW told "that its competitors were a distant second in market realism and in keeping costs down." (Ros Davidson. A New Start. *Windpower Monthly* (June 1996) 4.)

39 Philip J. DiVirgilio. Interview November 3, 1995.

40 The feeling on the part of the renewables companies that their interests were secondary within the Independent Energy Producers Association (IEP) to that of the gasfire industry, was one of the reasons to start CEERT in 1990. Interview V. John White (Executive Director of CEERT). October 17, 1995.

41 A few years before the bankruptcy, the President and CEO of U.S. Windpower, asked himself "What will happen to renewable energy sources as the global energy economy shifts from a regulated to a, command-and-control structure to one that features competition and privatization?" His answer was "The prevailing view is that renewables will wither away in these new, open markets – unable to compete. Once again the conventional wisdom is often stated, widely accepted – and wrong. We see the move to privatization and an open, and competitive energy economy as reasons for optimism, not pessimism." (Alderson 1994: 1)

42 Glenn Ikemoto quoted in Serchuk (1995: 235).

43 Paul Gipe illustrates this difference in business culture. "If typical European businessmen are sitting around a table with a big pile of money on it, everyone would take his share and if there is some money left over, it would be divided up. But, with American firms like U.S. Windpower the situation would be quite different. In that case, all these businesspeople are sitting around, and Gerald Alderson (former President and CEO of U.S. Windpower) would go 'This is mine! Thank you very much and have a nice day.'" (Interview November 2, 1995)

44 See Kjær (1988) and § 3.4.5.

45 Ejvind Larsen. Super Bowl som samfundsmodel for SF (The Super Bowl as a model of society for the Socialist People's Party. *Information*. January 26-27, 1991. Quoted in Borish (1991: 261).

46 The Danes had ample experience with civilian technology projects. The nuclear power program at Risø was an exception to this rule. However, while NASA was super confident after the Apollo project and adored by the American people, Risø experienced a real identity crisis as caused by the antinuclear movement and the resulting change in public opinion.

47 This was not merely a positive choice. The Danes had initially hoped to buy either ready-made blades from American manufacturers or computer design codes. For lack of financial resources, the Danes were forced to rely on their own technological resources. Helge Petersen. Interview March 27, 1995.

48 Helge Petersen. Interview March 27, 1995.

49 According to Beck (1997: 103) "*Subpolitics* is distinguished from politics in that (a) agents *outside* the political or corporatist system are allowed to appear on the stage of social design (this group includes professional and occupational groups, the technical intelligentsia in companies, research institutions and management, skilled workers, citizen's initiatives, the public sphere, and so on), and (b) not only social and collective agents, but *individuals* as well compete with the latter and each other for the emerging power to shape politics."

50 Between 1974 and 1990, Karnøe (1991: 248) discerns five stages within the development of the

Danish wind turbine industry: a grassroots period (1974-1979), an early industrialization phase (1980-1983), a forced industrialization phase (1984-1985), a time of crisis (1986-1988), and a period of stable growth (1989-1990).

51 Participation on the local level represented an essential second order element with the policy belief system of the cooperative coalition.

52 The German sociologist Ferdinand Tönnes distinguishes between "*Gemeinschaft* and *Gesellschaft* – that is, between a traditional, small-scale, face-to-face *community* resting on a universal sense of solidarity and a modern, rationalistic, impersonal *society* resting on self-interest." (Putnam 1992: 114)

53 In anticipation of possible developments within the business pole of this network, the Danish government had taken two main actions. First, in 1976, the ATV-committee had recommended the establishment of a test facility for small wind turbines as part of the government's WPP. Second, in August 1977, the cooperative coalition had stipulated for the introduction of an investment subsidy on renewable energy technology the moment a sufficient level of industrial activity would justify the use of such a market stimulation instrument. In 1978, the Test Station for Smaller Turbines was actually set up at Risø National Laboratory, and the market subsidy law came into force in the autumn of 1979.

54 FDV and Ebbe Johansen (1982). Danish Windpower. In: Proceedings Wind Energy Expo 1982 and National Conference AWEA. Amarillo, Texas. October 24-27, 1982. pp. 62-64.

55 At the time the ATV-committee proposed to set up a test facility, it was expected that such a Test Station would mainly be dealing with those kind of stand-alone heat producing wind turbines (ATV 1976: 35). Supporting wind turbines that worked independently from the grid was not directly at odds with the established coalition's preference for a centralized electricity generation system. When the Test Station started, however, power producing wind turbines that could be connected to the public grid were becoming increasingly popular.

56 Because the capacity of the turbine was related to the electricity consumption of the owner, and the price per kilowatt-hour of a medium-sized wind turbine was substantially lower than that of a small one, it was more economic to collectively buy and own a turbine.

57 Some folk high schools even set up courses like 'Wind Energy: Theory and Practice.' Steven M. Borish (1991: ch. 11) describes his personal experiences with attending that course at the progressive folk high school in Kolding.

58 This was also promoted by the fact that these small firms simply neither had the financial nor the technical resources to follow a more sophisticated and expensive design-by-extrapolation approach.

59 The grid-connected electricity generating cooperative wind turbine took the place of the wind rose as the most popular turbine. From a political point of view one could say that whereas the wind rose was a pure or typical grassroots technology, the power producing wind turbines that could be connected to the grid represented an ideologically hybrid form of technology. On the one hand, this hybrid supported local involvement and decentralized decision-making, and thus strengthened the beliefs of the cooperative coalition. On the other hand, it made wind turbine owners dependent on the existing centralized grid, or better, the existing technological system, which embodied the political ideas of the established coalition. The appearance of a hybrid technology, therefore, was a double-edged phenomenon. For the pragmatic sides of both coalitions the grid-connected wind turbine represented a potential compromise or even synthesis. In contrast, for sectarians such a compromise represented a threat to their firm belief in either a completely centralized or decentralized electricity system. The main course of the development, however, was guided by pragmatic coalitional behavior. The pragmatic sides of both coalitions came to back the development of the electricity-producing grid-connected (medium-sized) wind turbine. This hybrid technological concept elicited the political cooperation between the two coalitions.

60 The fact that the government policy kept the Danish market small enticed companies like Bonus, Nordtank and Vestas to look for new markets.
61 A journey to California in 1985, sponsored by the Technology Council, to interest new firms in marketing abroad (Karnøe 1991: 219) illustrates that Danish policy makers were not aware of the market situation in California. A 'positive' outcome of this trip would have increased the ongoing strong competition among Danish wind turbine manufacturers. As seen, this would have lowered the profits of the wind industry that already were decreasing to a dangerously low level. This well-meant, but counterproductive effort illustrates the ignorance of the Danish government with respect to wind turbine market in the United States.
62 Cf. §§ 8.2.2 – 8.2.3.
63 Cf. § 6.3.3.
64 Cf. § 6.4.2.
65 Cf. § 5.2.
66 Along the same line of reasoning, Hajer (1993: 56) claims "a story line provides the narrative that allows the scientist, environmentalist, politician, etc., to illustrate where his or her work fits into the jig saw."
67 The three investor-owned utilities' investment in conservation grew from $ 31 million in 1980 to just over $ 107 million in 1985 and achieved energy savings of about 12 billion kWh (CPUC 1993: 72).
68 The overcapacity was due to the fact that together with third-party generators nuclear plants had come on line (see e.g. § 2.4.3).
69 In § 5.1.7, we showed that the BRPU proceeding can be seen as a complex political compromise within and between the sustainable-grow-and-build alliance and the regulated competition alliance.
70 Jan Smutny-Jones. Director IEP Association. Interview October 31, 1995.
71 In 1993, cogeneration power plants provided some of the least expensive electricity serving the State of California, costing ratepayers 3-6 ¢/kWh (IEP 1993: 2). This compares to average electricity rates for PG&E and SCE of over 10 ¢/kWh. This high cost is mainly due to fact that electricity from California's nuclear power plants is extremely high. For example, in the same year, electricity from the nuclear power plant Diablo Canyon cost 11.4 ¢/kWh (Ferguson 1993: 1).
72 At the time, the radical side of the public interest coalition and the pragmatic wing of the libertarian coalition formed the so-called regulated competition alliance.
73 Ironically, it was exactly the same rapid innovation process that was nurtured within the flexible/cooperative windenergy network in Denmark that caused the California export experience to become a financial disaster. On the regional, low-competitive, and gradually developing Danish home market, cooperative R&D worked and prepared the Danes from a technological perspective – not from a managerial point of view! – for the export market. On the rapidly growing, highly competitive California export market, Danish firms came to compete each other on the basis of product technology. In this case, technological cooperation overheated the process of creative destruction and almost caused the elimination of the Danish wind turbine industry.

10 The Intelligence of Innovation

1 The section's title refers to Charles E. Lindblom's (1965) classical study *The Intelligence of Democracy*.
2 Since wind energy is becoming more competitive this condition is weakening. At this moment, however, wind energy still needs some governmental support.
3 This explains the title of Chapter 3 'Denmark: Living Wind Energy.'

4 Another way of putting this is that in Denmark the isolation of the hard side of populism caused the inclusion of the soft side of populism. In California, the inclusion and dominance of the hard side of populism led to the exclusion of the soft side of populism in the early 1980s.
5 Quoted in Graham (1994: 72).
6 This partly explains the title of Chapter 2 'California: common sense wind energy.'
7 When we follow Touraine's (1997: 189) argument that "democratic politics attempts to promote intervention so as to protect the weak from domination of the strong," we must notice that on the energy policy subsystem level the stimulation of an entrepreneurial mode of innovation should be regarded as highly democratic.
8 For example, since 1982, the independent power industry supplied about fifty-six percent of all new generation brought on line in California (CPUC 1993: 123).
9 Jan Smutny-Jones. Executive Director IEP. Interview Oct 31, 1995.
10 Text from the title track of the CD *De Onderstroom* (*The Undercurrent*) by Stef Bos (1997). In English: "The undercurrent, that no one sees, defines the direction in every field."
11 Beck's words echo the insights of Marx. In the revolution year 1848, Marx expressed in *The Communist Manifesto* that "The executive of the modern State is but a committee for managing the common affairs of the whole bourgeoisie." (McLellan 1971: 192)
12 Italics added.
13 Widening the scope of public innovation policies brings along the need for a more sophisticated view on what defines the success of innovation. The network approach makes it clear that the success of, for example, wind energy innovation policy cannot simply be determined by the number of megawatts installed per year. One could easily draw up a list of qualitative and quantitative measures of success for the development of each network pole. With respect to the science and technical poles one might, for example, think of indicators like financial commitment to R&D, levels of skills of personnel, quality of the education system, innovation rate, product range, advancement of the quality control system, expectations with regard to the long-term viability of the utilised technological concepts, et cetera. With respect to the production pole, the amount of turnover and profit rates, quality of the management of manufacturing, change in market position, viability of the market strategy may be usable indicators for innovative success (cf. Dodgson and Bessant 1996). The user will probably see the cost of wind power, the noise wind turbines make, quality of service and maintenance as important criteria for success. Such a list of criteria originates from the beliefs of the various players involved in the innovation network.
14 With respect to joint innovation learning, all that glitters is not gold. High expectations in defiance of facts or a lack of knowledge with respect to the capability of other innovation actors is probably one of the main forces that drives innovation. Ignorance, therefore, often plays a vital role in innovation, especially during its early phase. However, as the innovation process carries on, too high expectations will be thwarted, and ignorance or a lack of information will readily prove to be counterproductive. It is exactly the interactive nature of innovation that, once it is going, requires joint innovation learning of all involved actors.
15 See Wildavsky's quotation at the top of this subsection.
16 The issue of the democratic monitoring of innovation will be treated in somewhat larger detail in the next subsection.
17 In general, one could argue that too strong regulation combined with too weak stimulation hampers techno-economic innovation. In this situation, public policy is likely to become symbolic action and, as a result, its democratic quality will diminish. Too strong stimulation combined with too weak regulation will probably lead to a lot of market activity. The democratic and techno-economic quality of the resulting innovation will be highly uncertain. Policy makers, in fact, hand in the primacy of politics, and give away democratic guidance to the market. It is important to strike the right balance between government stimulation and regulation.

11 Wind Energy Futuribles

1 The Progressive conservation movement that emerged in the late nineteenth century in reaction to reckless exploitation of America's natural resources represented the first wave. The Sierra Club and the National Audubon Society were established during that period. Dunlap and Mertig (1992: 1-3) name the 'wilderness movement' during the 1950s the second wave of conservationism. For the sake of convenience, Dowie (1995: 106) leaves out that episode and points to the publication of Rachel Carson's *Silent Spring* (1962) as the beginning of the second wave (which logically represents the third wave within the terminology of Dunlap and Mertig).

2 Lori Jablonski. Interview Oct 17, 1995.

3 Carl Weinberg. Interview Oct 27, 1995.

4 *Idem.*

5 These are the five elements of the new energy paradigm as identified by Lee, Ball, and Tabors in their book *Energy Aftermath* (1990).

6 See the statement of Ejvin Beuse at the top of this section.

7 Tage Dræbye, was MP for RV in the early 1980s, and is currently the chairman of the Technical Committee for Wind Power under the Energy Agency (*Energistyrelsens faglige udvalg for vindkraft*). Interview Apr 7, 1995.

8 One of the largest joint projects is a wind farm near Avedøre Holme near Copenhagen. The project consists of twelve 300-kW turbines, of which half is owned by the wind turbine cooperative Avedøre Vindkraft I/S. The expected yearly electricity production of these six turbines is about 3,500,000 kWh (Hagensen 1992).

9 For example, the average Danish household consumes about 2,000 kWh. Since shareholders are allowed only to have shares up to 150 percent of their own electricity use, the wind turbine cooperative described in the above note, needs more than one thousand members. Thus, an equal share within a similar project with 1-MW turbines will already require more than three thousand participants.

10 Formerly, only people who lived within either the community in which the wind turbine is sited, or within the neighbouring community were allowed to buy a wind turbine share. Nowadays, also people who have lived at least two years within the last ten years within the community, or neighbouring community in which the wind turbine is sited are allowed to own wind shares. Moreover, nowadays every adult (and not like before every household) may buy wind shares that correspond to a yearly power production of 30.000 kWh.

11 The former chairman of FDV, Birger T. Madsen (1993) drew this historical analogy on the occasion of DV's fifteenth anniversary.

12 Helge Petersen. Interview Mar 27, 1995.

Samenvatting

Wind der verandering
Een vergelijkende studie van de politiek van windenergie-innovatie in Californië en Denemarken

Inleiding

Dit boek is een bescheiden zoektocht naar mogelijke antwoorden op de vraag: Hoe kunnen beleidsmakers techno-economische innovatie in democratisch gelegitimeerde banen leiden? Deze vraag wordt behandeld op basis van een gedetailleerde en omvattende vergelijkende studie naar de rol die beleidsmakers hebben gespeeld in de ontwikkeling van windenergie in Californië en Denemarken.

Het boek bestaat uit vier delen. Deel Een *Het verhaal van windenergie* geeft een uitgebreide beschrijving van de ontwikkeling van windenergie in Californië en Denemarken. Deel twee *Het politieke spel* geeft een analyse van het politieke debat rondom windenergie en plaatst dit in een historische perspectief. Deel drie *Het innovatiespel* analyseert de interactie tussen het politieke veld, de technologische wereld en het bedrijfsleven op het gebied van windenergie. Deel vier *Conclusies* beziet de betekenis van de bevindingen op het gebied van windenergie voor de algemene uitdaging van politieke sturing van techno-economische innovatieprocessen. Het boek eindigt met een discussie over de toekomst van windenergie in Californië en Denemarken.

Deel een: Het verhaal van windenergie

De twee oliecrises in de jaren zeventig brachten een heftige politieke discussie op gang over de toekomst van het energievoorzieningssysteem. Met betrekking tot de elektriciteitssector zagen sommigen de oplossing in het gebruik van kernenergie. Anderen wezen op het belang van energiebesparing en stonden het gebruik van alternatieve energiebronnen voor. In Californië en Denemarken werd de kernenergieoptie na vele jaren van maatschappelijk debat op een zijspoor gezet. Alternatieve elektriciteitsopwekkers, waaronder windmolens, kregen in beide staten een zekere politieke steun. Zowel de aard van deze politieke steun als ook de aard van de resulterende techno-economische ontwikkeling verschilden sterk.

Californië: 'Common Sense' windenergie

In Californië kon windenergie profiteren van een verzameling van beleidsinstrumenten, die erop gericht was marktwerking te introduceren binnen de monopolistische elektriciteitssector. In de Verenigde Staten brak de federale overheid in 1977 een lans voor de private opwekking van elektriciteit door de invoering van PURPA. Deze wet verplichtte de gevestigde elektriciteitsbedrijven voor een redelijk bedrag stroom in te kopen van onafhankelijke elektriciteitsproducenten. De Californische overheid pakte, via het opstellen van een aantal standaard contracten, de implementatie van PURPA voortvarend aan. Dit beleid leidde echter pas tot een succes na de invoering in september 1983 van het zogenaamde ISO4 contract dat aanbieders van elektriciteit voor een periode van tien jaar inkomsten uit elektriciteitsproductie garandeerde.

Dit had een hause aan decentrale projecten tot gevolg. In twee jaar werd voor meer dan 10.000 MW aan contracten getekend. Het leeuwendeel kwam voor rekening van wamte-kracht centrales. Gestimuleerd door gunstige belastingaftrekmogelijkheden op federaal en staatsniveau namen windprojecten ongeveer een vijfde deel voor hun rekening. Commerciële ontwikkelaars plaatsten, gegroepeerd in grote windparken, duizenden windmolens op de heuvels in Californië. Deze onstuimige ontwikkeling leidde tot overcapaciteit binnen de energievoorziening en derhalve werd besloten tot de afschaffing van dit ISO4 contract in april 1985. Aan het einde van dat jaar werd tevens het nationale belastingkrediet opgeheven. Een overgangsregeling zorgde er voor dat het merendeel van de reeds getekende contracten nog steeds in aanmerking kwam voor dit krediet. Zodoende kon tachtig procent van de projecten, die elektriciteit leverden op basis van het ISO4 contract, ook na 1985 worden gerealiseerd. Pas in het begin van de jaren negentig werden de laatste projecten voltooid. In totaal zo'n 1700 MW aan windvermogen.

Sinds de afschaffing van de standaard contracten werd op een laag pitje gediscussieerd over de vraag op welke wijze nieuw te bouwen vermogen op een controleerbare manier aanbesteed zou moeten worden. CEERT speelt binnen deze discussie een belangrijke rol. De lobby-activiteiten van deze coalitie van milieuorganisaties en ontwikkelaars van duurzame energie leidden in 1991 tot een wet die stelde dat een bepaald gedeelte van het nieuwe vermogen diende te bestaan uit duurzame energiebronnen. Tijdens het eerste biedproces in 1994 werd hiervoor een vijfde deel gereserveerd. De windparkontwikkelaars maakten volledig gebruik van dit aanbod. In een rechtszaak, aangespannen door de zuidelijke elektriciteitsbedrijven SCE en SDG&E, besliste het machtige federale bureau, de *Federal Energy Regulatory Commission*, dat de wijze waarop het biedproces (bekend onder het acronym BRPU) was opgezet in strijd was met de nationale wetgeving.

Omdat inmiddels ook het debat rondom de herstructurering van de elektriciteitssector was opgelaaid stierf de BRPU een langzame dood en de gouden bergen, die lange tijd voor de windindustrie aan de horizon gloorden, verdwenen als sneeuw voor de zon. De invoering van 'volledige' marktwerking binnen de elektriciteitssector eind 1995 legde de markt voor windenergie in Californië volledig stil en verplichtte de industrie tot het vinden van nieuwe wegen.

Denemarken: Levende windenergie

In Denemarken onstond in de tweede helft van de jaren zeventig, in nauwe samenhang met een levendige alternatieve-energiebeweging waarin stevig geëxperimenteerd werd met het zelf bouwen en gebruiken van windmolens, een klein industrieel potentieel op het gebied van windenergie. Om deze bedrijvigheid te stimuleren voerde de overheid in

1979 een investeringssubsidie in op duurzame elektriciteitsgeneratoren. De regeling was aanleiding voor bedrijven als Bonus, Vestas en Nordtank windturbines te gaan assembleren. Zij verkochten hun molens vooral aan buurtgroepen, georganiseerd in windmolencoöperaties. In 1982 zakte de thuismarkt in en gingen de fabrikanten naarstig op zoek naar nieuwe markten en vonden deze in Californië. De export naar deze staat groeide van 40 turbines in 1982 naar een maximum van 3000 turbines in 1985.

In 1984 trok de thuismarkt sterk aan, nadat de elektriciteitsbedrijven (DEF) – onder druk van de politiek – een 10-jarige overeenkomst hadden gesloten met de particuliere windmoleneigenaren. Deze regelde goede teruglevertarieven en hief tevens grotendeels de bestaande rigide particuliere plaatsings- en productiebeperkingen op. Dit laatste bood institutionele beleggers de mogelijkheid gesubsidieerd windparken op te zetten. Verrast door deze ontwikkeling sloot de overheid in 1985 in het diepste geheim een convenant met DEF. De productie- en plaatsingseisen werden weer verscherpt om de speculatieve ontwikkeling van windparken tegen te gaan. Ook verplichtte deze zogenaamde '100-MW afspraak' de elektriciteitssector in de komende vijf jaar 100 MW aan windvermogen te installeren.

Na 1986 leidde het wegvallen van de Californische markt tot zware concurrentie op de windturbinemarkt. Slechts één bedrijf ging niet failliet, anderen werden overgenomen of gingen verder met nieuw kapitaal. De herstructurering van de industrie, de snel teruglopende subsidie en het vooruitzicht op een nieuwe markt voor grote molens, dwongen de windturbine-industrie enerzijds tot herstructurering en anderzijds tot een snelle opschaling van de windmolens. In Denemarken werden tussen 1985 en 1991 ongeveer 300 molens per jaar geplaatst. Het gemiddelde turbinevermogen nam in deze periode toe van 80 tot 200 kW.

In april 1990 publiceerde de regering de nota *Energie 2000* waarin het streven naar een duurzaam energievoorzieningssysteem centraal stond. Als onderdeel van dit plan werd een tweede 100-MW overeenkomst gesloten. Desondanks stagneerde de markt voor windturbines van elektricteitsbedrijven in de eerste helft van de jaren negentig. Lokaal verzet tegen windparken en gebrek aan ruimte speelden de elektriciteitssector parten. Ook de markt voor particuliere windmolens zakte in deze periode in, ondanks de Windturbine Wet van 1992, waarin de opvolging van de 1984-overeenkomst geregeld werd. De stagnerende thuismarkt leverde echter geen echte problemen meer op voor de sterk op de exportgerichte Deense industrie.

Deel twee: Het politieke spel

Aan de hand van een theoretische raamwerk voor beleidsverandering, -traditie en beleidsgericht leren (hoofdstuk 4) wordt het politieke debat rondom windenergie geanalyseerd en in een historisch perspectief geplaatst. Het scala aan argumenten wordt met behulp van de methode voor politieke evaluatie van Fischer bestudeerd. Deze methode onderscheidt eerste orde argumentaties, die betrekking hebben op de inhoud en effectiviteit van beleid, en tweede orde argumentaties – de 'harde kern' van de beleidstheorie – die betrekking hebben op de normatieve rechtvaardiging van het beleid. Het beleidsproces wordt geconceptualiseerd in termen van interacterende *ideologische coalities*, waarvan de leden gemeenschappelijke harde kern beleidsopvattingen hebben. Deze harde kern opvattingen komen niet uit de lucht vallen, maar passen binnen zogenaamde *ideologische tradities*, die hun oorsprong hebben in de specifieke politieke geschiedenis van een land.

Om zich te handhaven onder steeds veranderende (politieke) omstandigheden, om politiek relevant te blijven en zodoende de eigen opvattingen om te kunnen zetten in beleid, zijn ideologische coalities veroordeeld tot een eeuwigdurend proces van *beleidsgericht leren*. Daar harde kern opvattingen niet leiden tot eensluidende beleidsvoorstellen, kan binnen één coalitie verschil van mening bestaan met betrekking tot de preciese invulling van het beleid. Om de diversiteit in gedrag binnen een coalitie te beschrijven maken we een onderscheid tussen *sektariërs* en *pragmatici*. Dit onderscheid hangt nauw samen met het onderscheid tussen twee vormen van beleidsgericht leren. Een *competitieve vorm*, die er op gericht is de eigen positie binnen het debat te verstevigen door het versterken van de eigen beleidsopvattingen of het verzwakken van de argumenten van de tegenstander. Daarnaast een *coöperatieve vorm*, die gericht is op het integreren van verschillende visies binnen het beleid. Daar waar sektariërs de nadruk leggen op competitieve vormen van leren, staan pragmatici vaak open voor coöperatieve vormen van leren tussen coalities. Wanneer er sprake is van leren tussen coalities spreken we van een *alliantie*. Indien er sprake is van een langdurige coöperatieve leerrelatie spreken we van een *beleidsgerichte leeralliantie*. Dergelijke leerallianties vormen zich rondom *beleidsverhalen* (*story lines* of *policy narratives*) en construeren deze tevens.

Beleidsverandering en beleidsgericht leren

Vier ideologische coalities speelden een rol binnen het energiedebat in Californië: de *corporate capitalist coalition*, de *public interest coalition*, de *libertarian coalition* en de *communitarian coalition*. De beleidstheorieën van deze coalities staan samengevat in Figuur 5-2 (p. 121) en 5-3 (p. 122). De wisselwerking tussen deze coalities bepaalde de dynamiek van het Californische debat. Deze is schematisch weergegeven in Figuur 5-1 (p. 118).

De *public interest coalition*, bestaande uit voornamelijk consumenten- en milieuorganisaties, zette zich – vanuit een gevoel van wantrouwen ten opzichte van de politieke en economische rol van grote bedrijven – in voor de economische en sociale belangen van de gewone burger. Techno-economische innovatie diende ingezet te worden voor het algemene belang. De *corporate capitalist coalition*, waarbinnen de grote Californische elektriciteitsbedrijven een prominente rol speelden, zagen daarentegen grote bedrijven als de onbetwiste aanjagers van techno-economische en maatschappelijke vooruitgang. Een alliantie van deze twee coalities domineerde het energiedebat vanaf de Tweede Wereldoorlog tot en met het midden van de jaren zestig. De basis voor deze alliantie was een beleid gericht op groei van de productie en het gebruik van elektriciteit, die gekoppeld aan technologische vernieuwing, leidde tot lagere stroomtarieven voor de burger en goede winstmarges voor de gereguleerde monopolistische elektriciteitsbedrijven.

De energiecrises, de strijd rondom kernenergie en een verhoogd milieubewustzijn zette bovengenoemde alliantie onder druk en bood ruimte aan de alternatieve visies van de *libertarian coalition* en de *communitarian coalition*. De *libertarian coalition*, bestaande uit onder andere onafhankelijke stroomproducenten, zag de vrije markt als de perfecte wijze om welzijn te creëren en te verdelen, beschouwde de monopolistische marktstructuur van de elektriciteitssector als de oorzaak van het energieprobleem en wilde deze openen voor kleine ondernemers. De *communitarian coalition* tenslotte, geworteld in de *appropriate technology* beweging, zag de monopoliepositie van de grote elektriciteitsbedrijven vooral als een democratisch probleem. Deze coalitie geloofde in directe democratie en zag lokale productie van elektriciteit als een manier om politieke macht te decentraliseren en gemeenschapszin te stimuleren. De opkomst van deze nieuwe visies leidde tot een tweede alliantie.

Beide allianties bleven tot het midden van de jaren negentig naast elkaar bestaan en vormden ieder een basis voor coöperatief leren tussen de deelnemende coalities. De *corporate capitalist coalition* en de *public interest coalition* vormden een leeralliantie rondom het beleidsverhaal van ecologische modernisering. De 'ecologische modernisering'-alliantie dacht dat via juiste regulering de bestaande elektriciteitssector in staat zou zijn om op den duur goedkoop en betrouwbaar schone stroom te leveren. De andere alliantie (bestaande uit de *libertarian coalition*, de *communitarian coalition* en een deel van de *public interest coalition*) was gebaseerd op het 'ondernemers'-beleidsverhaal (*entrepreneurial modernization*), waarbinnen gepleit werd voor deregulering en marktwerking.

In het midden van de jaren negentig viel de 'ecologische modernisering'-alliantie uit elkaar doordat de *corporate capitalist coalition* zich aansloot bij de *libertarian coalition*. Door dit nieuwe verbond domineerde de 'ondernemers'-alliantie het energiebeleid en werd de weg naar deregulering van de Californische elektriciteitsproductiemarkt geopend.

In Denemarken speelden slechts twee coalities een relevante rol: de *established coalition* en de *cooperative coalition*. Figuur 5-5 (p. 135) vat de beleidsopvattingen van deze coalities samen. De dynamiek van het Deense debat is schematisch weergegeven in Figuur 5-4 (p. 133).

De *established coalition* voelde zich de grondlegger en de beschermer van de Deense welvaartstaat (het zogenaamde Deense model). Economische groei en technologische vooruitgang diende de welvaartstaat te faciliteren. Elektriciteit werd beschouwd als de motor van economische groei en als een welvaartverhogend product. De *cooperative coalition* voorzag grenzen aan de economische groei. Ze stond een ecologisch duurzame maatschappij voor. Daar waar de één het bestaande representatieve politieke stelsel en een centraal gereguleerd elektriciteitssysteem ondersteunde, was de ander voor een actieve lokale democratie en een gedecentraliseerd systeem.

Binnen het Deense energiedebat vormden de *established* en *cooperative coalition* vanaf het einde van de jaren zeventig een stabiele leeralliantie rondom het beleidsverhaal van ecologische modernisering, waarbinnen duurzame ontwikkeling – als strategie om economische groei te bestendigen – werd omhelst.

Beleidstraditie en beleidsgericht leren

In hoofdstuk 6 wordt onderzocht op welke wijze de mix van nationale politieke culturen beleidsgericht leren binnen het energiebeleid mogelijk heeft gemaakt en beperkt heeft. Nadat de politieke tradities binnen Californië en Denemarken zijn beschreven, wordt geconcludeerd dat de harde kern opvattingen van de ideologische coalities die de afgelopen drie decennia binnen het energiebeleid een relevante rol speelden in sterke mate verbonden zijn met de onderscheiden politieke tradities. Ideologische coalities blijken vertegenwoordigers van politieke tradities.

De ideologie van de *communitarian coalition* is verbonden met de ideeën van Jefferson en het ideaal van onafhankelijkheid en individuele verantwoordelijkheid. De *libertarian coalition* is geïnspireerd door het klassieke liberalisme van de tweede helft van de 19de eeuw. Binnen beide 'klassieke' Amerikaanse politieke tradities verkondigt men de stelling dat minder overheid beter is. De *corporate capitalist coalition* is ideologisch verbonden met het institutionele grootkapitalisme, dat aan het begin van deze eeuw opkomt. Bezorgheid over de politieke en economische macht van de grote bedrijven leidde in Amerika tot een sterke opleving van het consumentenbewustzijn. De *public interest coalition* past binnen deze traditie van het opkomen voor het belang van de consument.

Pas sinds de tweede helft van de jaren zestig wordt het Deense model van de welvaartstaat, dat centraal staat binnen het gedachtengoed van de *established coalition*, breed ondersteund. Dit model bracht verscheidene politieke stromingen (conservatisme, liberalisme en socialisme) samen en ontdeed ze van hun oorspronkelijke bedoelingen. De ideeën van de *cooperative coalition* zijn te herleiden tot de filosofie achter de succesvolle agrarische coöperatieve beweging aan het einde van de 19de eeuw. Deze filosofie was sterk geënt op Grundtvig's inzichten omtrent de relatie tussen leken en experts, tussen solidariteit en persoonlijke vrijheid, en democratie als dialoog.

Nadat ideologische coalities zijn onthuld als dragers van politieke tradities, wordt bezien op welke wijze de afgelopen drie decennia verschuivingen zijn opgetreden binnen de nationale mix van tradities. Zowel in Californië als in Denemarken groeide na de Tweede Wereldoorlog langzaam een brede consensus rondom het concept van de welvaartstaat en het nut van overheidsbemoeienis met de economie. In Denemarken belichaamde de *established coalition* deze naoorlogse consensus. In Amerika gaf de alliantie tussen de *corporate capitalist coalition* en de *public interest coalition* hieraan vorm. Deze alliantie vertegenwoordigde het zogenaamde *liberal establishment* en bepaalde het politieke klimaat in de jaren zestig. Gedurende de jaren zeventig slonk het publieke vertrouwen in het establishment sterk. Dit bood ruimte voor de opkomst van twee vormen van populisme. De Californische *communitarian coalition* en de Deense *cooperative coalition* vertegenwoordigden de zogenaamde zachte kant van het populisme, die voor meer democratie pleitte. De *libertarian coalition* vertegenwoordigde de harde kant van het Amerikaanse populisme, die pleitte voor minder overheidsbemoeienis. Terwijl in Californië de zachte kant van het populisme alras zijn aanhang verloor, werd de harde kant onthaald als typisch Amerikaans en ging geleidelijk het politieke klimaat domineren. Hierdoor veranderde de politieke cultuur in Amerika sterk en werd het *liberal establishment* vervangen door het *conservative establishment*. In Denemarken gebeurde totaal iets anders. De harde kant van het populisme werd als zijnde on-Deens in de ban gedaan, terwijl de zachte kant door het establishment werd gezien als een kritisch correctief om de eigen crisis te bezweren en de Deense maatschappij te moderniseren.

Uit bovenstaande volgt dat de twee Californische leerallianties de *liberal* en *conservative establishment* vertegenwoordigde binnen het energiebeleidssubsysteem. Leerallianties kunnen getypeerd worden als zogenaamde *establishment allianties*. Het conservatieve politieke klimaat bemoeilijkte beleidsgericht leren binnen de *liberal establishment alliance* en ondersteunde leren binnen de *conservative establishment alliance*, die tenslotte in de jaren negentig binnen het energiebeleid dominant werd. De Deense politieke cultuur bood slechts plaats voor één leeralliantie. Leren binnen deze alliantie werd ondersteund door een goedgezinde, stabiele politieke omgeving.

Deel drie: Het innovatiespel

Aan de hand van een theoretisch raamwerk voor innovatie, innovatietraditie en innovatiegericht leren (hoofdstuk 7) wordt het windenergie-innovatieproces geanalyseerd vanuit een historisch perspectief. Dit geschiedt aan de hand van het innovatienetwerkconcept. Innovatienetwerken worden beschreven door middel van vijf polen: een wetenschaps-, een technologie-, een productie-, een consumptie- en een politieke pool. Wetenschappers en technologen spelen een dominante rol binnen de wetenschaps- en technologiepolen, ondernemers en managers zijn de belangrijkste spelers binnen de productiepool, gebruikers spelen een rol binnen de consumptiepool en beleidsmakers bezetten de politieke pool. Binnen een innovatienetwerk zijn zodoende allerlei typen actoren actief, met verschillende belangen en (professionele) visies en inzichten. Voor succesvolle innovatie is het nodig dat alle polen binnen een netwerk ontwikkeld zijn en dat de activiteiten binnen de verschillende innovatiepolen bij elkaar aansluiten. Het completeren en convergeren van een netwerk via beleidsmaatregelen vereist gezamenlijk innovatiegericht leren (*joint innovation learning*) tussen beleidsmakers en andere innovatieactoren.

Nationale innovatietradities

In hoofdstuk 8 wordt eerst een beschrijving gegeven van de gezamenlijke historische ontwikkeling van techno-economische en politieke tradities in Amerika en Denemarken. Deze beschrijving laat zien dat er langdurige relaties bestaan tussen specifieke politieke en techno-economische tradities. Techno-economische tradities zijn afhankelijk van de steun van politieke tradities en omgekeerd. Om deze wederzijdse afhankelijkheid aan te geven wordt de term *innovatietraditie* ingevoerd. Verbonden met bepaalde innovatietradities wordt in Californië en Denemarken een aantal ideaaltypische *politiek gewenste windenergie-innovatienetwerken* (WEIN) onderscheiden.

De *liberal establishment alliance* in Californië stond een zogenaamd *civilian/corporate WEIN* voor. Deze alliantie wilde de ontwikkeling van windenergiecentrales bestaande uit zeer grote multimegawatt windmolens, die een bijdrage moesten leveren aan het bestaande gecentraliseerde elektriciteitssysteem. Hoog-technologische kennis – aanwezig binnen nationale onderzoeksinstituten, universiteiten en onderzoeksafdelingen van grote bedrijven – diende aangewend te worden voor het maken van dergelijke grote molens. Deze molens dienden door grote bedrijven (uit de vliegtuigindustrie) geproduceerd te worden en door de monopolistische elektriciteitsbedrijven gebruikt te worden. De *libertarian coalition* trok de monopoliepositie van de grote elektriciteitsbedrijven in twijfel en wilde dat het midden- en kleinbedrijf gebruik zou maken van nieuwe kleinschalige alternatieve elektriciteitsproductietechnieken, zoals warmte/krachtinstallaties en windmolens, om een positie te verwerven binnen de elektriciteitsproductiesector. Deze coalitie stond een *entrepreneurial WEIN* voor, waarin onafhankelijke elektriciteitsproducenten de belangrijkste gebruikers van windenergietechnologie waren. Hoogtechnologische, kleine of middel-grote bedrijven in coöperatie met universiteitsgroepen werden in staat geacht om deze technologie op korte termijn te commercialiseren. De *communitarian coalition* tenslotte prefereerde zogenaamde lokale *communal WEIN's*. Windenergie werd als middel gezien om de controle over de productie van elektriciteit te decentraliseren. Boeren en kleine gemeenschappen zouden windmolens, gemaakt door vakmensen binnen kleine, lokale, niet-winstgeoriënteerde bedrijfjes, gaan gebruiken voor de lokale stroomproductie.

De ideeën van de Deense *established coalition* rondom de implementatie van windenergie leken sterk op die van de *liberal establishment* in Amerika. Een consortium van grote Deense bedrijven zou windmolens bouwen voor windturbineparken van de elektriciteitsbedrijven. Het nationale energieonderzoekscentrum in Risø – in samenwerking met de Technische Universiteit van Denemarken – zou de kennis leveren voor het bouwen van grote multimegawatt molens. Het door de *established coalition* gewenste netwerk wordt *flexible WEIN* genoemd. Evenals de *communitarian coalition* in Californië hoopte de *cooperative coalition* dat er her en der lokale windenergieinnovatienetwerken zouden ontstaan. Coöperaties zouden windmolens gaan produceren voor gemeenschappen of consumentencoöperaties. Dit ideaaltypische netwerk noemen we het *cooperative WEIN*.

Windenergie-innovatie ontrafeld

In hoofdstuk 9 wordt de dynamiek van windenergieinnovatie onder de loep genomen door zowel de interactie van beleidsmakers met andere innovatieactoren als de wisselwerking tussen verschillende *WEIN's* te onderzoeken.

Tijdens de ambtsperiode van gouverneur Brown werden de bovengenoemde drie typen *WEIN's* simultaan gestimuleerd (zie Figuur 9-2, p. 222). De inspanningen van de Californische overheid om een *corporate/civilian* en *communal WEIN* op te zetten mislukte vanwege een gebrek aan politieke steun, gebrek aan steun uit het bedrijfsleven, gebrek aan vraag (zowel van private gebruikers als elektriciteitsbedrijven) en gebrek aan betrouwbare en efficiënte technologie. Met andere woorden, ondanks de inspanningen bleven de polen van deze twee *WEIN's* onderontwikkeld.

Als resultaat van een verzameling van maatregelen (PURPA, belastingkredieten, Wind Resource Center) – oorspronkelijk bedoeld om andere technologieën (zonneboilers en warmte/kracht-gascentrales) en innovatienetwerken te stimuleren – ontwikkelde het *entrepreneurial WEIN* zich wel onverwacht sterk. Ondernemers ontdekten dat de bonte set van beleidsinstrumenten de mogelijkheid bood om, door particuliere beleggers gefinancierde, windparken op te zetten. De *entrepreneurial innovation* aanpak was reeds met succes beproefd op het gebied van warmte/kracht-gascentrales en werd omarmd door de regering Brown. De sterke groei van de vraag leidde echter tot veel (soms frauduleuze) windparkontwikkelaars en veel zwakke fabrikanten, die veel armzalige windmolens maakten. Een belangrijke reden was het gebrek aan betrouwbare productinformatie. Het Californische beleid leidde tot een ontwikkeling waarin de bestaande zwakke technologie vervangen werd door nieuwe armzalige technologie. Dit proces wordt gekarakteriseerd als '*destructieve creatie*.' In de markt ontwikkelden zich twee strategieën om dit proces te omzeilen. U.S. Windpower volgde de strategie van verticale integratie. Door zelf windmolens te ontwikkelen en deze te gebruiken in de eigen windparken maakte het bedrijf zich immuun voor de afstemmingsproblemen binnen het *entrepreneurial WEIN*. Het kopen van Deense windmolens, waarvan wel betrouwbare informatie aanwezig was, was de tweede strategie die ervoor zorgde dat het innovatieproces binnen het *entrepreneurial WEIN* zichzelf niet opblies.

In het midden van de jaren tachtig verdween het reguleringsregime, dat geleid had tot de onstuimige groei van het aantal windmolens in Californië. Belastingkredieten op het gebied van windenergie en het lucratieve lange termijn standaardcontract werden afgeschaft. Terwijl beleidsmakers zich gingen beraden over de wijze waarop nieuw te installeren vermogen op een controleerbare manier aanbesteed zou moeten worden, verliet U.S. Windpower zijn strategie van verticale integratie en begon met de ontwikkeling van een nieuwe windturbine speciaal voor de grote elektriciteitsbedrijven. Het bedrijf verschoof zijn activiteiten aldus van het *entrepreneurial WEIN* (zoals voorgestaan door de naar 'ondernemers'-leeralliantie) naar het *corporate WEIN* (zoals voorgestaan door de 'ecologische modernisering'-alliantie). In het begin van de jaren negentig was er veel activiteit binnen de techniek- en bedrijvenpool van dit nieuwe *coporate WEIN*. Op dat moment was de overheid klaar met de voorbereidingen van het nieuwe aanbestedingsproces. Dit sloot niet aan bij de ontwikkelingen van het *corporate WEIN*, maar bood het *entrepreneurial WEIN* mogelijkheden voor een verse start. Doordat het aanbestedingsproces een mislukking werd en het herstructureringsdebat de interesse in windenergie sterk deed verminderen, bleef de windenergie-industrie in het midden van de jaren negentig met lege handen staan.

De ontwikkeling van windenergie in Denemarken werd bepaald door de interactie tussen het *flexible WEIN* en het *cooperative WEIN* (zie Figuur 9-3, p. 240).

In de jaren zeventig concentreerde het beleid zich op het tot stand brengen van een *flexible WEIN*. Het Deense windenergieprogramma richtte zich zowel op de ontwikkeling van, als op de wisselwerking tussen de verschillende polen. Zo werd het programma gecoördineerd door de toekomstige gebruiker, de elektriciteitsbedrijven, en werd geprobeerd een consortium van grote bedrijven te interesseren voor de windturbinemarkt. Dit laatste mislukte en bedreigde in het begin van de jaren tachtig het voortbestaan van het *flexible WEIN*. De overheid initiëerde hierop de oprichting van het bedrijf DWT om windturbines voor de elektriciteitssector te ontwikkelen. Tegen die tijd was er al een bloeiende windturbine-industrie ontstaan, die middelgrote molens produceerde voor de thuismarkt, maar vooral voor de export naar Californië. Deze industrie had zijn oorsprong binnen het *cooperative WEIN*.

De *cooperative coalition* probeerde niet alleen het energiebeleid te beïnvloeden, maar probeerde zijn ideeën ook direct in praktijk te brengen door het opzetten van een *cooperative* WEIN. Dit gebeurde onder andere door gemeenschappelijke zelfbouwexperimenten, het organiseren van bijeenkomsten om technologische kennis uit te wisselen, de oprichting van een vereniging voor windmoleneigenaren en – niet te vergeten – het kopen, vaak om ideoloische motieven, van een windmolen. De actoren die zich inzetten voor de ontwikkeling van het *cooperative* WEIN worden 'ecologische practici' (*ecological practitioners*) genoemd.

Tot aan het einde van de jaren zeventig werd de ontwikkeling van het *cooperative* WEIN bepaald door de filosofie van de *cooperative coalition*. Vanaf die tijd begon de overheid zich met deze ontwikkeling bezig houden en gingen de ideeën van de *established coalition* de ontwikkeling van het netwerk mede bepalen. De oprichting van het Test Station voor Kleine Windmolens, de instelling van een investeringssubsidie en de invoering van een nieuwe set van plaatsings- en consumptiecriteria, leidde in het begin van de jaren tachtig tot de ontwikkeling van een nieuw type innovatienetwerk, het zogenaamde *flexible/cooperative* WEIN, met nieuwe typen van actoren en relaties. In de productiepool traden – later succesvolle – bedrijven binnen en windcoöperaties gingen de gebruikerspool typeren. De gecontroleerde stimulatie van windenergie zorgde ervoor dat de samenhang tussen de verschillende activiteiten binnen het nieuwe netwerk behouden bleef. Hoewel de gereguleerde thuismarkt een ideale voedingsbodem vormde voor de ontwikkeling van technologie, bood zij te weinig voor een bloeiende industrie. Gesteund door de overheid ging men op zoek naar exportmarkten en vond deze in Californië. Dit leidde tot een exponentiële groei van de Deense windmolenproductie in de eerste helft van de jaren tachtig. De technologie ontwikkelde zich ook onstuimig.

Het succes van het *flexible/cooperative* WEIN bood de *established coalition* de mogelijkheid om het zeer zwakke *flexible* WEIN te versterken. Men realiseerde zich namelijk dat de productiepool van de *flexible/cooperative* WEIN een rol kon spelen binnen het *flexible* WEIN. In 1985, creëerde de regering, door middel van de '100-MW' afspraak, een gegarandeerde afzetmarkt voor windturbines voor de elektriciteitssector. Dit stimuleerde de fabrikanten hun strategie te veranderen in de richting van het ontwikkelen en maken van grote windturbines voor de elektriciteitsbedrijven. De bedrijven hadden echter vooralsnog hun handen vol aan de export. In steeds sneller tempo vervingen nieuwe efficiëntere Deense turbines 'oude' types op de Californische markt. Dit oververhitte proces van '*creatieve destructie*' verkleinde de winstmarges sterk en veroorzaakte, tesamen met het inzakken van de export, het failliet van bijna alle Deense fabrikanten. Het vertrouwen in Deense technologie zorgde echter voor een snelle herstructurering van de industrie, hetgeen er voor zorgde dat de opgebouwde kennis en ervaring behouden bleef. Het vooruitzicht van een nieuwe (Europese) markt voor grote molens was ook aantrekkelijk voor de geldschieters. De nieuwe bedrijven professionaliseerden hun productieproces en zette nieuwe R&D afdelingen op, die in staat waren om de wetenschappelijke kennis, ontwikkeld binnen het *flexible* WEIN, te gebruiken bij het ontwerpen van nieuwe megawatt windturbines voor de elektriciteitssector. Het innovatienetwerk veranderde zo in het *flexible* WEIN dat de *established coalition* in de jaren zeventig voor ogen had gehad. Echter, als gevolg van beleidsgericht leren met de *cooperative coalition* rondom het beleidsverhaal van ecologische modernisering had de *established coalition* inmiddels het concept van duurzame ontwikkeling omarmd. Om aan te geven dat de politieke visie achter dit type WEIN is veranderd, wordt van een *flexible ecological WEIN* gesproken.

In Denemarken ging een langdurig leerproces rondom het beleidsverhaal van ecologische modernisering hand in hand met innovatiegericht leren op het gebied van windenergie, waarbij de 'grassroots' technologie werd aangepast aan de normen van de elektriciteitssector en een internationaal toonaangevende windturbineindustrie ontstond. De Deense casus laat zien dat ecologische modernisering niet slechts een beleidsverhaal, maar meeromvattend, een *innovatieverhaal* is. Daarom wordt gesproken van het innovatieverhaal van *ecologische innovatie*. Windenergie-innovatie is het symbool geworden van deze manier van innoveren.

In Californië werden twee competitieve beleidsverhalen onderscheiden. Het leren rondom ecologische modernisering werd slechts in geringe mate ondersteund door innovatieactiviteit. Er was duidelijk een gebrek aan innovatiegericht leren tussen beleidsmakers, managers uit de elektriciteitssector en technologen. Het 'ondernemers'-beleidsverhaal werd sterk ondersteund door techno-economische ontwikkeling. De activiteiten van de onafhankelijke stroomproducenten pasten perfect binnen dit beleidsverhaal. Het 'ondernemers'-verhaal blijkt eveneens een breed innovatieverhaal. Het boegbeeld van dit innovatieverhaal werd gevormd door de kleinschalige warmte/kracht-gascentrales, die economisch konden concurreren met de grote centrales van de elektriciteitsbedrijven. Dit gold niet voor windparken. In tegenstelling tot de Deense windturbineindustrie, die het symbool van ecologische innovatie was, was de Californische windparkindustrie het lelijke eendje binnen het 'ondernemers'-innovatieverhaal.

Deel vier: Conclusies

De intelligentie van innovatie

Het concluderende hoofdstuk 10 zet alle bevindingen nog eens op een rijtje en beantwoordt vervolgens de centrale vraagstelling. Om techno-economische innovatie in een democratisch gelegitimeerde richting te leiden dienen beleidsmakers serieus aandacht te besteden aan drie met elkaar verbonden processen (zie Figuur 10-1, p. 283).

Ten eerste dient het democratische gehalte van het politieke debat rondom innovatie gewaarborgd te worden. Politiek pluralisme wordt alom genoemd als de essentie van democratische besluitvorming. In het geval ideologische coalities op een vijandige manier ten opzichte van elkaar staan, blijft politiek pluralisme echter een leeg begrip. De gevalsstudies laten zien dat integrerend gedrag de basis vormt voor lange termijn beleidsgerichte leerprocessen tussen ideologische coalities, die nodig zijn om gezamenlijk nieuwe politieke betekenis te geven aan techno-economische innovatie. Kortom, integrerend coalitiegedrag effectueert, op constructieve wijze, politiek pluralisme en is daarom een belangrijke voorwaarde voor de kwaliteit van het democratische debat rondom techno-economische ontwikkelingen.

Ten tweede dient het democratische gehalte van het implementatieproces geoptimaliseerd te worden. De gevalsstudies laten zien dat de ontwikkeling van windmolens niet gesimplificeerd kan of mag worden tot een technisch probleem. Bij de invoering van windenergie zijn talloze aspecten, actoren en visies aan de orde. Implementatie van innovatie vereist zodoende een continu leerproces tussen beleidsmakers en andere actoren binnen het innovatienetwerk. De studie toont aan dat er verschillende politiek geprefereerde innovatienetwerken zijn. Als politiek pluralisme de essentie van democratische besluitvorming is, dient men dit principe niet slechts binnen het politieke debat, maar ook binnen de implementatie van innovatiebeleid centraal te stellen en effectief te maken. Politiek pluralisme leidt tot innovatie pluralisme, wat kan leiden tot een politiek diverse set van innovatienetwerken of, als resultaat van lange termijn beleidsgericht leren tussen coalities, tot een innovatienetwerk waarbinnen verschillende politieke waarden geïntegreerd zijn (zie de Deense casus).

Ten derde dienen beleidsmakers toezicht te houden op de politieke betekenis van techno-economische innovatie. Omdat techno-economische innovatie een relatief autonoom proces is, legt ook optimale politieke sturing de richting van deze ontwikkeling niet volledig vast. De beleidsmaker kan op ieder moment een beentje gelicht worden. Met andere woorden, terwijl de beleidsmaker denkt het techno-economisch proces naar links te sturen, gaat het de andere kant uit. Bijvoorbeeld in Denemarken gaf in 1985 het beleid gericht op de stimulatie van het lokale, particuliere gebruik van windenergie, institutionele beleggers de kans om op een lucratieve wijze windparken aan de noord-west kust van Jutland te ontwikkelen. Deense beleidsmakers grepen echter snel in en maakten deze politiek ongewenste ontwikkeling onmogelijk. Dit voorbeeld geeft aan dat techno-economische ontwikkeling de beleidsmakers, juist vanwege haar politieke implicaties, continu uitdaagt. Het is aan de beleidsmaker zich steeds uitgedaagd te voelen om techno-economische ontwikkelingen politiek te interpreteren en deze aan te passen aan de maatschappelijke wensen en politieke voorkeuren.

Windenergietoekomstbeelden

Bij wijze van nawoord biedt hoofdstuk 11 een beredeneerde blik in de toekomst van windenergie. De uitdaging in Californië is om binnen een gedereguleerde competitieve elektriciteitssector kansen te creëren voor duurzame vormen van elektriciteitsopwekking. Een door de *communitarian coalition* voorgestane manier is het aanbieden van groene stroom aan groepen en individuele burgers. De *public interest coalition* pleit voor het instellen van een wettelijke bepaling, die garandeert dat een bepaald minimum percentage van de verkochte elektriciteit in Californië wordt gegenereerd door alternatieve energiebronnen. Een combinatie van beide oplossingsrichtingen ligt voor de hand.

In Denemarken is het ideaal van een duurzame elektriciteitsvoorziening wijd verspreid. Hierdoor ligt een geleidelijke, maar gestage, uitbreiding van het gebruik van windmolens (al dan niet in zee) in de toekomst in het verschiet. Grootschaligheid zet het coöperatieve bezit en beheer van windmolens echter onder druk. Pragmatici binnen de *cooperative coalition* zien de verdere institutionalisering van de coöperatieve beweging rondom windenergie als een oplossing voor dit dilemma. Anderen bepleiten een toekomst waarin op grote schaal windmolens bij of op het huis geplaatst worden.

References

Achterhuis, Hans (1988) *Het Rijk der Schaarste: Van Thomas Hobbes tot Michel Foucault.* Baarn: Ambo.

— (1997) 'Lang Leve de Grondwet van de Techniek'. *Technisch Weekblad* 15-10-1997.

Achterhuis, Hans, Philip Brey, René Munnik, Martijntje Smits, Pieter Tijms, Peter-Paul Verbeek (1997) *Van Stoommachine tot Cyborg: Denken over Techniek in de Nieuwe Wereld.* Amsterdam: Ambo.

Acs, Zoltan J. and David B. Audretsch (1993) 'Innovation and Firm Size: The New Learning', in Mark Dodgson and Roy Rothwell [eds.]. *Small Firms and Innovation: The External Influences.* pp. 23-35. Geneva: Inderscience Enterprises.

Alderson, Gerarld R. (1995) 'Competition, Privatization and Renewables: Wind as the Test Case', in Leonard S. Hyman [ed.]. *The Privatization of Public Utilities.* Vienna, Va: Public Utilities Reports.

Asmus, Peter (1993) 'Saving Energy Becomes Company Policy. Utilities and the Environment – Profit from Efficiency'. *The Amicus Journal* Winter, pp. 38-42.

ATV, Danish Academy for Technical Sciences (May 1975) *Translation from the Danish: Vindkraft: Oversigtsrapport fra Vindenergiudvalget under Akademiet for de Tekniske Videnskaber.*

— (January 1976) *Vindkraft 2: Forslag til Handlingsprogram.*

Ball, R.H., *et al.* (September 1972) *California's Electricity Quandary: II. Planning for Power Plant Siting.* Santa Monica, CA: Rand Corporation.

Bachrach, Peter, and Morton S. Baratz (1970) *Power and Poverty: Theory and Practice.* New York: Oxford Univ. Press.

Bardach, Eugene (1977) *The Implementation Game: What Happens After a Bill Becomes Law.* Cambridge, Mass.: MIT Press.

Beck, Ulrich (1992) *Risk Society: Towards a New Modernity.* London: Sage.

— (1997) *The Reinvention of Politics: Rethinking Modernity in the Global Social Order.* Cambridge, UK: Polity Press.

Bell, Charles G., and Charles M. Price (1980) *California Government Today: Politics of Reform.* Homewood, IL: Dorsey Press.

Bell, Daniel (1996) 'The Protestant Ethic'. *World Policy Journal* Fall, pp. 35-39.

Bendure, Glenda, and Ned Friary (1996) *Denmark: A Lonely Planet Travel Survival Kit.* Hawthorn, Australia: Lonely Planet.

Benestad, Olav, Lars Brinck, Lennart Emborg, Bjarne Juul-Kristensen, Alexander Kristiansen, Niels I. Meyer, Kyösti Pulliainen, Eivind Selvig, and Anssi Vaittinen (1993) 'Energy and Environment: Nordic Energy Scenarios for 2010 and 2030'. *Energy Policy* December, pp. 1225-1236.

Bernstein, Richard J. (1976) *The Restructuring of Social and Political Theory. London*: Methuen & Co.

Beuse, Ejvin (1993) 'OVE-Landsmøde. Stigende utålmodighed'. *Vedvarende Energi & Miljø* No. 3, pp. 22-23.

Bijker, Wiebe E., Thomas P. Hughes, and Trevor Pinch [eds.] (1987) *The Social Construction of Technological Systems: New Directions in the Sociology and History of Technology.* Cambridge, Mass.: MIT Press.

Bijker, Wiebe E., and John Law [eds.] (1992) *Shaping Technology/Building Society: Studies in Sociotechnical Change*. Cambridge, MA: MIT Press.

Bijker, Wiebe E. (1993) 'Do Not Despair: There Is Life after Constructivism'. *Science, Technology, & Human Values* Vol. 18, No. 1, pp. 113-138.

Blegaa, Sussanne, Frede Hvelplund, Johannes Jensen, Lars Josehsen, Hans Linderoth, Niels I. Meyer, Niels P. Balling, and Bent Sørensen (October 1976) *Skitse til Alternativ Energiplan for Danmark*. Copenhagen: OOA, OVE.

Blumer, Herbert (1969) *Symbolic Interactionism: Perspective and Method*. Englewood Cliffs: Prentice Hall.

Boes, Jan (1989) 'Economy and Development: A Capital Viewpoint', in Riedijk, Willem, Jan Boes, and Wim Ravesteijn [eds.]. *Appropriate Technology in Industrialized Countries*. pp. 35-71. Delft: Delft Univ. Press.

Boon, Louis (1983) *De List der Wetenschap: Variatie en Selectie: Vooruitgang Zonder Rationaliteit*. Baarn: Ambo.

Borish, Steven M. (1991). *The Land of the Living: The Danish Folk High Schools and Denmark's Non-Violent Path to Modernization*. Nevada City, CA: Blue Dolphin.

Bos, Stef (1997) *De Onderstroom* (Compact Disc). Antwerpen: HKM Records.

Briggs, John (1992) *Fractals: The Patterns of Chaos: Discovering a New Aesthetic of Art, Science, and Nature*. London, New York: Simon & Schuster.

Broch, Hermann (1959) *Massenpsychologie*. Zurich.

Brunsted, Birgit *et al.* (November 1979) *OOA's Folkepjece: Danmark Uden Atomkraft*. Copenhagen: Organisationen til Oplysning om Atomkraft.

Bryson, John E. (1980) 'New Directions for Utilities', in: Vaugn Nelson [ed.]. Proceedings of the National Conference, Summer 1980, AWEA, Pittsburgh, Penn, June 8-11, pp. 68-70. Washington, DC: AWEA.

Burn, Duncan (1978) *Nuclear Power and the Energy Crisis*. London: Macmillian.

Burnham, Laurie [ed.] (1992) *Renewable Energy: Sources for Fuels and Electricity*. London: Island.

Bush, Vannevar (1945) *Science: The Endless Frontier*. Washington, DC: U.S. Government.

Callon, M., P. Laredo, V. Rabeharisो, T. Gonard, and T. Leray (1992) 'The Management and Evaluation of Technological Programs and the Dynamics of Techno-Economic Networks: The Case of the AFME'. *Research Policy* No. 21, pp. 215-236.

Caracostas, Paraskevas, and Ugur Muldur (1998) *Society: The Endless Frontier: A European Vision of Research and Innovation Policies for the 21st Century*. Brussels: European Commission.

Carpenter, Stanley R. (1992) 'Instrumentalists and Expressivists: Ambiguous Links Between Technology and Democracy', in Langdon Winner [ed.]. *Democracy in a Technological Society*. pp. 161-174. Dordrecht: Kluwer Academic. Philosophy and Technology, Volume 9

Carson, Rachel (1962) *Silent Spring*. Boston, MA: Houghton, Mifflin.

Cassedy, Edward S., and Peter Z. Grossman (1990) *Introduction to Energy: Resources, Technology, and Society*. Cambridge, UK: Cambridge Univ. Press.

Cassell, Philip [ed.] (1993) *The Giddens Reader*. London: Macmillan.

Cavallo, Alfred J., Susan M. Hock, and Don R. Smith (1992) 'Wind Energy: Technology and Economics', in Laurie Burnham [ed.]. *Renewable Energy. Sources for Fuels and Electricity*. pp. 121-156. London: Island Press.

CEC, California Energy Commission (1977) *1977 Biennial Report: California Energy Trends and Choices. Summary*. Sacramento, CA: CEC.

— (1979) *1979 Biennial Report*. Sacramento, CA: CEC.

— (1980) *Wind Energy Program: Progress Report*. Sacramento, CA: CEC.

— (January 1981) *Decade of the Sun: Program Plan for the Maximum Implementation of Solar Energy Through 1990. Final Report*. Sacramento, CA: CEC.

CEC, (April 1985) *Wind Energy Development in California. Status Report.* Sacramento, CA: CEC.
— (January 1990) *Report of the Statewide Collaborative Process: An Energy Efficiency Blueprint for California.* Sacramento, CA: CEC. Internal Report.
— (December 1992) *Results from the Wind Project Performance Reporting System. 1991 Annual Report.* Sacramento, CA: CEC.
CEERT, Center for Energy Efficiency and Renewable Technologies (1994) 'Special BRPU Edition'. *Coalition Energy News* Winter, pp. 1-9.
Chandler, A. D. (1977) *The Visible Hand: The Managerial Revolution in American Business.* Cambridge, UK: Cambridge Univ. Press.
Child, John (1995) 'Follett: Constructive Conflict', in Pauline Graham [ed.]. *Mary Parker Follett – Prophet of Management: A Celebration of Writitngs from the 1920s.* pp. 87-95. Boston, Mass.: Harvard Business School Press.
Christensen, Jørgen Grønnegård, and Peter Munk Christiansen (1992) *Offentlig Forvaltning i Danmark: Forvaltning og Omgivelser.* Herning: Systime.
Clark, Wilson (1974) *Energy for Survival: The Alternative to Extinction.* Garden City, NY: Anchor Press.
Cohen, Maurice J. (1997) 'Risk Society and Ecological Modernisation: Alternative Visions for Post-Industrial Nations'. *Futures* Vol. 29, No. 2, pp. 105-119.
Constant II, Edward W. (1980) *The Origins of the Turbojet Revolution.* Baltimore: Johns Hopkins Univ. Press.
CPUC, California Public Utilities Commission (February 1993) *California's Electric Industry: Perspectives on the Past, Strategies for the Future.* (*The Yellow Book.*) San Francisco, CA: CPUC, Division of Strategic Planning.
— (April 1994a) *Order Instituting Rulemaking (R.94-04-031) on the Commission's Proposed Policies Governing Restructuring California's Electric Service Industry and Reforming Regulation, and Order Instituing Investigation (I.94-04-032) on the Commission's Proposed Policies Governing Restructuring California's Electric Service Industry and Reforming Regulation.* (*The Blue Book.*) April 20, 1994. San Francisco, CA: CPUC.
— (August 1994b) *CPUC Handbook on the Regulation of California's Public Utilities and Transportation Companies.* San Francisco, CA: Public Affairs Office.
Cronberg, T., and K. H. Sørensen (1995) 'Similar Concerns, Different Styles? A Note on European Approaches to the Social Shaping of Technology', in COST Social Sciences A4, Proceedings of the COST A4 in Ruvaslahti, Finland, January 13-14, 1994.

Dam, Poul (1983) *N. F. S. Grundtvig.* Copenhagen: Ministry of Foreign Affairs.
Dannemand Andersen, Per (1990) 'Licensing Procedures and Technological Innovation', in J. Allesch [ed.]. *Consulting in Innovation: Practice – Methods – Perspectives.* pp. 157-168. Amsterdam: Elsevier Science Publishers.
— (1993) *En Analyse af den Teknologiske Innovation i Dansk Vindmølleindustri – Herunder Prøvestationen for Vindmøllers Dobbeltrolle som Forskningsinstitution og Godkendende Myndighed.* København: Samfundslitteratur.
Davidson, Ros (1988) 'Settled out of Court'. *WindPower Monthly* No. 8, p.11.
— (1989a) 'Shaping up for a New Future'. *WindPower Monthly* No. 5, pp. 14-16.
— (1989b) 'Danish Firm Bows out of California'. *WindPower Monthly* No. 7, p. 17.
— (1989c) 'Think Tank Pioneers Set out to Produce a World Beater'. *WindPower Monthly* No. 11, p. 11.
— (1990a) 'Protestors Raging over SERI Choice'. *WindPower Monthly* No. 3, p. 17.
— (1990b) 'Fayette in Trouble Again'. *WindPower Monthly* No. 10, p. 15.
— (1996) 'Kenetech Files for Bankruptcy Protection'. *WindPower Monthly* No. 6, pp. 20-21.

DEA, Danish Energy Agency (March 1979) *Energiforskning. Handelsministeriets Energiforskingsprogram.*
DEF, Association of Danish Electric Utilities (1971a-1995a). *Annual Reports.*
— (1973b) *Uddrag af Vindkraftudvalgets Betænkning af 1962 (Betækningens hovedafsnit).*
— (August 18, 1976b) *Tilslutning af Vindmøller: Mindre, Private Dieselanlæg og Lignende til Elproduktion.* Cirk. Nr. S 33/1976. (*DEF's Provisional 1976-Guidelines*)
— (May 11, 1984b) *Draft for Agreement Concerning Settlement with and Connection of Wind Power Plants Producing Electricity.* (*The 1984-agreement*)
— (1994b) *Electricity Supply in Denmark.*
DEFU, Research Association of Danish Electric Utilities (February 1979). *Handelsministeriets og Elværkernes Vindkraftprogram.* Lyngby: DEFU.
Dewey, John (1910) *How We Think.* Boston, MA: D.C. Heath.
Dickson, David (1974) *Alternative Technology and the Politics of Technical Change.* Glasgow: Fontana/Collins.
Divone, Louis (1975) 'Workshop Introduction', in Proceedings of the Second Workshop on Wind Energy Conversion Systems. Washington, DC, June 9-11, 1975, p.1. Washington, DC: NSF/ERDA.
Doctor, R. D., *et al.* (September 1972) *California's Electricity Quandary: III. Slowing the Growth Rate.* Santa Monica, CA: Rand Corporation.
Dodgson, Mark, and John Bessant (1996) *Effective Innovation Policy: A New Approach.* London: International Thomson Business Press.
Dorf, Richard C. (1978) *Energy, Resources, & Policy.* Reading, MA: Addison-Wesley.
Dosi, Giovanni, Christopher Freeman, Richard Nelson, Gerald Silverberg, and Luc Soete [eds.] (1988) *Technical Change and Economic Theory.* London: Pinter.
Douglas, Mary and Aaron Wildavsky (1982) *Risk and Culture: An Essay on the Selection of Technological and Environmental Dangers.* Berkeley: Univ. of California Press.
Dowie, Mark (1995) *Losing Ground: American Environmentalism at the Close of the Twentieth Century.* Cambridge, Mass., London, England: MIT Press.
Drager, Kerry (1979) 'Despite Promotion by the Brown Administration and a Substantial State Income-Tax Credit, There Has Not Been Rapid Progress toward a Solar California'. *California Journal* November, pp. 393-395.
Drejer, Anders (1996) 'Frameworks for the Management of Technology: Towards a Contingent Approach'. *Technology Analysis & Strategic Management* Vol. 8, No.1, pp. 9-20.
Dror, Yehezkial (1973) *Public Policy-Making Reexamined.* Bedfordshire, England: Leonard Hill Books.
Dunlap, Riley E., and Angela G. Mertig [eds.] (1992) *American Environmentalism: The U.S. Environmental Movement, 1970-1990.* Philadelphia: Taylor and Francis.

Eberg, Jan, Rinie van Est, and Henk van de Graaf [eds.] (1996) *Leren met Beleid: Beleidsverandering en Beleidsgericht Leren bij Nimby-, Milieu- en Technologiebeleid.* Amsterdam: Spinhuis.
Edson, Karen K. (1983) 'California's Wind Industry: The Sacaramento Perspective', in Proceedings Wind Energy Expo '83 and National Conference, AWEA. San Francisco, CA, October 17-18, pp. 221-229. Washington, DC: AWEA.
Edquist, Charles and Bengt-Åke Lundvall (1993) 'Comparing the Danish and Swedish Systems of Innovation', in Richard R. Nelson [ed.]. *National Innovation Systems: A Comparative Analysis.* pp. 265-298. New York, Oxford: Oxford Univ. Press.
Elkraft (March 1993a) *Plan-92/93: Elkraft's Development Plan for 1992 and 1993.*
— (1993b) *Annual Report 1993.*

Ellul, Jacques (1964 [1954]) *The Technological Society*. New York: Vintage Books.
Elmore, R. F. (1980) 'Backward Mapping: Implementation Research and Policy Decisions'. *Political Science Quarterly* No. 94, pp. 601-616.
EM, Ministry of Energy (November 1981) *Energiplan 81*.
— (December 1985) *Aftale Mellem Energiministeriet og Elsam, Elkraft, Danske Elværkers Forening om Øget Anvendelse af Vedvarende og Indenlandske Energikilder i Elproduktionen*. (*The First 100-MW Agreement*)
— (March 13, 1990) *Rapport om Elværkernes Vindkraftudbygning. København: Energiministeriet*, Arbejdsgruppen Vedrørende Elværkernes Vindkraftudbygning.
— (April 1990) *Energy 2000. A Plan of Action for Sustainable Development*. (*The Green Plan*) København: Energimisteriet.
Energistyrelsen (Juni 1993) *Rapport fra Styregruppe om Udskiftning af Gamle Vindmøller.*
— (December 1994) *Vindmøller i Danske Farvande – En Undersøgelse af de Visuelle Forhold ved Opstilling af Vindmøller på Havet*. Miljø- og Energiministeriets udvalg om Havbaserende Vindmøller.
Ergas, Henry (1986) *Does Technology Policy Matter?* CEPS Papers No. 29. Brussels: Centre for European Policy Studies.
Evans, Nigel, and Chris Hope (1984) *Nuclear Power: Future, Costs and Benefits*. Cambridge, UK: Cambridge Univ. Press.

FDV, Foreningen af Danske Vindmøllefabrikanter (August 1994) *Perspektiv 2004: En Vurdering af Dansk Vindmølleindustris Muligheder frem til År 2004*.
Federal Energy Administration (November 1974) *Project Independence Blueprint. Final Task Force Report. Solar Energy*.
Fenn, Scott (1984) *America's Electric Utilities: Under Siege and in Transition*. New York: Praeger.
Ferguson, Paul R., and Glenys J. Ferguson (1994) *Industrial Economics: Issues and Perspectives*. London: Macmillan.
Ferguson, Rich (1993) 'Should California Retire Its Nuclear Power Plants?' *Coalition Energy News* Fall, pp. 1-3.
Feynman, Richard (1988) *What Do You Care What Other People Think?* New York: W. W. Norton.
Fischer, Frank (1980) *Politics, Values and Public Policy: The Problem of Methodology*. Boulder, Colo: Westview.
— (1985) 'Critical Evaluation of Public Policy: A Methodological Case Study'. Paper presented at the 1985 meeting of the International Political Science Association (IPSA) in Paris, July 15-20.
— (1994) *Evaluating Public Policy*. Boulder, Colo: Westview.
Fischer, Jean (1979) 'Mit Plusenergi-Hus', in Inger Christensen, Niels I. Meyer, and Ole Thyssen. Energi [eds.]. *Skriftrækken Krise og Utopi*. København: Gyldendal.
Flavin, Christopher (July 1981) *Wind Power: A Turning Point*. Washington, DC: Worldwatch Institute. (Worldwatch Paper 45)
Follett, Mary Parker (1918) *The New State: Group Organization, the Solution for Popular Government*. New York: Longmans, Green.
— (1924) *Creative Experience*. New York: Longmans, Green.
Fonsmark, Henning (1991) *Historien om den Danske Utopi: Et Idépolitisk Essay om Danskernes Velfærdsdemokrati*. København: Gyldendal.
Frank, Ellen (1978) 'The Whiz Kid Energycologist'. *New Times* August 21, pp. 37-46.
Freeman, Christopher, and Bengt-Åke Lundvall [eds.] (1988) *Small Countries Facing the Technological Revolution*. London and New York: Pinter.
Fuchs, Doris Andrea (1995) 'Energy Restructuring in California: Environmental Implications'.

Paper presented at the Euroconference "Sustainability Dimensions of Energy Policy." Roskilde, Denmark. March 13-24, 1995.
Fukuyama, Francis (1989) 'The End of History?' *The National Interest* No. 16, pp. 3-18.

Galbraith, John Kenneth (1963) *The Affluent Society*. New York: NAL/Dutton.
Gantenbein, Douglas (1994) 'Running with the Wind'. *Alaska Airlines Magazine*. September, pp. 80-84.
Giddens, Anthony (1976) *New Rules of Sociological Method: A Positive Critique of Interpretative Sociologies*. London: Hutchinson; New York: Harper & Row.
— (1979) *Central Problems in Social Theory: Action, Structure and Contradiction in Social Analysis*. London: Macmillian.
— (1982) *Profiles and Critiques in Social Theory*. London: Macmillan.
— (1984) *The Constitution of Society: Outline of a Theory of Structuration*. Cambridge, UK: Polity Press.
— (1990) *The Consequences of Modernity*. Cambridge, UK: Polity Press.
Giedion, Siegfried (1948) *Mechanization Takes Command*. New York: Oxford Univ. Press.
Gipe, Paul (1983) *Wind Energy: How to Use It*. Harrisburg, PA: Stackpole Books.
— (1995) *Wind Energy Comes of Age*. New York: John Wiley & Sons.
Glaser, Barney, and Anselm Strauss (1967) *The Discovery of Grounded Theory*. Chicago: Aldine.
Godtfredsen, Finn, Peter Hjuler Jensen, and Poul Erik Morthorst (1993) *Wind Energy in Denmark: Development in Wind Turbine Technology and Economics since 1990*. Risø: Risø National Laboratory. (Internal Paper)
Graham, Pauline [ed.] (1995) *Mary Parker Follett – Prophet of Management: A Celebration of Writings from the 1920s*. Boston, Mass: Harvard Business School Press.
Grastrup, Henning, Asger Jensen, Mogens Johansson, B. Maribo Pedersen, and Vagn E. Rasmussen (1977) *Rejserapport fra Vindkraft-Studiebesøg i USA 2.-10. Maj 1977*. Lyngby: DEFU.
Grastrup, Henning, and Poul Nielsen (1990) *Large-Scale Wind Turbines. In: Wind Energy in Denmark. Research and Technological Development, 1990*. Ministry of Energy, Danish Energy Agency.
Grin, John, and Henk van de Graaf (1994) 'Handelingstheorieën en Beïnvloeding in Netwerken: Ongelijksoortige Rationaliteiten en Congruente Betekenissen'. *Beleidswetenschap* No. 4, pp. 349-366.
— (1996) 'Technology Assessment as Learning'. *Science, Technology, & Human Values* No. 1, pp. 72-99.
Grin, John, and Rob Hoppe (1997) 'Towards a Theory of the Policy Process: Problems, Promises and Prospects of the ACF'. Paper presented at the ARIPS Meeting 'Seminar on Policy learning and the Use of the Advocacy Coalition Framework (ACF).' Amsterdam, February 4, 1997.
Grin, John, Henk van de Graaf, and Rob Hoppe (1997) *Technology Assessment through Interaction. A Guide*. The Hague: Rathenau Institute. (Working Document Rathenau Institute, W57)
Griswold, A. Whitney (1948) *Farming and Democracy*. New York.
Gundelach, Peter (1993) 'Danskernes Særpræg'. *Sociologi* No. 2, pp. 6-9.

Habermas, Jürgen (1970) *Toward a Rational Society*. Boston: Beacon.
Hagensen, Flemming (1992) 'Vindkraft i Københavns Sydhavn'. *VEinformation* No. 49, pp. 6-7.
Hajer, Maarten A. (1993) *The Politics of Environmental Discourse: A Study of the Acid Rain Controversy in Great Britain and the Netherlands*. Oxford: Univ. of Oxford.
Hall, Stuart, and Bram Gieben [eds.] (1992) *Formations of Modernity*. Open Univ.: Polity Press, Blackwell.
Halligan, Julie (October 1993) *Biennial Resource Plan Update: A Primer*. San Francisco, CA:

CPUC, Commission Advisory and Compliance Division, Environmental and Energy Advisory Branch.
Hansen, Preben, and Flemming A. Larsen (1983) *Factsheet Denmark: N. F. S. Grundtvig.* Copenhagen: Danish Ministry of Foreign Affairs.
Hansen, Jacob L. (1993) 'Samarbedje med Folketinget på Sidelinien'. *Naturlig Energi* No. 9, p. 7.
Harbers, Hans (1997) 'Technologie Is Politiek met Andere Middelen'. *Tijdschrift voor Wetenschap, Technologie en Samenleving* No. 5, pp. 158-161.
Harding, Jim (1983) 'PG&E Signs Big Wind Contract'. *Soft Energy Notes* February/March, p. 143.
Hart, John [ed.] (1984) *The New Book of California Tomorrow.* Los Altos, CA: William Kaufmann.
Hartz, Louis (1955) *The Liberal Tradition in America: An Interpretation of American Political Thought since the Revolution.* New York: Harcourt, Brace & World.
Harvey, David (1990) *The Condition of Modernity: An Enquiry into the Origins of Cultural Change.* Cambridge, Mass. & Oxford, UK: Blackwell.
Harvey, Richard B. (1985) *The Dynamics of California Government and Politics. Second Edition.* Belmont, CA: Wadsworth.
Heclo, Hugh (1974) *Social Policy in Britain and Sweden.* New Haven: Yale Univ. Press.
Heerding, A. (1980) *Geschiedenis van de N.V. Philips' Gloeilampen-Fabrieken: Deel 1.* Den Haag: Philips.
Heymann, Matthias (1992) *Why Were the Danes the Best? Social Determinants of Wind Turbine Development in Denmark, West Germany, and the USA.* Stuttgart: Univ. of Stuttgart – Institute for Energy Economics and the Rational Use of Energy (IER). (Unpublished manuscript)
— (1995) *Die Geschichte der Windenergienutzung 1890-1990.* Frankfurt: Campus.
Hill, Gladwin (1968) *Dancing Bear: An Inside Look at California Politics.* New York: World.
Himmelstrup, Per, Kate Hegelund, and Helle Askgaard (eds.) (1992) *Discover Denmark: On Denmark and the Danes; Past, Present and Future.* Copenhagen: The Danish Cultural Institute/Systime.
Hirsh, Richard F. (1989) *Technology and Transformation in the American Electric Utility Industry.* Cambridge, UK: Cambridge Univ. Press.
HM, Ministry of Trade (May 1976) *Dansk Energipolitik 1976.*
Hofstadter, Richard (1955) *The Age of Reform.* New York: Vintage Books.
Hoppe, Rob, Henk van de Graaf, and Asje van Dijk (1987) 'Implementation Research and Policy Design: Problem Tractability, Policy Theory, and Feasibility Testing'. *International Review of Administrative Sciences* No. 53, pp. 581-604.
Hoppe, Rob, and Aat Peterse (1993) *Handling Frozen Fire: Political Culture and Risk Management.* Boulder, San Francisco, Oxford: Westview.
Horn, David G. (1994) 'Book Review of Shaping Technology/Building Society, by Wiebe E. Bijker and John Law [eds.] (1992)'. *Science, Technology, & Human Values* Vol. 19, No. 3, pp. 386-388.
Huber, Joseph (1985) *Die Regenbogengesellschaft: Ökologie und Sozialpolitik.* Frankfurt am Main: Fischer.
Hughes, Thomas P. (1983) *Networks of Power: Electrification in Western Society, 1880-1930.* Baltimore, London: John Hopkins Univ. Press.
— (1986) 'The Seamless Web: Technology, Science, Etcetera, Etcetera'. *Social Studies of Science* No. 16, pp. 281-292.
Hvelplund, Frede, Klaus Illum, Johannes Jensen, Niels I. Meyer, Jørgen S. Nørgård, and Bent Sørensen (1983) *Energi for Fremtiden: Alternative Energiplan 1983.*
Hyman, Leonard S. (1988) *America's Electric Utilities: Past, Present and Future.* Arlington, Va: Public Utilities Reports.
Høeg, Peter (1996) *The History of Danish Dreams.* London: Harvill Press.

IEEE Power Engineering Society (1974) *Credibility and the Public Trust. What Can the Power Industry and the Media Do?* Proceedings Summer Meeting and Energy Resources Conference. Anaheim, CA. Piscataway: IEEE.

IEP, Independent Energy Producers Association (October 1993) *Independent Power: A California Success Story*. Sacramento, CA: IEP.

Ihde, Don (1990) *Technology and the Lifeworld*. Bloomington, Minneapolis: Indiana Univ. Press.

— (1993) *Postphenomenology*. Evanston, IL: Northwestern Univ. Press.

IRP Project Group (April 1993) *Situation Analysis for Integrated Resource Planning in the Danish Electric Utilities*. IRP-project, Distribution Companies in East and West Denmark, ELSAM, and ELKRAFT.

Jablonski, Lori (1995) 'Competition? Only If You Lose!' *Coalition Energy News* Spring, pp. 6-7.

Jamison, Andrew, and Jeppe Læssøe (1990) 'The Making of the New Environmentalism in Denmark', in Andrew Jamison, Ron Eyerman, Jacqueline Cramer, and Jeppe Læssøe. *The Making of the New Environmental Consciousness: A Comparative Study of the Environmental Movements in Sweden, Denmark and the Netherlands*. pp. 66-120. Edinburgh: Edinburgh Univ. Press.

Jensen, Peter Hjuler, Ture Falbe-Hansen, and Kurt Mortensen (1990) 'Offshore Wind Power in Denmark', in *Wind Energy in Denmark. Research and Technological Development, 1990*. Ministry of Energy, Danish Energy Agency.

Jespersen, Kim Dyre (1991) 'Vindeby: The First Offshore Wind Farm in the Danish Wind Power Program', in Proceedings Windpower '91. AWEA, DOE, SERI, Palm Springs, CA, September 24-27.

— (1992) 'Vindeby Offshore Wind Farm: The First Experiences', in Proceedings Special Topic Conference '92, EWEA. Herning, Denmark. September 8-11, 1992.

Johansson, Mogens (December 1975) *Vindkraftforskning i USA og Canada: Referat af Studierejse til USA og Canada, 17-28 oktober 1975*. Lyngby: DEFU.

Johnson, Huey D. (1979) 'Wind Energy: A Sound Alternative for California's Energy Future', in Vaughn Nelson [ed.]. Proceedings of the National Conference, Spring 1979, AWEA. San Francisco, CA, April 16-19, pp. 48-51. Washington, DC: AWEA.

Juul, Johannes (1961) 'Design of Wind Power Plants in Denmark', in U.N. Conference on New Sources of Energy. Vol. 7, pp. 229-239. Rome: United Nations.

— (1961) 'Economy and Operation of Wind Power Plants', in U.N. Conference on New Sources of Energy. Vol. 7, pp. 399-408. Rome: United Nations.

Jørgensen, Ulrik, and Peter Karnøe (1995) 'The Danish Wind-Turbine Story: Technical Solutions to Political Visions?', in Arie Rip, Thomas J. Misa and Johan Schot [eds.]. *Managing Technology in Society. The Approach of Constructive Technology Assessment*. pp. 57-82. London, New York: Pinter.

Kaplan, Thomas J. (1986) 'The Narrative Structure of Policy Analysis'. *Journal of Policy Analysis and Management*. Vol. 5, No. 5, pp. 761-778.

Karnøe, Peter (1991) *Dansk Vindmølleindustri: En Overraskende International Succes: Om Innovationer, Industriudvikling og Teknologipolitik*. København: Samfundslitteratur.

— (1993) *Approaches to Innovation in Modern Wind Energy Technology: Technology Policies, Science, Engineers and Craft Traditions*. Stanford, CA: Stanford Univ. – Center for Economic Policy Research. (CEPR Publication No. 334)

— (1995) 'Institutional Interpretations and Explanations of Differences in American and Danish Approaches to Innovation', in W. Richard Scott and Søren Christensen [eds.]. *The Institutional Construction of Organizations: International and Longitudinal Studies*. pp. 243- 276. Thousand Oaks, London, New Delhi: Sage.

Karnøe, Peter (1996) 'The Social Process of Competence Building'. *International Journal of Technology Management* Vol. 11, No. 7/8, pp. 770-789.

Kenetech (1993) *Annual Report 1993: Energy That Makes a Difference*. San Francisco, CA: Kenetech.

Kickert, W.J.M. (1993) 'Scheiding tussen Beleid en Uitvoering in Denemarken'. *Bestuurskunde* No. 7, pp. 330-340.

Kingdon, John (1984) *Agendas, Alternatives and Public Policies*. Boston, Toronto: Little, Brown.

Kingsford, P.W. (1969) *Electrical Engineering: A History of the Men and the Ideas*. New York: St. Martin Press.

Kjær, Tyge (September 1988) *Strukturanalyse af Dansk Vindmølleindustri. Udarbejdet for Foreningen af Danske Vindmøllefabrikanter*. Roskilde: Institut for Miljø, Teknologi & Samfund.

Koeppl, Gerard W. (1982) *Putnam's Power from the Wind. Second Edition*. New York: Van Nostrand Reinhold.

Krieger, Martin H. (1981) *Advice and Planning*. Philadelphia: Temple Univ. Press.

Kristensen, Peer Hull (1989) 'Denmark's Concealed Production Culture: Its Socio-Historical Construction and Dynamics at Work', in Finn Borum and Peer Hull Kristensen [eds.]. *Technological Innovation and Organizational Change: Danish Patterns of Knowledge, Networks and Culture*. pp. 165-188. Gylling, DK: Narayana.

— (1995) 'Denmark: Many Small Worlds', in Arnaldo Bagnasco and Charles F. Sabel [eds.]. *Small and Medium-Size Enterprises*. London, England: Pinter.

Lakatos, Imre (1976) 'Falsification and the Methodology of Scientific Research', in I. Lakatos and A. Musgrave [eds.]. *Criticism and the Growth of Knowledge*. pp. 97-196. Cambridge, UK: Cambridge Univ. Press.

Lamar (1986) 'Gone With the Wind'. *Time*, January 20.

Lamarre, Leslie (1992) 'A Growth Market in Windpower'. *EPRI Journal*, December, pp. 4-15.

Lambright, W. Henry, Michael Crow, and Ralph Shangraw (1988) 'National Projects in Civilian Technology', in J. David Roessner [ed.]. *Government Innovation Policy: Design, Implementation, Evaluation*. Houndsmills, Basingstoke, Hampshire: Macmillan Press.

Lamott, Kenneth (1971) *Anti-California: Report from Our First Parafascist State*. Boston: Little, Brown.

Larédo, Philippe, and Philippe Mustar (1996) 'The Techno-Economic Network: A Socio-Economic Approach to State Intervention in Innovation', in Rod Coombs, Albert Richards, Pier Paolo Saviotti, and Vivien Walsh [eds.]. *Technological Collaboration: The Dynamics of Cooperation in Industrial Innovation*. pp. 143-164. Cheltenham, UK, Brookfield, VT: Edward Elgar.

Larsen, Ejvind (1983) *Det Levende Ord*. København: Rosinante.

Lawson, Max (1992) 'Profiles of Educators. N.F.S. Grundtvig (1783-1872)'. *Prospects* Vol. XXII, No. 1, pp. 113-119.

Leary, Mary Ellen (1980) 'California's Public Utilities Commission Has Determined That the Construction of New Power Plants in the State Would Be "Too Financially and Environmentally Expensive for Californians." Backed by Sudden Support from Utility Companies, the CPUC is Bucking the Desires of the Federal Government and Is Taking Bold Steps to Assure That California's Electricity Needs for Future Years Will Be Provided through Energy Conservation'. *California Journal*, June, pp. 219-223.

Lee, Eugene C. [ed.] (1966) *The California Governmental Process: Problems and Issues*. Boston, MA and Toronto: Little, Brown.

Lee, Thomas H., Ben C. Ball, Jr., and Richard D. Tabors (1990) *Energy Aftermath: How We Can*

Learn From the Blunders of the Past and Create a Hopeful Energy Future. Boston: Harvard Business School Press.

Lemaire, T. (1986) *De Indiaan in Ons Bewustzijn.* Baarn: Ambo.

Leonard, George B. (1962) 'California'. *Look*, September, pp. 30-31. (Reprinted as 'Forces Shaping the Future', in Eugene C. Lee [ed.] (1966). *The California Governmental Process: Problems and Issues.* pp. 11-13. Boston: Little, Brown.)

Lerner, James (July 1978) *Wind-Electric Power: A Renewable Energy Source for California.* California Energy Commission, Solar Office, Wind Energy Program.

Lerner, James, and Mat Ginosar (1979) 'A Proposed Wind Energy Program for the State of California', in Vaughn Nelson [ed.]. Proceedings of the National Conference, Fall 1978, AWEA, Cape Cod, MA, September 25-27, pp. 132-142. Washington, DC: AWEA.

Lindblom, Charles E. (1965) *The Intelligence of Democracy: Decision-Making through Mutual Adjustment.* New York: Free Press.

Lindsey, Quentin W. (1988) 'Technological Innovation Policy: The Role of the States', in J. David Roessner [ed.]. *Government Innovation Policy: Design, Implementation, Evaluation.* Houndsmills, Basingstoke, Hampshire: Macmillan Press.

Lintsen, Harry W., M.S.C. Bakker, E. Homburg, D. van Lente, J.W. Schot, and G.P.J. Verbong [eds.] (1993) *Geschiedenis van de Techniek in Nederland: De Wording van een Moderne Samenleving, 1800-1890. Deel III: Katoen – Gas, Licht en Elektriciteit – Bouwen.* Zutphen: Walburg.

Lipset, Seymour Martin (1996) *American Exceptionalism: A Double-Edged Sword.* New York, London: W.W. Norton.

Lundvall, Bengt-Åke (1988) 'Innovation As an Interactive Process: From User-Producer Interaction to the National System of Innovation', in Giovanni Dosi, Christopher Freeman, Richard Nelson, Gerald Silverberg, and Luc Soete [eds.]. *Technical Change and Economic Theory.* London: Pinter.

Lynn, Laurence E. (1981) *Managing the Public's Business. The Job of the Government Executive.* New York: Basic Books.

— (1987) *Managing Public Policy.* Boston, Toronto: Little, Brown.

Madsen, Birger T. (1988) 'Too Much Haste'. *WindPower Monthly* No. 4, pp. 8-9.

— (1991) 'The Danish Wind Power Industry 1991: Product and Market Development', in Proceedings Windpower '91, AWEA, DOE, SERI. Palm Springs, CA, September 24-27, pp. 82-87.

— (1993) 'Tiden inde for DV's Professionalisering'. *Naturlig Energi* No. 9, pp. 8-9.

— (1996) 'Dansk Vindkraft Har 60% af Verdensmarkedet'. *Vindformation* No. 6, pp. 1-11.

Madsen, Birger T., and Asbjørn Bjerre (1993) 'Mulighederne i Udskifting, Energipolitisk Set?' *Naturlig Energi* No. 11, p. 11.

Madsen, Birger T., Flemming Rasmussen, and Peter Hauge Madsen (1990) 'Technology Development for Commercial Wind Turbines', in *Wind Energy in Denmark. Research and technological development, 1990.* Ministry of Energy, Danish Energy Agency.

Majone, Giandomenico (1980) 'Policies As Theories'. *Omega* No. 8, pp. 151-162.

— (1989) *Evidence, Argument, and Persuasion in the Policy Process.* New Haven, London: Yale Univ. Press.

March, Frederic, Edward H. Dlott, Donald H. Korn, Frederick R. Madio, Robert C. McArthur, and William A. Vachon (1982) *Wind Power for the Electric Utility Industry. Policy Incentives for Fuel Conservation.* Lexington, MA: D.C. Heath.

March, James G., and Johan P. Olsen (1989) *Rediscovering Institutions: The Organizational Basis of Politics.* New York: The Free Press.

Maribo Pedersen, B. (1990) 'Wind Energy in Denmark', in *Wind Energy in Denmark. Research and Technological Development, 1990.* Ministry of Energy, Danish Energy Agency.
Mazmanian, Dan (1992) 'Toward a New Energy Paradigm', in John J. Kirlin [ed.]. *California Policy Choices. Vol. 8*, pp. 203-215. Los Angeles, Sacramento, Washington, DC: Univ. of Southern California, School of Public Administration.
McCarthy, T.A. (1973) 'A Theory of Communicative Competence'. *Philosophy of the Social Sciences* No. 3, pp. 135-156.
McCormack, Mike (1975) 'Guest Speech', in Proceedings of the Second Workshop on Wind Energy Conversion Systems. Washington, DC, June 9-11, pp. 72-78. Washington, DC: NSF. NSF-RA-N-75-050.
McLellan, D. (1971) *The Thought of Karl Marx.* London: Macmillan.
Meyer, Niels I., K. Helveg Petersen, and Villy Sørensen (1981) *Revolt From the Center.* London, Boston: Marion Boyars.
Meyer, Niels I., Olav Benestad, Lennart Emborg, and Ervind Selvig (1993) 'Sustainable Energy Scenarios for the Scandinavian Countries'. *Renewable Energy* Vol. 3, No. 2/3, pp. 127-136.
Miller, Albert, and Richard Simon (1978) *Wind Power Potential in California.* California Energy Commission, Solar Office, Wind Energy Program.
Miller, Kenneth E. (1996) *Friends and Rivals: Coalition Politics in Denmark, 1901-1995.* Lanham, New York, London: Univ. Press of America.
Mirviss, Alan H. (1983) 'Tax Credits Under Fire'. *Soft Energy Notes* No. 2, pp. 2-3.
Mitchell, W.J.T. [ed.] (1981) *On Narrative.* Chicago: Univ. of Chicago Press.
MM, Ministry of Environmental Affairs (June 1991) *Betænkning I fra Vindmølleplaceringsudvalget. 100 MW i Elværksregi inden 1994.* Miljøministeriet, Planstyrelsen.
— (February 1992) *Betænkning II fra Vindmølleplaceringsudvalget.* Miljøministeriet, Planstyrelsen.
Mooz, W.E., and C.C. Mow (September 1972) *California's Electricity Quandary: I. Estimating Future Demand.* Santa Monica, CA: Rand Corporation.
Morgan, Gareth (1986) *Images of Organization.* Beverly Hills, Newbury Park, London, New Delhi: Sage.
Mowery, David C. and Nathan Rosenberg (1993) 'The U.S. National Innovation System', in Richard R. Nelson [ed.]. *National Innovation Systems: A Comparative Analysis.* pp. 29-75. New York, Oxford: Oxford Univ. Press.
Mowry, George (1951) *The California Progressives.* Berkeley, CA: Univ. of California Press.
Mumford, Lewis (1964) 'Authoritarian and Democratic Technics'. *Technology and Culture* No. 5, pp. 1-18.
Munro, John F. (1993) 'California Water Politics: Explaining Policy Change in a Cognitively Polarized Subsystem', in Paul A. Sabatier and Hank C. Jenkins-Smith [eds.]. *Policy Change and Learning. An Advocacy Coalition Approach.* pp. 105-127. Boulder, San Francisco, Oxford: Westview.
Mørup-Petersen, Vilhelm and Søren Pedersen (1992) 'Vindmølleprojekter i Samarbejde mellem NESA og Private Vindmølle'. Paper presented at the EWEA-conference *The Potential of Wind Farms.*

Nagel, Stuart S. (1987) 'Series Editor's Introduction', in Frank Fischer and John Forester [eds.]. *Confronting Values in Policy Analysis: The Politics of Criteria.* pp. 7-9. Newbury Park, CA, Beverly Hills, CA, London, New Dehli: Sage.
Nalder, Ian A. (1993) 'Ignoring Limits to Growth'. *IEEE Technology and Society Magazine*, Summer, pp. 24-28.
Nelson, Richard R. [ed.] (1993) *National Innovation Systems: A Comparative Analysis.* Oxford, New York: Oxford Univ. Press.

Nelson, Richard R. and Nathan Rosenberg (1993) 'Technological Innovation and National Systems', in Richard R. Nelson [ed.]. *National Innovation Systems: A Comparative Analysis*. pp. 3-22. New York, Oxford: Oxford Univ. Press.

Nielsen, T.W. (1984) 'På Vej mod en Folkelig Teknologi', in F. Slumstrup [ed.] (1984). *Højskolen til Debat*. København: Vindrose.

Noordegraaf, M. and M.M. Teeuw (1997) 'Reciprociteit en Integratie: Het Realistisch Idealisme van Mary Parker Follett'. *Bestuurskunde* No. 2, pp. 67-76.

Nordtank Energy Group (October 17, 1995) *Offering Circular 1995*. Copenhagen: Alfred Berg Bank A/S.

NRC, Nuclear Regulatory Commission (1975) *Reactor Safety Study: An Assessment of Accident Risks in U.S. Commercial Nuclear Power Plants*. Washington, DC: NRC. (WASH-1400)

NSF/NASA Solar Energy Panel (December 1972) *An Assessment of Solar Energy as a National Energy Resource*.

NSF/NASA, National Science Foundation/National Aeronautics and Space Administration (December 1973) *Wind Energy Conversion Systems: Workshop Proceedings*. Washington, DC: NSF. (U.S. Department of Commerce. National Technical Information Service. PB-231 341)

Nørgård, Jørgen S., and Bente L. Christensen (1993) 'Towards Sustainable Energy Welfare'. *Perspectives in Energy* Vol. 2, pp. 313-332.

Nørgård, Jørgen S., P.S. Nielsen, and J. Viegand (1994) 'Low Electricity Europe', in Anibal T. de Almeida, Arthur H. Rosenfeld, Jacques Roturier, and Jørgen S. Nørgård [eds.]. *Integrated Electricity Resource Planning*. pp. 399-417. Dordrecht: Kluwer Academic.

OAT, Office of Appropriate Technology (1979) *Present Value: Constructing a Sustainable Future*.

— (1981) *Working Together: Self-Reliance in California's Communities*.

Olse, Ole Jess (1995) 'The Danish Electricity Supply Industry and the European Internal Market'. Roskilde Univeristy, Department of Economics and Planning. (Internal Paper)

Paris, Ellen (1984) 'The Great Windmill Tax Dodge'. *Forbes*, March 12, pp. 39- 40.

Park, Jack, and John Obermeier (1982) *Common Sense Wind Energy*. Sacramento, CA: State of California, OAT/CEC.

Petersen, Flemming, Jytte Thorndahl, and Birgitte Wistoft (1993) *Skrifter fra Elmuseet (4): Som Vinden Blæser*. Elmuseet.

Petersen, Henrik Kragelund, and Per Dannemand Andersen (1995) *Wind Turbine Characteristics and Costs, 1982-1995*. Roskilde, DK: Risø. (Risø-R-837, EN).

Pfeffer, J., and G.R. Salancik (1978) *The External Control of Organizations*. New York: Harper & Row.

Piore, Michael J., and Charles F. Sabel (1984) *The Second Industrial Divide: Possibilities for Prosperity*. New York: Basic Books.

Pirsig, Robert M. (1991) *Lila: Een Onderzoek naar Zeden*. Amsterdam: Bert Bakker.

Plummer, James L., Eugene Oatman, and Pradeep K. Gupta (1985) *Strategic Management and Planning for Electric Utilities*. Englewood Cliffs, New Jersey: Prentice-Hall.

President's Materials Policy Commission (*The Paley Commission*) (1952) *Resources for Freedom. Vol. 1: Foundations for Growth and Security*. Washington, DC: U.S. Government Printing Office.

Pressman, Jeffrey L., and Aaron Wildavsky (1979, 2nd ed.) *Implementation: How Great Expectations in Washington Are Dashed in Oakland; Or, Why it's Amazing that Federal Programs Work at All, This Being a Saga of the Economic Development Administration as Told by Two Sympathetic Observers Who Seek to Build Morals on a Foundation of Ruined Hopes*. Berkeley, Los Angeles, London: Univ. of California.

Pursell, Carroll (1993) 'Presidential Address: The Rise and Fall of the Appropriate Technology Movement in the United States, 1965-1985'. *Human Technology* No. 5, pp. 629-637.

Quade, E.S. (1975) *Analysis for Public Decisions.* New York: Elsevier.

Rasmussen, Bent and Flemming Øster (1990) 'Power Production from the Wind', in *Wind Energy in Denmark: Research and Technological Development, 1990.* Copenhagen: Ministry of Energy, Danish Energy Agency.

Rasmussen, Jørgen (1982) *Energien til Magten: En Analyse af de Økonomiske, Politiske og Organisatoriske Forhold Omkring de Danske Elektricitetsselskabers Udvikling fra 1890 og til i Dag.* Aalborg: Aalborg Universitetsforlag. (Serien om Energi nr. 5)

Ravesteijn, Wim (1989) 'Technology and Culture: Impulse towards a Cultural-Anthropological Approach to Appropriate Technology', in Willem Riedijk, Jan Boes, and Wim Ravesteijn [eds.]. *Appropriate Technology in Industrialized Countries.* pp. 3-34. Delft: Delft Univ. Press.

Rawls, John (1971) *A Theory of Justice.* Cambridge, Mass.: Harvard Univ. Press

Reich, Robert B. (1983) *The Next American Frontier.* New York: Times Books.

Rein, Martin and Donald A. Schön (1986) 'Frame-Reflective Policy Discourse'. *Beleidsanalyse* No. 4, pp. 4-18.

Reinke, Stefan (1980) 'The Pieces in Huey Johnson's Chess Game: Guide to Who Does What among Major Environmental Organizations'. *California Journal* February, pp. 87-89.

Reitsma, H.A., and J.M.G. Kleinpenning (1991) *The Third World in Perspective.* Assen, Maastricht: Van Gorcum.

Riedijk, Willem (1986) *Technology for Liberation. Appropriate Technology for New Employment.* Delft: Delft Univ. Press.

Riedijk, Willem, Jan Boes, and Wim Ravesteijn [eds.] (1989) *Appropriate Technology in Industrialized Countries.* Delft: Delft Univ. Press.

Rios, Delia M. (1981) 'California's Love Affair with Small Hydropower Plants'. *California Journal*, February, pp. 81-82.

Risø (1983) *Risø Mellem Fortid og Fremtid: Risøs 25 Års Jubilæum den 6. Juni 1983.* Roskilde: Forsøgsanlæg Risø.

Rittel, H.W.J. and M.M. Webber (1973) 'Dilemmas in a General Theory of Planning'. *Policy Sciences* No. 4, pp. 155-169.

Roberts, Marc J., and Jeremy S. Bluhm (1981) *The Choices of Power: Utilities Face the Environmental Challenge.* Cambridge, Mass.: Harvard Univ. Press.

Robertson, David B., and Dennis R. Judd (1989) *The Development of American Public Policy: The Structure of Policy Restraint.* Glenview, IL, Boston, London: Scott, Foresman.

Robinson, John Bridger (1982) 'Apples and Horned Toads: On the Framework-Determined Nature of the Energy Debate'. *Policy Sciences* No. 15, pp. 23-45.

Roe, David (1984) *Dynamos and Virgins.* New York: Random House.

Roe, Emery (1994) *Narrative Policy Analysis: Theory and Practice.* Durham, London: Duke Univ. Press.

Rosenberg, Nathan (1982) *Inside the Black Box: Technologies and Economics.* Cambridge, UK: Cambridge Univ. Press.

Rothwell, Roy, and Mark Dodgson (1993) 'Technology-based MSEs: Their Role in Industrial and Economic Change', in Mark Dodgson and Roy Rothwell [eds.]. *Small Firms and Innovation: The External Influences.* (A special publication of the *International Journal of Technology Management* Vol. 8, No. 5/6)

Roussel, Philip A., Kamal N. Saad, and Tamara J. Erickson (1991) *Third Generation R&D: Managing the Link to Corporate Strategy.* Arthur D. Little.

Rubin, Hal (1979) 'Guide to Nuclear Power in California'. *California Journal*, June, pp. 200-207.

Sabatier, Paul A. (1987) 'Knowledge, Policy-Oriented Learning and Policy Change'. *Knowledge: Creation, Diffusion, Utilization* 8 (June), pp. 649-692.

— (1993) 'Policy Change over a Decade or More', in Paul A. Sabatier, and Hank C. Jenkins-Smith [eds.]. *Policy Change and Learning: An Advocacy Coalition Approach.* pp. 13-39. Boulder, San Francisco, Oxford: Westview Press.

Sabatier, Paul A., and Hank C. Jenkins-Smith [eds.] (1993) *Policy Change and Learning: An Advocacy Coalition Approach.* Boulder, San Francisco, Oxford: Westview.

— (April 1997, 2nd revision). 'The Advocacy Coalition Framework: An Assessment', final version to be published in Paul A. Sabatier [ed.]. *Theories of the Policy Process.* Boulder, Co.: Westview. Forthcoming.

Sabel, Charles F. (1982) *Work and Politics: The Division of Labor in Industry.* Cambridge, UK: Cambridge Univ. Press.

Savage, J. A. (1993) 'Deregulation Derby'. *California Journal*, December, pp. 33-35.

Schlesinger, Jr., Arthur M. (1986) *The Cycles of American History.* London: André Deutsch.

Schmidt, Erik Ib (1983) 'Omstilling og Fornyelse', *in Mellem Fortid og Fremtid. Risøs 25 Års Jubilæum den 6. Juni 1983.* Risø: Forsøgsanlæg Risø.

Schulte Nordholt, J. W. (1963) *Amerika: De Weg van een Wereldmacht.* Baarn: In den Toren.

Schumacher, E. F. (1989 [1973]) *Small is Beautiful: Economics as if People Mattered.* London: Blond & Briggs.

Schumpeter, J. A. (1980 [1934]) *The Theory of Economic Development.* London: Oxford Univ. Press.

Schwarz, Michiel, and Michael Thompson (1990) *Divided We Stand: Redefining Politics, Technology and Social Choice.* New York, London: Harvester Wheatsheaf.

Schön, Donald A. (1983) *The Reflective Practitioner: How Professionals Think in Action.* New York: Basic Books.

Schön, Donald A. and Martin Rein (1994) *Frame Reflection: Toward the Resolution of Intractable Policy Controversies.* New York: Basic Books.

Sclove, Richard E. (1995) *Democracy and Technology.* New York, London: The Guilford Press.

SEAS (December 19, 1987). *SEAS 75 År: Udgivet i Anledning af SEAS A/S 75 Års Jubilæm den 19. December 1987.* Haslev, DK: SEAS.

Sederberg, P. C. (1984) *The Politics of Meaning: Power and Explanation in the Construction of Social Reality.* Tucson, Ariz.: Univ. of Arizona Press.

Serchuk, Adam Harris (1995) *Federal Giants and Wind Energy Entrepreneurs: Utility-Scale Windpower in America, 1970-1990.* Blacksburg, VA: Virginia Polytechnic Institute and State Univ. (Dissertation).

Sidenius, Niels Chr. (1989) *Dansk Industri Politik: Nye Løsninger På Gamle Problemer?* Randers: Hovedland.

Simmons, Harvey (1973) 'System Dynamics and Technocracy', in H. S. D. Cole, Christopher Freeman, Marie Jahoda, and K. L. R. Pavitt [eds.]. *Thinking About The Future: A Critique of The Limits To Growth.* London: Sussex Univ. Press.

Slaughter, Sheila (1993) 'Beyond Basic Science: Research University Presidents' Narratives of Science Policy'. *Science, Technology, & Human Values* Vol. 18, No. 3, pp. 278-302.

Slichter, Katy (1982) 'Farming the Trade Winds: Windfarms, Ltd. Generates Electricity for Oahu'. *Soft Energy Notes*, May/June, pp. 48-49.

Slumstrup, Finn (1984) *Factsheet Denmark: The Danish Folk High School.* Copenhagen: Ministry of Foreign Affairs.

Smit, W. A. (1989) 'Defence Technology Assessment and the Control of Emerging Technologies', in M. L. A. ter Borg and W. A. Smit [eds.]. *Non-Provocative Defence as a Principle of Arms Reductions and its Implications for Assessing Defence Technologies.* pp. 61-77. Amsterdam: VU Univ. Press.

Smits, Martijntje (1997) 'Langdon Winner: Technologie als Schaduwconstructie', in Hans Achterhuis *et al. Van Stoommachine tot Cyborg: Denken over Techniek in de Nieuwe Wereld.* pp. 93-115. Amsterdam: Ambo.

Smits, Ruud, and Jos Leyten (1991) *Technology Assessment: Waakhond of Speurhond? Naar een Integraal Technologiebeleid.* Zeist: Kerckebosch.

Snell, Tee Loftin (1974) *The Wild Shores: America's Beginnings.* The National Geographic Society.

SolarCal Council (January 1979) *Toward a Solar California: The SolarCal Council Action Program.* Sacramento, CA: State of California.

Starr, Chauncey, and Richard Rudman (1973) 'Parameters of Technological Growth'. *Science* No. 182, pp. 358-364.

Starr, Kevin (1973) *Americans and the California Dream 1850-1915.* New York, Oxford: Oxford Univ. Press.

Starss, Thomas A. (1988) 'Legislative Incentives and Energy Technologies: Government's Role in the Development of the California Wind Energy Industry'. *Ecology Law Quarterly* Vol. 15, pp. 103-158.

Steinbruner, John D. (1974) *The Cybernetic Theory of Decision.* Princeton: Princeton Univ. Press.

Stiesdal, Henrik (1990) 'Denmark in Focus: Product Development and Marketing Opportunities for Danish Wind Power in the 1990s'. *WindPower Monthly* No. 12, pp. 26-28.

Strauss, Anselm, and Juliet Corbin (1990) *Basics of Qualitative Research: Grounded Theory Procedures and Techniques.* London: Sage.

Stryker, Sean D. (1993) 'Knowledge and Power in the Students for a Democratic Society, 1960-1970'. *Berkeley Journal of Sociology* Vol. 38, pp. 89-138.

Sørensen, Bent (1989) *Blegdamsvej 17. Et Essay om Fysik og Virkelighed.* Nysyn/Munksgaard

Taggart, Paul (1995) 'New Populist Parties in Western Europe'. *West European Politics* No. 1, pp. 34-51.

Tennyson, George P. (1978) 'Wind Energy: An Accelerating Program', in Vaughn Nelson [ed.]. Proceedings of the National Conference, AWEA. Amarillo, TX, March 1-5, 1978, pp. 153-161. Canyon, TX: Alternative Energy Institute, West Texas State University.

— (1979a) 'The Federal Wind Energy Program: An Overview', in Vaughn Nelson [ed.]. Proceedings of the National Conference, Fall 1978, AWEA. Cape Cod, MA, September 25-27, pp. 33-37. Washington, DC: AWEA.

— (1979b) 'The Federal Wind Energy Program: An Overview', in Vaughn Nelson [ed.]. Proceedings of the National Conference, Spring 1979, AWEA. San Francisco, CA. April 16-19, pp. 1-11. Washington, DC: AWEA.

Thayer, Jr., Robert L., and Carla M. Freeman (1987) 'Altamont: Public Perceptions of a Wind Energy Landscape'. *Landscape and Urban Planning* No. 14, pp. 379-398.

Thayer, Jr., Robert L. (1988) 'The Aesthetics of Wind Energy in the United States: Case Studies in Public Perception', in W. Palz [ed.]. Proceedings of an International Conference. European Community Wind Energy Conference. Herning, Denmark, June 6-10, pp. 470-476.

— (1994) *Gray World, Green Heart: Technology, Nature, and the Sustainable Landscape.* New York: John Wiley & Sons.

Thompson, Michael, Richard Ellis, and Aaron Wildavsky (1990) *Cultural Theory.* Boulder, San Francisco, & Oxford: Westview.

Thorndal, Jytte (1993) 'Gedsermøllen: En Dansk Nyskabelse'. *Fabrik og Bolig. Det Industrielle Miljø i Danmark* No. 2, pp. 3-19.

Touraine, Alain (1997) *What Is Democracy?* Boulder, Colo: Westview.

Tvind (1976) *Lad 100 Møller Blomstre.* Ultborg: Skipper Klement.

Tyler, Cameron (1982) 'Windfarms Muster Megawatts'. *Soft Energy Notes*, September/October, p. 107.

UCS, Union of Concerned Scientists (1993) *1993 Annual Report: 1969-1994, 25 Years*. Cambridge, Mass: UCS.

Van Beek, Jos (1981) 'Stormtest voor Deense Windmolens'. *Duurzame Energie* No. 3, p. 19.

Van de Graaf, Henk, and Rob Hoppe (1989) *Beleid en Politiek: Een Inleiding in de Beleidswetenschap en de Beleidskunde*. Bussum: Couthinho

Van der Wouden, H.C. (1990) *De Dynamiek van Beleid: Onbetaalde Arbeid en Theorieën over Beleid*. Amsterdam.

Van Dijk, H. (1991) 'Rede bij de Verlening van het Eredoctoraat aan Dr. Richard von Weizsäcker', in *Von Weizsäcker Doctor Honnoris Causa Erasmus Universiteit Rotterdam*. Utrecht: Lemma.

Van Gunsteren, Herman (1994) *Culturen van Besturen*. Meppel, Amsterdam: Boom.

Van Muijen, Marie-Louise (1993) *Better Safe Than Provocative: A Policy Science Perspective on the West European Non-Provocative Defence Debate in the 1980s*. Amsterdam: VU Univ. Press.

Van Rossem, Maarten (1991) *De Verenigde Staten in de Twintigste Eeuw*. Den Haag: SDU.

Vestergaard, Frede (1993) 'Vindmøller i Modvind'. *Weekendavisen*, September 3-9.

Von Hippel, Erich (1976) 'The Dominant Role of Users in the Scientific Instrument Innovation Process'. *Research Policy* No. 5, pp. 212-239.

— (1988) *The Sources of Innovation*. New York: Oxford Univ. Press.

Walters, Dan (1983) 'Tax Loopholes Lose Exemption from Regular Legislative Debate. A New Attitude Toward Special-Interest Breaks'. *California Journal*, July, pp. 255-256.

Washburn Rubin, Pat (1981) 'Moving Uncertainlly into a New Era', in John Hart [ed.]. *The New Book on California Tomorrow: Reflections and Projections from the Golden State*. pp. 291-295. Los Altos, CA: William Kaufman.

Weinberg, Carl J., and Robert H. Williams (1990) 'Energy From the Sun'. *Scientific American* Vol. 263, No. 3, pp. 146-155.

Wildavsky, Aaron (1979) *Speaking Truth to Power: The Art and Craft of Policy Analysis*. Boston: Little, Brown and Company.

Willett, Cynthia (1980) 'Expensive Drilling, Government Red Tape: Why the State Has Been Slow to Develop Geothermal Energy'. *California Journal*, May, pp. 204-205.

Williams, W.A. (1961) *The Contours of American History*. New York: World.

Winner, Langdon (1986) *The Whale and the Reactor: A Search for Limits in an Age of High Technology*. Chicago: Chicago Univ. Press.

— [ed.] (1992) *Democracy in a Technological Society. Philosophy and Technology, Vol. 9*. Dordrecht: Kluwer Academic.

Wistoft, Birgitte, Flemming Petersen, and Jytte Thorndahl (1991) *Elektricitetens Aarhundrede: Dansk Elforsynings Historie. Bind 1: 1891-1940*. Copenhagen: Danske Elværkers Forening.

Wistoft, Birgitte, Jytte Thorndahl, and Flemming Petersen (1992) *Elektricitetens Aarhundrede: Dansk Elforsynings Historie. Bind 2: 1940-1991*. Copenhagen: Danske Elværkers Forening.

Yinger, Robert J. (1983) 'The Status of Wind Turbine Development in Southern California Edison's Service Territory', in Proceedings Wind Energy Expo '83 and National Conference, AWEA. San Francisco, CA. October 17-19, pp. 276-282. Washington, DC: AWEA.

Østergaard, Iben (1993) '1 MW Mølle på Avedøre Holme'. *VEinformation* No. 54, pp. 12-14.

List of Interviews

California

Aitken, Donald W. (Nov 8, 1995; San Francisco Airport) In 1991, Aitken was invited to become a staff member of the Union of Concerned Scientists (UCS). Since then he has worked as Senior Energy Analyst within UCS's renewable energy program.

Boyd, Hap (Nov 7, 1995; Telephone) Since the early 1980s, Hap Boyd worked as lawyer and political lobbyist for U.S. Windpower. In 995, he started to work with Zond Energy Systems.

Coe, Virginia A. (Nov 17, 1995; San Francisco) In the early 1980s, Gigi Coe worked as the Deputy Director of the Office of Appropriate Technology. At the moment, she is Director of the CPUC's Division of Strategic Planning that got the electric restructuring debate going. She is co-author of the *Yellow Book* (CPUC 1993).

Dasovich, Jeffrey. (Nov 17, 1995; San Francisco) Senior Policy Analyst at the CPUC's Division of Strategic Planning. Dasovich is one of the authors of the *Yellow Book* (CPUC 1993).

DiVirgilio, Philip J. (Nov 3, 1995; Tehachapi) Director Wind Turbine Marketing at Zond Energy Systems, Inc. In 1994, DiVirgilio was hired by Zond to work on the utility wind energy market. Before that he worked for the gas cogeneration industry.

Ferguson, Rich (Oct 17, 1995; Sacramento) Rich Ferguson works for the California division of the Sierra Club. He represents this environmental organization within the board of CEERT.

Gipe, Paul (Nov 2 & 3, 1995; Bakersfield) Paul Gipe is an internationally recognized authority on wind energy. He has an environmental background. As a college student, he worked on the strip-mining issue. Looking for alternative energy sources, Paul Gipe became interested in wind energy. During the California wind boom, he came to work with Zond as a maintenance mechanic. As a result, he has hands-on experience in nearly every aspect of the technology. Gipe has served as the West Coast representative of AWEA and as Executive Director of the Kern Wind Energy Association (cf. Gipe 1983, 1995).

Hamrin, Jan (Oct 25, 1995; San Francisco) Jan Hamrin is affectionately called the 'den mother' of renewable resources. She has a PhD in ecology and public policy. In the 1970s, she ran solar programs for the CEC. In 1981, she founded the IEP, of which she was Executive Director from 1982 to 1990. Since then, she works at Hansen, McOuat & Hamrin, Inc.

Jablonski, Lori (Oct 17, 1995; Sacramento) Lori Jablonski studied political science and was active within the environmental movement in Hawaii. She started to work at CEERT in 1990.

Kirshner, Dan (Oct 11, 1995; Oakland) Dan Kirshner has BA degrees in Economics and Physics. Since 1976, he has worked at the Environmental Defense Fund's California office. During the 1970s, he developed innovative integrated resource planning software, known as Elfin. He is the EDF representative within the board of CEERT.

Loyola, Juanito (Jul 28, 1993; Sacramento) Associate Energy Specialist, who works for the CEC, Energy Technology Development Division, Research and Development Office. In the 1990s, she worked as Program Manager of the Wind Performance Reporting System (WPRS).

McFarland, Jan (Oct 23, 1995; Telephone) After she studied political science, she came to work for CEC Chairman Richard Maullin in 1977-1978. After this she worked with Huey D. Johnson (1979), the State Secretary for Resources. Before joining the law firm Skadden, Arps, Slate, Meagher & Flom in 1987, she advised CEC Chairman Chuck Imbrecht. At the end of the 1980s, she took the initiative to establish CEERT. Since mid-December of 1993, she works as a senior policy advisor for the Environmental Protection Agency. Her primary duties include monitoring global climate change policies and the environmental impacts of federal energy policy.

Miller, Eric (Oct 26, 1995; San Francisco) Eric Miller has been in the energy business since he was in college. At that time, he helped to build the Rocky Mountain Institute, headed by Amory Lovins, and followed a graduate program at Dartmouth College (New Hampshire, NE) run by the author of *Limits to Growth*, Dennis Meadows. After his study, he became involved in the acid rain legislation that was passed in 1992, and put a limit on sulphur oxide emissions. Since 1990, he works at U.S. Windpower, which at the time was staffing up its marketing group. At the time of the interview, Miller was responsible for business development in the Western United States.

Obermeier, John (Jul 26, 1993; Davis) Physical engineer, who worked for two years at the Office of Appropriate Technology (OAT). As OAT's Wind Energy Specialist, Obermeier was instrumental in getting the first PURPA wind energy contracts signed in the service territory of PG&E and SCE. Since 1981, he works as an independent consultant, owning his own company OTECH Engineering in Davis.

Rader, Nancy (Nov 11, 1995; Berkeley) After studying political science, Nancy Rader started to work for consumer and environmental advocacy groups (so-called Ralph Nader groups). Now she is an independent Renewable Energy Consultant. Since 1988, she almost exclusively worked for AWEA.

Smith, Don R. (Oct 25, 1995; Oakland) Physical engineer, who has worked for years as a steady consultant for PG&E Wind Energy Research.

Smutny-Jones, Jan (Oct 31, 1995; Sacramento) Jan Smutny-Jones is General Council of the Independent Energy Producers (IEP) since 1987. Since 1990, he is Executive Director of this trade association.

Starrs, Thomas A. (Oct 26, 1995; Walnut Creek) Energy scholar, who in the second half of the 1980s studied the government's role in the development of the California wind energy industry (cf. Starrs 1988).

Thayer, Jr., Robert L. (Oct 18, 1995; Davis) Robert L. Thayer, Jr. is Professor and Head of the Landscape Architecture Program at the University of California, Davis. His work and writings emphasize resource conservation strategies for land planning. In relation to this, he has studied the public response to wind energy development in California (cf. Thayer and Freeman 1987, Thayer 1988, 1994).

Thomson, Thomas W. (Nov 9, 1995; San Francisco) Mechanical engineer, who has worked on the CPUC's alternative energy program (esp. QF contract issues) since 1984. He is now Supervisor of the Resource Advisory Section of the CPUC's Commission Advisory and Compliance Division.

Weinberg, Carl (Oct 27, 1995; Walnut Creek) Engineer who started up the R&D group of the PG&E company in the early 1980s. When this group became formalized in the mid-1980s, Weinberg became head of the R&D department. In this job, he became closely involved with the development of private wind farms at Altamont Pass. He left PG&E a few years ago to become a private consultant.

White, V. John (Oct 17, 1995; Sacramento) V. John White is Executive Director of CEERT. In the 1980s, he worked as a legislative advocate for the independent power industry.

Denmark

Andersen, Jan (Mar 30, 1995; Roskilde) Andersen is Associate Professor at the Department of Environment, Technology and Social Studies of Roskilde University. During the 1980s, he worked for the Energy Agency (1979-1986) and the Ministry of Energy.

Dannemand Andersen, Per (Nov 3, 1993; Risø) Per Dannemand Andersen works at the Risø Test Station. He wrote his PhD dissertation about the Test Station's role within the technical innovation in the Danish wind turbine industry (Dannemand Andersen 1990, 1993).

Dræbye, Tage (April 7, 1995; Copenhagen) Academic engineer who works as a private consultant. He has a political background. Dræbye was MP for the *Radikale Venstre* from 1979 to 1984. Chairman of the Energy Agency's Technical Committee for Wind Energy (*Energystyrelsens Faglige Udvalg for Vindkraft*) that was established in 1991.

Fog, Niels (Nov 8, 1993; Copenhagen) Academic engineer, who from 1990 to 1994 dealt with wind energy issues within the Association of Danish Electric Utilities (DEF).

Godske, Bjørn (Mar 29, 1995; Ballerup) Electrical engineer, who works at the Planning Department of the power company Elkraft.

Godtfredsen, Finn (Nov 3, 1993; Risø) Economist, who works at the Risø Test Station for Wind Turbines (cf. Godtfredsen, Hjuler Jensen, and Morthorst 1993).

Hansen, Tove Kjær (Mar 29, 1995; Ballerup) Hansen works for the Section Environmental & Licensing of Elkraft's Planning Department. For years, she has been much involved in the physical planning of utility-owned wind farms.

Hjuler Jensen, Peter (Nov 15, 1993; Telephone) Mechanical engineer, who has worked at the Risø Test Station for Wind Turbines since it started in 1978. In 1993, Hjuler Jensen became Head of the Test Station.

Karnøe, Peter (Nov 5 & 12, 1993; Copenhagen) Associate Professor at the Institute of Organization and Industrial Sociology of the Copenhagen Business School. Peter Karnøe wrote his dissertation on the development of the Danish wind turbine industry (Karnøe 1991). In 1992, he stayed at the Department of Economics of Stanford University in California, to study and compare American and Danish approaches to innovation in wind turbine technology (Karnøe 1993, 1995, 1996).

Kjær, Tyge (Mrt 31, 1993; Copenhagen) Tyge Kjær is Associate Professor at the Department of Environment, Technology and Social Studies of Roskilde University. At the end of the 1980s, he was hired by FDV to study the downfall of the Danish wind turbine industry after (and during) the California export boom (cf. Kjær 1988).

Krohn, Søren (Apr 7, 1995; Copenhagen) Director General of the Danish Wind Turbine Manufacturers Association (DVF).

Lemming, Jørgen (Nov 2, 1993 & Apr 5, 1995; Copenhagen) Mechanical engineer, who successively worked for the Technical University of Denmark, the Ministry of Energy (planning and implementation of energy R&D programs), the Danish Technological Institute, and DK-TEKNIK (external coordinator of governmental energy R&D programs). Since 1988, Lemming works as Head of the Wind Energy Section of the Renewable Energy Division of the Danish Energy Agency.

Maribo Pedersen, B. (Apr 6, 1995; Lyngby) Senior Lecturer at the Department of Fluid Mechanics at the Technical University of Denmark. Has been involved in wind energy, in particular the Danish Wind Power Program (WPP), since the 1970s. He was a member of the ATV Wind Energy Committee, and is a member of the Energy Agency's Technical Committee for Wind Energy (*Energystyrelsens Faglige Udvalg for Vindkraft*, est. 1991).

Meyer, Niels I. (Nov 9, 1993; Lyngby) Professor of Physics at the Physics Department's Energy Group of the Technical University of Denmark. Since the early 1970s, Meyer has promoted renewable energy in numerous ways. In the 1970s, he initiated the ATV Wind Energy Committee, the above mentioned Energy Group, and the writing of the Alternative Energy Plan (Blegaa *et al.* 1976). He was one of the authors of the book *Revolt from the Center* (Meyer, Petersen, Sørensen 1981). From 1977 to 1985, Meyer was a member of the Technology Council (*Teknologirådet*). From 1980 to 1992, he was a member of the Energy Research Committee (*Energiforskningsudvalg*) of the Ministry of Energy. In 1983, Meyer became Chairman of the newly established Steering Group for Renewable Energy. When in 1991, the steering group was institutionalized as the Renewable Energy Council (*VE-Rådet*) under the Energy Agency, Niels Meyer remained its Chairman.

Nielsen, Torben Glar (Apr 5, 1995; Haslev) Torben Glar Nielsen is the Technical Manager at the Lolland & Falster power company SEAS. Until 1994, SEAS has been a shareholder of VESTAS.

Petersen, Helge (Mar 27, 1995; Copenhagen) As a member of the second ATV Wind Energy Committee, Helge Petersen proposed the idea of a Test Station for Small Wind Turbines. In 1978, he became the first Head of the Risø Test Station. He was the designer of the blades of the famous windmill at the Tvind Folk High Schools. He is a member of the Technical Committee for Wind Energy (*Energystyrelsens Faglige Udvalg for Vindkraft*, est. 1991).

Riishøj, Søren (Apr 4, 1995; Copenhagen) From 1981 to 1994, Søren Riishøj was Member of Parliament and Energy Specialist for the Socialist People's Party (SF). During the last years, he chaired the Permanent Energy Committee of the *Folketing*.

Svenson, Jan (Mar 29, 1995; Ballerup) Jan Svenson works as a Wind Energy Specialist for Elkraft Consult Ltd. In 1984, he was one of the founders of the wind turbine manufacturer Danwin.

Sørensen, Bent (Nov 11, 1993; Roskilde) Before becoming a Professor of Physics at the Roskilde University in 1980, Bent Sørensen worked at the Nordic Institute for Theoretical Nuclear Physics and the Niels Bohr Institute (1967-1980). Inspired by Henry W. Kendall (Henry W. Kendall is Professor of Physics at MIT. In the 1960s, Kendall was among the first physicists who came to seriously criticize nuclear power technology. He has been on the board of UCS since 1973. In 1990, Kendall received the Nobel price for Physics.) at MIT, he became engaged within the anti-nuclear power struggle in Denmark. Sørensen was one of the initiators of the Alternative Energy Plans (Blegaa *et al.* 1976, Hvelplund *et al.* 1983).

Index